ORDINARY DIFFERE

ORDINARY DIFFERENTIAL EQUATIONS

SECOND EDITION

Purna Chandra Biswal
Assistant Professor of Mathematics
Parala Maharaja Engineering College
Berhampur, Odisha

PHI Learning Private Limited
New Delhi-110001
2012

₹ 295.00

ORDINARY DIFFERENTIAL EQUATIONS, Second Edition
Purna Chandra Biswal

ISBN-978-81-203-4622-2

Second Printing (Second Edition) . . . . **September, 2012**

Published by Asoke K. Ghosh, PHI Learning Private Limited, M-97, Connaught Circus, New Delhi-110001 and Printed by Mudrak, 30-A, Patparganj, Delhi-110091.

To

My Parents

Jagannath Biswal

and

Nishamani Biswal

Contents

Preface

Encouraging feedback received from the teachers and students motivated me to publish this second edition. While all the features of the first edition have been retained in this edition, two new sections, one on methods of regrouping and another on independent functions are incorporated to broaden the coverage. In addition, numerous problems and chapter-end exercises with hints are also included in this edition.

Any suggestion for improvement of the text is welcome.

Purna Chandra Biswal

Preface to the First Edition

This textbook has been developed over a period of many years and meets the needs of a one-term undergraduate course in ordinary differential equations. No specific prerequisite except for the knowledge of basic calculus is required to understand the content of this textbook. In this book, I have presented the derivation of the formula for each type of differential equation, along with standard examples worked out according to the derived formulae, to make the book precise. The notation used in this textbook is very commonly used by mathematicians.

A large number of judiciously selected worked-out examples has made the book simple and lucid for classroom instruction. These examples often convey interesting logical principles. Numerous exercises, along with hints for solution in some cases, have been included at the end of each section to gauge the students' understanding and grasp of the theory.

I am grateful to the Director, Prof. Sangram Mudali and the Dean, Dr. Ajit Kumar Panda, both of National Institute of Science & Technology (NIST), for extending all possible cooperation, help and support in the preparation of this textbook. I am thankful to Mrs. Sunanda Samal for her continuous and unfailing encouragement throughout the period of writing this book.

I am also indebted to all those who contributed to the writing of this textbook in various ways, and to the staff of PHI Learning for their efforts in publishing this book within a short period of time.

I would appreciate receiving by e-mail at purnabiswal@rediffmail.com any helpful suggestion for improving the presentation of the material. Nonetheless I fervently hope that this book fulfils the requirement for an accessible textbook suitable for undergraduate mathematics and engineering courses of universities all over India.

Purna Chandra Biswal

Chapter 1

Introduction

The mathematical formulation of many problems in science, engineering and economics gives rise to differential equations. An equation involving one dependent variable and its derivatives with respect to one or more independent variables is called a *differential equation.* Many of the general laws of nature in physics, chemistry, biology and astronomy find their most natural expression in the language of differential equations. Applications also abound in mathematics itself, especially in geometry, and in engineering, economics and many other fields of applied sciences. In this book, we have represented $y(x)$ as the dependent variable and x as the independent variable. Again, $\frac{dy}{dx}$ is represented by y'.

The following are some of the examples of differential equations:

(1) $y' + xy = x^2$

(2) $y''' + x^4y'' + y' + y = \sin x$

(3) $x^2y'''y' + 2xy'' = (x^2 + y^2)y^2$

(4) $\frac{\partial y}{\partial t} + \left(\frac{\partial y}{\partial x}\right)^2 = 4xt$

In (4), the dependent variable y is a function of two independent variables x and t. Again, $\frac{\partial y}{\partial t}$ and $\frac{\partial y}{\partial x}$ are called *partial derivatives* of the dependent variable y with respect to t and x respectively.

1.1 Classification of Differential Equations

Differential equations are classified as either *ordinary differential equations* or *partial differential equations,* depending on whether one or more independent variables are involved. In the abovementioned set of differential equations, only (4) is a partial differential equation. Differential equations are also classified according to the *order* or *degree* of the highest order derivatives present in the differential equation. A differential equation is said to be of order n if the order of the highest order derivative is n. A differential equation is said to be of degree n if the degree (power) of the highest order derivative is n.

In the abovementioned set of differential equations, (1) and (4) are of first order, and (2) and (3) are of third order. Only (4) is of second degree.

A differential equation is called *linear* if and only if the following conditions are satisfied:

1. Every dependent variable and derivative involved occurs to the first degree only.
2. Product of any form of dependent variable is absent.

A differential equation is called *nonlinear* if it fails to be linear. In the following differential equations, the first two are linear and the last four are nonlinear:

(1) $y'' + x^2y' + xy = \sin x$

(2) $\frac{\partial y}{\partial t} - x\frac{\partial^2 y}{\partial x^2} = 0$

(3) $y'' + (y')^3 + y = 0$

(4) $y' + y^2 = 0$

(5) $y'' + yy' = x^2$

(6) $\frac{\partial^2 y}{\partial x^2} + \frac{\partial y}{\partial x}\frac{\partial y}{\partial t} + \frac{\partial^2 y}{\partial t^2} = 0$

Equations given in (3) and (4) are nonlinear because the degree of the derivative y' in (3) is three and that of the dependent variable y is two in (4). Again, (5) is nonlinear because of the presence of term yy', which is the product of the dependent variable y and the first derivative y' and (6) is also nonlinear because of the occurrence of $\frac{\partial y}{\partial x}\frac{\partial y}{\partial t}$.

A differential equation can be classified as either *homogeneous* or *nonhomogeneous* depending on whether that differential equation does not contain or contains some additive term containing only independent variable respectively. In the above set of differential equations, only (1) and (5) are nonhomogeneous differential equations, whereas other four are homogeneous differential equations.

Exercises 1.1

1. Find the order of each of the following differential equations:
   (a) $y' + xy = e^x$
   (b) $y'' + (y')^2 = \cos x$
   (c) $y'' + y'y = x$
   (d) $xy'' - y = 0$
   (e) $x^2y' + y^2 = 0$
   (f) $y''' + 3(y')^2 + y = 0$
   (g) $(y')^3 = ((y')^2 + 1)^{\frac{1}{2}}$
   (h) $x\left(\frac{\partial y}{\partial x}\right) + t\left(\frac{\partial y}{\partial t}\right) = x^2t$
   (i) $t\left(\frac{\partial y}{\partial x}\right) + x\left(\frac{\partial y}{\partial t}\right) = t^2x$

(j) $x^2\left(\frac{\partial^2 y}{\partial x^2}\right)+y\left(\frac{\partial y}{\partial t}\right)=0$

(k) $\frac{\partial^2 y}{\partial x^2}+\frac{\partial y}{\partial x}+yt=0$

2. Find the degree of each of the following differential equations:

   (a) $y'+xy=e^x$
   (b) $y''+(y')^2=\cos x$
   (c) $y''+y'y=x$
   (d) $xy''-y=0$
   (e) $x^2y'+y^2=0$
   (f) $y'''+3(y')^2+y=0$
   (g) $(y')^3=((y')^2+1)^{\frac{1}{2}}$
   (h) $x\left(\frac{\partial y}{\partial x}\right)+t\left(\frac{\partial y}{\partial t}\right)=x^2t$
   (i) $t\left(\frac{\partial y}{\partial x}\right)+x\left(\frac{\partial y}{\partial t}\right)=t^2x$
   (j) $x^2\left(\frac{\partial^2 y}{\partial x^2}\right)+y\left(\frac{\partial y}{\partial t}\right)=0$
   (k) $\frac{\partial^2 y}{\partial x^2}+\frac{\partial y}{\partial x}+yt=0$

3. Classify whether the following differential equations are ordinary or partial:

   (a) $y'+xy=e^x$
   (b) $y''+(y')^2=\cos x$
   (c) $y''+y'y=x$
   (d) $xy''-y=0$
   (e) $x^2y'+y^2=0$
   (f) $y'''+3(y')^2+y=0$
   (g) $(y')^3=((y')^2+1)^{\frac{1}{2}}$
   (h) $x\left(\frac{\partial y}{\partial x}\right)+t\left(\frac{\partial y}{\partial t}\right)=x^2t$
   (i) $t\left(\frac{\partial y}{\partial x}\right)+x\left(\frac{\partial y}{\partial t}\right)=t^2x$
   (j) $x^2\left(\frac{\partial^2 y}{\partial x^2}\right)+y\left(\frac{\partial y}{\partial t}\right)=0$
   (k) $\frac{\partial^2 y}{\partial x^2}+\frac{\partial y}{\partial x}+yt=0$

4. Classify whether the following differential equations are linear or nonlinear:

   (a) $y'+xy=e^x$
   (b) $y''+(y')^2=\cos x$
   (c) $y''+y'y=x$

(d) $xy'' - y = 0$

(e) $x^2y' + y^2 = 0$

(f) $y''' + 3(y')^2 + y = 0$

(g) $(y')^3 = ((y')^2 + 1)^{\frac{1}{2}}$

(h) $x\left(\frac{\partial y}{\partial x}\right) + t\left(\frac{\partial y}{\partial t}\right) = x^2t$

(i) $t\left(\frac{\partial y}{\partial x}\right) + x\left(\frac{\partial y}{\partial t}\right) = t^2x$

(j) $x^2\left(\frac{\partial^2 y}{\partial x^2}\right) + y\left(\frac{\partial y}{\partial t}\right) = 0$

(k) $\frac{\partial^2 y}{\partial x^2} + \frac{\partial y}{\partial x} + yt = 0$

5. Classify whether the following differential equations are homogeneous or nonhomogeneous:

(a) $y' + xy = e^x$

(b) $y'' + (y')^2 = \cos x$

(c) $y'' + y'y = x$

(d) $xy'' - y = 0$

(e) $x^2y' + y^2 = 0$

(f) $y''' + 3(y')^2 + y = 0$

(g) $(y')^3 = ((y')^2 + 1)^{\frac{1}{2}}$

(h) $x\left(\frac{\partial y}{\partial x}\right) + t\left(\frac{\partial y}{\partial t}\right) = x^2t$

(i) $t\left(\frac{\partial y}{\partial x}\right) + x\left(\frac{\partial y}{\partial t}\right) = t^2x$

(j) $x^2\left(\frac{\partial^2 y}{\partial x^2}\right) + y\left(\frac{\partial y}{\partial t}\right) = 0$

(k) $\frac{\partial^2 y}{\partial x^2} + \frac{\partial y}{\partial x} + yt = 0$

1.2 Formulation of Differential Equations

Let $y' + ay = 0$ be any differential equation, and $u(x) = e^{-ax}$ be any dependent variable. It is easy to verify that when $u(x)$ is substituted in the left side of the differential equation $y' + ay = 0$, we produce the right hand side. In other words, some expression for a dependent variables $u(x)$ is called a *solution* of some differential equation if and only if when that expression for the dependent variable is substituted in the left hand side of that differential equation, we produce the corresponding right hand side. In general, there are two different types of questions that come to one's mind, and they are:

1. How to formulate a differential equation when a dependent variable is given?
2. How to find a dependent variable when a differential equation is given?

In this section, we are going to discuss the first question. The second question will be discussed from Chapter 2 onwards. The general principles for formulating a differential equation, when the dependent variable is given, are:

1. Find a relation between the dependent variable and the possible derivatives of the dependent variable.
2. Eliminate the parameter between the dependent variable and the possible derivatives of the dependent variable.

EXAMPLE 1.1 Find a differential equation corresponding to the dependent variable $y(x) = cx + c^2$.

Solution: One can get a differential equation by eliminating the constant c present in the dependent variable expression. Now, we have $y = cx + c^2$ and $y' = c$. Again, we have $(y')^2 = c^2$. Now, consider the three expressions $y = cx + c^2$, $xy' = cx$ and $(y')^2 = c^2$. In other words, the required differential equation is

$$\begin{aligned}(y')^2 + xy' - y &= c^2 + cx - (cx + c^2)\\ &= 0\end{aligned}$$

EXAMPLE 1.2 Find a differential equation corresponding to the expression

$$\phi(x, y) = x^2 + y^2 = a^2$$

Solution: It is easy to verify that

$$y = (a^2 - x^2)^{\frac{1}{2}}$$

and

$$\begin{aligned}y' &= -\frac{x}{(a^2 - x^2)^{\frac{1}{2}}}\\ &= -\frac{x}{y}\end{aligned}$$

The corresponding differential equation can be expressed in the form

$$x\,\mathrm{d}x + y\,\mathrm{d}y = 0$$

EXAMPLE 1.3 Find the differential equation corresponding to set of all concentric circles with the centre at the origin.

Solution: Clearly, the equation of any circle with the centre at the origin is $x^2 + y^2 = r^2$. In other words, the corresponding differential equation is $x + yy' = 0$.

EXAMPLE 1.4 Find the differential equation corresponding to the set of all circles tangent to the x-axis at the origin.

Solution: Clearly, the equation of the circle is $x^2 + y^2 = 2by$. Again, the differentiation of $x^2 + y^2 = 2by$ gives $x + (y - b)y' = 0$. Eliminating the parameter b between $x^2 + y^2 = 2by$ and $x + (y - b)y' = 0$, we have $(x^2 - y^2)y' = 2xy$.

EXAMPLE 1.5 Find the differential equation corresponding to the set of all circles tangent to the y-axis at the origin.

Solution: Clearly, the equation of the circle is $x^2 + y^2 = 2bx$. Again, the differentiation of $x^2 + y^2 = 2bx$ gives $x + yy' = b$. Eliminating the parameter b between $x^2 + y^2 = 2bx$ and $x + yy' = b$, we have $2xyy' = y^2 - x^2$.

Exercises 1.2

1. Show that the differential equation corresponding to the dependent variable $y(x) = e^{ax}$ is not unique.
2. Show that the differential equation corresponding to the dependent variable $y(x) = \sin x$ is not unique.
3. Show that the differential equation corresponding to the dependent variable $y(x) = \cos x$ is not unique.
4. Find a differential equation corresponding to each of the following dependent variables:

   (a) $y(x) = \frac{x^2}{2} + x$

   (b) $y(x) = (x^2 + c)e^{-x}$

   (c) $y(x) = 2x + ae^x + be^{2x} + 3$

   (d) $y(x) = a(1 - \cos x)$

   (e) $y(x) = a \cos 3x + b \sin 3x$

   (f) $y(x) = x^r$

   (g) $y(x) = e^{ax}$

5. Find the differential equation by eliminating a, b and c from $y(x) = a\,e^x + b\,e^{2x} + c\,e^{-x}$.
6. Find the differential equation by eliminating a from $y(x) = ax + \frac{b}{a}$.
7. Find the differential equation by eliminating a and b from $y(x) = ae^{cx} \cos kx + be^{cx} \sin kx$.
8. Find the differential equation by eliminating a and b from the following family of curves:

   (a) $y(x) = ax + bx^2$

   (b) $y(x) = ae^{kx} + be^{-kx}$

   (c) $y(x) = a \sin kx + b \cos kx$

   (d) $y(x) = a + be^{-2x}$

   (e) $y(x) = ax + b \sin x$

   (f) $y(x) = ax + b \cos x$

   (g) $y(x) = ae^x + bxe^x$

   (h) $y(x) = ae^x + be^{-3x}$

   (i) $y(x) = ax + bx^{-1}$

   (j) $y(x) = ax^5 + bx^{-1}$

9. Find the differential equation by eliminating a and b from the following family of curves:
   (a) $y(x) = a + bx + cx^2$
   (b) $y(x) = ae^x + be^{-2x}$
   (c) $y(x) = a\cos(\ln x) + b\sin(\ln x)$
   (d) $x^3 - 3x^2y = a$
   (e) $y\sin x - xy^2 = c$
   (f) $y\ln y' + x(1 - y') = 0$
   (g) $y(x) = ae^{3x} + be^{-x}$
   (h) $y(x) = ae^x + be^{-3x}$
   (i) $y(x) = ax + bx^{-2}$
   (j) $y(x) = ax^3 + bx^{-1}$

10. Find the differential equations of the family of plane curves which are:
   (a) All circles of unit radius.
   (b) All confocal central conics.
   (c) All parabolas with foci at the origin and axes along the x-axis.
   (d) All parabolas with vertices at the origin and foci on the y-axis.
   (e) All parabolas having their axes parallel to the y-axis and touching the straight line $y = 3x$ at the origin.
   (f) All tangents to the parabola $y^2 = 2x$.
   (g) All straight lines at a unit distance from the origin.
   (h) All circles of fixed radius r with the centre on the x-axis.
   (i) All circles of fixed radius r with the centre on the y-axis.

Chapter 2

First Order Equations

The most general form of the first order differential equation is $y' = h(x, y)$. It becomes easy for determining the corresponding solution of any first order differential equation when the function $h(x, y)$ has the following form:

1. $h(x, y) = -p(x)y$
2. $h(x, y) = f(x)$
3. $h(x, y) = -p(x)y + f(x)$

2.1 First Order Homogeneous Equation

Suppose $y' + p(x)y = 0$ be any arbitrary first order ordinary differential equation. One can rewrite the given differential equation as

$$\frac{dy}{y} = -p(x)\,dx \tag{2.1}$$

This method is called *variable separable method* because variables are separated. Again, one can easily integrate Eq.(2.1). In other words, the solution of the ordinary differential equation $y' = -p(x)y$ is given by

$$\begin{aligned} y(x) &= e^{\ln y(x)} \\ &= e^{\int \frac{dy}{y} + c} \\ &= e^{\int -p(x)\,dx\, +\, c} \\ &= e^{c}e^{-\int p(x)\,dx} \\ &= ke^{-\int p(x)\,dx} \end{aligned} \tag{2.2}$$

The unknown k in the solution to the differential equation $y' + p(x)y = 0$, given in Eq.(2.2) can be determined if and only if one of the initial conditions $y(x_0)$ is given at $x = x_0$, which is discussed in Section 2.13.

EXAMPLE 2.1 Solve the first order homogeneous differential equation $y' + \cos(x)\, y = 0$.

Solution: Clearly, $p(x) = \cos(x)$. According to Eq.(2.2), the corresponding solution is given by

$$\begin{aligned} y(x) &= k\, e^{-\int p(x)\, dx} \\ &= k\, e^{-\int \cos\, x\, dx} \\ &= k\, e^{-\sin\, x} \end{aligned}$$

EXAMPLE 2.2 Solve the first order homogeneous differential equation $y' + \tan(x)\, y = 0$.

Solution: Clearly, $p(x) = \tan(x)$. According to Eq.(2.2), the corresponding solution is given by

$$\begin{aligned} y(x) &= k\, e^{-\int p(x)\, dx} \\ &= k\, e^{-\int \tan\, x\, dx} \\ &= k\, e^{\ln(\cos\, x)} \\ &= k \cos\, x \end{aligned}$$

EXAMPLE 2.3 Solve the first order homogeneous differential equation $y' - \cot(x)\, y = 0$.

Solution: Clearly, $p(x) = -\cot(x)$. According to Eq.(2.2), the corresponding solution is given by

$$\begin{aligned} y(x) &= k\, e^{-\int p(x)\, dx} \\ &= k\, e^{\int \cot\, x\, dx} \\ &= k\, e^{\ln(\sin\, x)} \\ &= k \sin\, x \end{aligned}$$

EXAMPLE 2.4 Solve the first order homogeneous differential equation $y' + x\, y = 0$.

Solution: Clearly, $p(x) = x$. According to Eq.(2.2), the corresponding solution is given by

$$\begin{aligned} y(x) &= k\, e^{-\int p(x)\, dx} \\ &= k\, e^{-\int x\, dx} \\ &= k\, e^{-\left(\frac{x^2}{2}\right)} \end{aligned}$$

EXAMPLE 2.5 Solve the first order homogeneous differential equation $xy' - y = 0$.

Solution: Clearly, $p(x) = -\frac{1}{x}$. According to Eq.(2.2), the corresponding solution is given by

$$\begin{aligned} y(x) &= k\, e^{-\int p(x)\, dx} \\ &= k\, e^{\int \left(\frac{1}{x}\right) dx} \\ &= k\, e^{\ln\, x} \\ &= kx \end{aligned}$$

EXAMPLE 2.6 Solve the first order homogeneous differential equation $\cos^2(x)\, y' - y = 0$.
Solution: Clearly, $p(x) = -\sec^2(x)$. According to Eq.(2.2), the corresponding solution is given by

$$\begin{aligned} y(x) &= k\, e^{-\int p(x)\, dx} \\ &= k\, e^{\int \sec^2 x\, dx} \\ &= k\, e^{\tan x} \end{aligned}$$

EXAMPLE 2.7 Solve the first order homogeneous differential equation $\sin^2(x)\, y' + y = 0$.
Solution: Clearly, $p(x) = \operatorname{cosec}^2(x)$. According to Eq.(2.2), the corresponding solution is given by

$$\begin{aligned} y(x) &= k\, e^{-\int p(x)\, dx} \\ &= k\, e^{-\int \operatorname{cosec}^2 x\, dx} \\ &= k\, e^{\cot\, x} \end{aligned}$$

Exercises 2.1

1. Solve the first order homogeneous differential equation $\cos(x)\, y' + \sin(x)\, y = 0$.
   HINT: According to Eq.(2.2), $y(x) = k \cos x$.
2. Solve the first order homogeneous differential equation $\cos^2(x)\, y' - y = 0$.
   HINT: According to Eq.(2.2), $y(x) = k\, e^{\tan\, x}$.
3. Solve the first order homogeneous differential equation $\sin^2(x)\, y' - y = 0$.
   HINT: According to Eq.(2.2), $y(x) = k\, e^{-\cot\, x}$.
4. Solve the first order homogeneous differential equation $x^2 y' + \sin\left(\frac{1}{x}\right)\, y = 0$.
   HINT: According to Eq.(2.2), $y(x) = k\, e^{-\cos\left(\frac{1}{x}\right)}$.
5. Solve the first order homogeneous differential equation $x^2 y' + \cos\left(\frac{1}{x}\right)\, y = 0$.
   HINT: According to Eq.(2.2), $y(x) = k\, e^{\sin\left(\frac{1}{x}\right)}$.
6. Solve the first order homogeneous differential equation $y' + x \cos(x^2)\, y = 0$.
   HINT: According to Eq.(2.2), $y(x) = k\, e^{-\frac{1}{2}\sin(x^2)}$.
7. Solve the first order homogeneous differential equation $y' + x \sin(x^2)\, y = 0$.
   HINT: According to Eq.(2.2), $y(x) = k\, e^{\frac{1}{2}\cos(x^2)}$.
8. Solve the first order homogeneous differential equation $y' + x \cos(x)\, y = 0$.
   HINT: According to Eq.(2.2), $y(x) = k\, e^{x \sin x + \cos x}$.
9. Solve the first order homogeneous differential equation $y' + x \sin(x)\, y = 0$.
   HINT: According to Eq.(2.2), $y(x) = k\, e^{\sin x - x \cos x}$.

10. Solve the first order homogeneous differential equation $(1-x)y' + y = 0$.
    HINT: According to Eq.(2.2), $y(x) = k(x-1)$.

11. Solve the first order homogeneous differential equation $y' - xy = 0$.
    HINT: According to Eq.(2.2), $y(x) = k\,e^{\frac{x^2}{2}}$.

12. Solve the first order homogeneous differential equation $y' + cy = 0$.
    HINT: According to Eq.(2.2), $y(x) = k\,e^{-cx}$.

13. Solve the first order homogeneous differential equation $y' + y = 0$.
    HINT: According to Eq.(2.2), $y(x) = k\,e^{-x}$.

2.2 First Order Nonhomogeneous Equation

Let $y' = f(x)$ be any arbitrary first order nonhomogeneous ordinary differential equation. One can rewrite the given differential equation as

$$\mathrm{d}y = f(x)\,\mathrm{d}x \tag{2.3}$$

This method is called *variables separable* because variables are separated. Again, one can easily integrate Eq.(2.3). In other words, the solution of the ordinary differential equation $y' = f(x)$ is given by

$$\begin{aligned} y(x) &= \int \mathrm{d}y \\ &= \int f(x)\,\mathrm{d}x + c \end{aligned} \tag{2.4}$$

The unknown c in the solution to the differential equation $y' = f(x)$ given in Eq.(2.4) can be determined if and only if one of the initial conditions $y(x_0)$ is given at $x = x_0$, which is discussed in Section 2.13.

EXAMPLE 2.8 Solve the first order nonhomogeneous differential equation $y' = x$.

Solution: Clearly, $f(x) = x$. According to Eq.(2.4), the corresponding solution is given by

$$\begin{aligned} y(x) &= \int f(x)\,\mathrm{d}x + c \\ &= \int x\,\mathrm{d}x + c \\ &= \frac{x^2}{2} + c \end{aligned}$$

EXAMPLE 2.9 Solve the first order nonhomogeneous differential equation $y' = x^2$.

Solution: Clearly, $f(x) = x^2$. According to Eq.(2.4), the corresponding solution is given by

$$\begin{aligned} y(x) &= \int f(x)\, \mathrm{d}x + c \\ &= \int x^2\, \mathrm{d}x + c \\ &= \frac{x^3}{3} + c \end{aligned}$$

EXAMPLE 2.10 Solve the first order nonhomogeneous differential equation $y' = \sin x$.

Solution: Clearly, $f(x) = \sin x$. According to Eq.(2.4), the corresponding solution is given by

$$\begin{aligned} y(x) &= \int f(x)\, \mathrm{d}x + c \\ &= \int \sin x\, \mathrm{d}x + c \\ &= -\cos x + c \end{aligned}$$

EXAMPLE 2.11 Solve the first order nonhomogeneous differential equation $y' = \cos x$.

Solution: Clearly, $f(x) = \cos x$. According to Eq.(2.4), the corresponding solution is given by

$$\begin{aligned} y(x) &= \int f(x)\, \mathrm{d}x + c \\ &= \int \cos x\, \mathrm{d}x + c \\ &= \sin x + c \end{aligned}$$

EXAMPLE 2.12 Solve the first order nonhomogeneous differential equation $y' = \sec^2 x$.

Solution: Clearly, $f(x) = \sec^2 x$. According to Eq.(2.4), the corresponding solution is given by

$$\begin{aligned} y(x) &= \int f(x)\, \mathrm{d}x + c \\ &= \int \sec^2 x\, \mathrm{d}x + c \\ &= \tan x + c \end{aligned}$$

EXAMPLE 2.13 Solve the first order nonhomogeneous differential equation $y' = \text{cosec}^2 x$.

Solution: Clearly, $f(x) = \text{cosec}^2 x$. According to Eq.(2.4), the corresponding solution is given by

$$\begin{aligned} y(x) &= \int f(x)\,\mathrm{d}x + c \\ &= \int \text{cosec}^2 x\,\mathrm{d}x + c \\ &= -\cot x + c \end{aligned}$$

Let $y' + p(x)y = f(x)$ be any arbitrary first order nonhomogeneous ordinary differential equation. If one can transfer the given equation into a nonhomogeneous differential equation of the form $u' = g(x)$, then the solution can be represented by the form given in Eq.(2.3). This can be done if and only if one multiplies the given equation $y' + p(x)y = f(x)$ by an unknown function $\mu(x)$, called *integrating factor* which satisfy the differential equation

$$\mu'(x) = \mu(x)p(x) \tag{2.5}$$

It is easy to verify that

$$\begin{aligned} f(x)\mu(x) &= \mu(x)\left(y' + p(x)y\right) \\ &= \mu(x)y' + \mu(x)p(x)y \\ &= \mu(x)y' + \mu'(x)y \\ &= (\mu(x)y(x))' \end{aligned} \tag{2.6}$$

Clearly, the unknown function $\mu(x)$ satisfies a first order homogeneous differential equation of the form given in Eq.(2.3). Again, the solution of Eq.(2.5) using Eq.(2.4) is given by

$$\mu(x) = e^{\int p(x)\,\mathrm{d}x} \tag{2.7}$$

The solution of the transferred nonhomogeneous equation $(\mu(x)y)' = \mu(x)f(x)$ according to Eq.(2.4) is given by

$$\begin{aligned} y(x) &= (\mu(x))^{-1}\left(\mu(x)y(x)\right) \\ &= (\mu(x))^{-1}\left(\int \mu(x)f(x)\,\mathrm{d}x\right) \\ &= e^{-\int p(x)\,\mathrm{d}x}\left(\int e^{\int p(x)\,\mathrm{d}x} f(x)\,\mathrm{d}x + c\right) \end{aligned} \tag{2.8}$$

The solution given in Eq.(2.8) is the most general solution for the first order linear differential equation. In particular, the solution given in Eq.(2.8) reduces to the solution given in Eq.(2.2) for $f(x) = 0$, and reduces to the solution given in Eq.(2.4) for $p(x) = 0$.

The unknown c in the solution to the differential equation $y' + p(x)y = f(x)$, given in Eq.(2.8) can be determined if and only if one of the initial condition $y(x_0)$ is given at $x = x_0$, which is discussed in Section 2.13. Again, the solution given in Eq.(2.8) can be represented without any arbitrary constant c by

$$y(x) = y(x_0)\, e^{-\int_{x_0}^{x} p(u)\, du} + e^{-\int p(x)\, dx} \int_{x_0}^{x} f(u)\, e^{\int p(u)\, du}\, du$$

EXAMPLE 2.14 Solve the first order nonhomogeneous equation $y' + x^2 y = x^2$.

Solution: Clearly, $p(x) = x^2$ and $f(x) = x^2$. According to Eq.(2.7), the corresponding integrating factor is given by

$$\begin{aligned} \mu(x) &= e^{\int x^2\, dx} \\ &= e^{\frac{x^3}{3}} \end{aligned}$$

Again, the solution of the given differential equation according to Eq.(2.8) is given by

$$\begin{aligned} y(x) &= e^{-\int x^2\, dx} \left(\int e^{\int x^2\, dx} x^2\, dx + c \right) \\ &= e^{-\frac{x^3}{3}} \left(\int x^2 e^{\frac{x^3}{3}}\, dx + c \right) \\ &= e^{-\frac{x^3}{3}} \left(e^{\frac{x^3}{3}} + c \right) \\ &= 1 + c\, e^{-\frac{x^3}{3}} \end{aligned}$$

EXAMPLE 2.15 Solve the first order nonhomogeneous equation $\cos^2(x) y' + y = 1$.

Solution: Clearly, $p(x) = \sec^2 x$ and $f(x) = \sec^2 x$. According to Eq.(2.7), the corresponding integrating factor is given by

$$\begin{aligned} \mu(x) &= e^{\int \sec^2 x\, dx} \\ &= e^{\tan x} \end{aligned}$$

Again, the solution of the given differential equation according to Eq.(2.8) is given by

$$\begin{aligned} y(x) &= e^{-\int \sec^2 x\, dx} \left(\int e^{\int \sec^2 x\, dx} \sec^2 x\, dx + c \right) \\ &= e^{-\tan x} \left(\int \sec^2 x\, e^{\tan x}\, dx + c \right) \\ &= e^{-\tan x} \left(e^{\tan x} + c \right) \\ &= 1 + c\, e^{-\tan x} \end{aligned}$$

EXAMPLE 2.16 Solve the first order nonhomogeneous equation $y' + 2xy = 2x^2$.

Solution: Clearly, $p(x) = 2x$ and $f(x) = 2x^2$. According to Eq.(2.7), the corresponding integrating factor is given by

$$\begin{aligned}\mu(x) &= e^{\int 2x\,\mathrm{d}x}\\ &= e^{x^2}\end{aligned}$$

Again, the solution of the given differential equation according to Eq.(2.8) is given by

$$\begin{aligned}y(x) &= e^{-\int 2x\,\mathrm{d}x}\left(\int e^{\int 2x\,\mathrm{d}x}2x^2\,\mathrm{d}x + c\right)\\ &= e^{-x^2}\left(\int 2x^2 e^{x^2}\,\mathrm{d}x + c\right)\\ &= e^{-x^2}\left(xe^{x^2} - \int e^{x^2}\,\mathrm{d}x + c\right)\\ &= x + c\,e^{-x^2} - e^{-x^2}\int e^{x^2}\,\mathrm{d}x\end{aligned}$$

EXAMPLE 2.17 Solve the first order nonhomogeneous equation $y' + y = xe^{-x}$.

Solution: Clearly, $p(x) = 1$ and $f(x) = xe^{-x}$. According to Eq.(2.7), the corresponding integrating factor is given by

$$\begin{aligned}\mu(x) &= e^{\int \mathrm{d}x}\\ &= e^{x}\end{aligned}$$

Again, the solution of the given differential equation according to Eq.(2.8) is given by

$$\begin{aligned}y(x) &= e^{-\int \mathrm{d}x}\left(\int e^{\int \mathrm{d}x}xe^{-x}\,\mathrm{d}x + c\right)\\ &= e^{-x}\left(\int x\,\mathrm{d}x + c\right)\\ &= e^{-x}\left(\frac{x^2}{2} + c\right)\\ &= \left(\frac{x^2}{2}\right)e^{-x} + c\,e^{-x}\end{aligned}$$

EXAMPLE 2.18 Solve the first order nonhomogeneous equation $y' - \frac{y}{x} = \frac{\ln x}{x}$.

Solution: Clearly, $p(x) = -\frac{1}{x}$ and $f(x) = \frac{\ln x}{x}$. According to Eq.(2.7), the corresponding integrating factor is given by

$$\begin{aligned}\mu(x) &= e^{-\int \frac{\mathrm{d}x}{x}}\\ &= \frac{1}{x}\end{aligned}$$

Again, the solution of the given differential equation according to Eq.(2.8) is given by

$$\begin{aligned}
y(x) &= e^{\int \frac{dx}{x}} \left(\int e^{-\int \frac{dx}{x}} \left(\frac{\ln x}{x} \right) dx + c \right) \\
&= x \left(\int \left(\frac{\ln x}{x^2} \right) dx + c \right) \\
&= x \left(-\frac{\ln x}{x} - \frac{1}{x} + c \right) \\
&= cx - \ln x - 1
\end{aligned}$$

EXAMPLE 2.19 Solve the first order nonhomogeneous equation $xy' - 2y = xe^x$.
Solution: Clearly, $p(x) = -\left(\frac{2}{x}\right)$ and $f(x) = e^x$. According to Eq.(2.7), the corresponding integrating factor is given by

$$\begin{aligned}
\mu(x) &= e^{\int -\left(\frac{2}{x}\right) dx} \\
&= e^{-2 \ln x} \\
&= e^{\ln(x^{-2})} \\
&= x^{-2}
\end{aligned}$$

Again, the solution of the given differential equation according to Eq.(2.8) is given by

$$\begin{aligned}
y(x) &= e^{-\int -\left(\frac{2}{x}\right) dx} \left(\int e^{\int -\left(\frac{2}{x}\right) dx} e^x \, dx + c \right) \\
&= x^2 \left(\int \left(\frac{e^x}{x^2} \right) dx + c \right) \\
&= x^2 \int \left(\frac{e^x}{x^2} \right) dx + cx^2
\end{aligned}$$

EXAMPLE 2.20 Solve the first order nonhomogeneous equation $y' = \frac{1}{e^y - 2x}$.
Solution: We can change the given differential equation in the form $x' + 2x = e^y$. Clearly, $p(y) = 2$ and $f(y) = e^y$. According to Eq.(2.7), the corresponding integrating factor is given by

$$\begin{aligned}
\mu(y) &= e^{\int 2 \, dy} \\
&= e^{2y}
\end{aligned}$$

Again, the solution of the given differential equation according to Eq.(2.8) is given by

$$\begin{aligned}
x(y) &= e^{-\int 2 \, dy} \left(\int e^{\int 2 \, dy} e^y \, dy + c \right) \\
&= e^{-2y} \left(\int e^{3y} \, dy + c \right) \\
&= e^{-2y} \left(\frac{e^{3y}}{3} + c \right) \\
&= \frac{e^y}{3} + ce^{-2y}
\end{aligned}$$

EXAMPLE 2.21 Solve the first order nonhomogeneous equation $x(x-1)y' + (2-x)y = 2x - 1$.

Solution: Clearly, $p(x) = \left(\frac{2-x}{x(x-1)}\right)$ and $f(x) = \left(\frac{2x-1}{x(x-1)}\right)$. According to Eq.(2.7), the corresponding integrating factor is given by

$$\begin{aligned}
\mu(x) &= e^{\int\left[\frac{2-x}{x(x-1)}\right]\mathrm{d}x} \\
&= e^{\int\left(\frac{1}{x-1}-\frac{2}{x}\right)\mathrm{d}x} \\
&= e^{\ln(x-1)-2\ln x} \\
&= e^{\ln\left(\frac{x-1}{x^2}\right)} \\
&= \frac{x-1}{x^2}
\end{aligned}$$

Again, the solution of the given differential equation according to Eq.(2.8) is given by

$$\begin{aligned}
y(x) &= e^{-\int\left(\frac{2-x}{x(x-1)}\right)\mathrm{d}x}\left(\int e^{\int\left(\frac{2-x}{x(x-1)}\right)\mathrm{d}x}\left(\frac{2x-1}{x(x-1)}\right)\mathrm{d}x + c\right) \\
&= \frac{x^2}{x-1}\left(\int\left(\frac{2x-1}{x^3}\right)\mathrm{d}x + c\right) \\
&= \frac{x^2}{x-1}\int\left(\frac{2}{x^2}-\frac{1}{x^3}\right)\mathrm{d}x + c\left(\frac{x^2}{x-1}\right) \\
&= \frac{x^2}{x-1}\left(-\frac{2}{x}+\frac{1}{2x^2}\right) + c\left(\frac{x^2}{x-1}\right) \\
&= \frac{1-4x}{2(x-1)} + c\left(\frac{x^2}{x-1}\right)
\end{aligned}$$

EXAMPLE 2.22 Solve the first order non-homogeneous equation $x(x-1)y' + (2-x)y = x^2(2x-1)$.

Solution: Clearly, $p(x) = \left(\frac{2-x}{x(x-1)}\right)$ and $f(x) = \left(\frac{x(2x-1)}{x-1}\right)$. According to Eq.(2.7), the corresponding integrating factor is given by

$$\begin{aligned}
\mu(x) &= e^{\int\left(\frac{2-x}{x(x-1)}\right)\mathrm{d}x} \\
&= e^{\int\left(\frac{1}{x-1}-\frac{2}{x}\right)\mathrm{d}x} \\
&= e^{\ln(x-1)-2\ln x} \\
&= e^{\ln\left(\frac{x-1}{x^2}\right)} \\
&= \frac{x-1}{x^2}
\end{aligned}$$

Again, the solution of the given differential equation according to Eq.(2.8) is given by

$$\begin{aligned}
y(x) &= e^{-\int\left(\frac{2-x}{x(x-1)}\right)\mathrm{d}x}\left(\int e^{\int\left(\frac{2-x}{x(x-1)}\right)\mathrm{d}x}\left(\frac{x(2x-1)}{x-1}\right)\mathrm{d}x + c\right)\\
&= \frac{x^2}{x-1}\left(\int\left(\frac{2x-1}{x}\right)\mathrm{d}x + c\right)\\
&= \frac{x^2}{x-1}\int\left(2-\frac{1}{x}\right)\mathrm{d}x + c\left(\frac{x^2}{x-1}\right)\\
&= \frac{x^2}{x-1}\left(2x-\ln x\right) + c\left(\frac{x^2}{x-1}\right)\\
&= \frac{2x^3}{x-1} - \ln x\left(\frac{x^2}{x-1}\right) + c\left(\frac{x^2}{x-1}\right)\\
&= \frac{2x^3}{x-1} + \frac{x^2}{x-1}(c-\ln x)
\end{aligned}$$

EXAMPLE 2.23 Solve the differential equation $y' + \sin(x)\,y = x\,e^{\cos x}$.

Solution: Clearly, $p(x) = \sin x$ and $f(x) = x\,e^{\cos x}$. Again, the solution of the given differential equation according to Eq.(2.8) is given by

$$\begin{aligned}
y(x) &= e^{-\int \sin x\,\mathrm{d}x}\left(\int x\,e^{\cos x}\,e^{\int \sin x\,\mathrm{d}x}\mathrm{d}x + c\right)\\
&= e^{\cos x}\left(\int x\,e^{\cos x}e^{-\cos x}\,\mathrm{d}x + c\right)\\
&= e^{\cos x}\left(\int x\,\mathrm{d}x + c\right)\\
&= e^{\cos x}\left(\frac{x^2}{2} + c\right)\\
&= \left(\frac{x^2}{2}\right)e^{\cos x} + c\,e^{\cos x}
\end{aligned}$$

EXAMPLE 2.24 Solve the differential equation $y' + \frac{y\ln y}{x-\ln y} = 0$.

Solution: In the given differential equation, y is a dependent variable and x is an independent variable. We can change the given differential equation into $x' + \frac{x-\ln y}{y\ln y} = 0$ in which x is a dependent variable and y is an independent variable. Clearly, we have $p(y) = \frac{1}{y\ln y}$ and $f(y) = \frac{1}{y}$. According to Eq.(2.7), the corresponding integrating factor is given by

$$\begin{aligned}
\mu(y) &= e^{\int\left(\frac{1}{y\ln y}\right)\mathrm{d}y}\\
&= e^{\ln(\ln y)}\\
&= \ln y
\end{aligned}$$

Again, the solution of the given differential equation according to Eq.(2.8) is given by

$$\begin{aligned}
x(y) &= e^{-\int\left(\frac{1}{y\ln y}\right)\mathrm{d}y}\left(\int e^{\int\left(\frac{1}{y\ln y}\right)\mathrm{d}y}\left(\frac{1}{y}\right)\mathrm{d}y + c\right)\\
&= \frac{1}{\ln y}\left(\int\left(\frac{\ln y}{y}\right)\mathrm{d}y + c\right)\\
&= \frac{1}{\ln y}\left(\frac{\ln^2 y}{2} + c\right)
\end{aligned}$$

EXAMPLE 2.25 Solve the equation $y' + \tan(x)\, y = \sin 2x$.

Solution: Clearly, $p(x) = \tan x$ and $f(x) = \sin 2x = 2\sin x\cos x$. Again, the solution of the given differential equation according to Eq.(2.8) is given by

$$\begin{aligned}
y(x) &= e^{-\int \tan x\,\mathrm{d}x}\left(\int 2\sin x\cos x\, e^{\int \tan x\,\mathrm{d}x}\,\mathrm{d}x + c\right)\\
&= \cos x\left(\int 2\sin x\,\mathrm{d}x + c\right)\\
&= \cos x\,(-2\cos x + c)\\
&= -2\cos^2 x + c\cos x
\end{aligned}$$

EXAMPLE 2.26 Solve the equation $y' + \frac{y}{1+x^2} = \frac{e^{\tan^{-1}(x)}}{1+x^2}$.

Solution: Clearly, $p(x) = \frac{1}{1+x^2}$ and $f(x) = \frac{e^{\tan^{-1}(x)}}{1+x^2}$. Again, the solution of the given differential equation according to Eq.(2.8) is given by

$$\begin{aligned}
y(x) &= e^{-\int\frac{\mathrm{d}x}{1+x^2}}\left(\int \frac{e^{\tan^{-1}(x)}}{1+x^2}\, e^{\int\frac{\mathrm{d}x}{1+x^2}}\,\mathrm{d}x + c\right)\\
&= e^{-\tan^{-1}(x)}\left(\int \frac{e^{2\tan^{-1}(x)}}{1+x^2}\,\mathrm{d}x + c\right)\\
&= e^{-\tan^{-1}(x)}\left(\frac{e^{2\tan^{-1}(x)}}{2} + c\right)\\
&= \frac{e^{\tan^{-1}(x)}}{2} + c\,e^{-\tan^{-1}(x)}
\end{aligned}$$

Exercises 2.2

1. Solve the differential equation $y' = f(x)$, where

   (a) $f(x) = (a^2 - x^2)^{\frac{1}{2}}$

HINT: It is easy to verify that

$$\begin{aligned}
y(x) &= \int f(x)\,\mathrm{d}x \\
&= \int \left(a^2 - x^2\right)^{\frac{1}{2}} \mathrm{d}x \\
&= x\left(a^2 - x^2\right)^{\frac{1}{2}} + \int \left(\frac{x^2}{\left(a^2 - x^2\right)^{\frac{1}{2}}}\right) \mathrm{d}x \\
&= x\left(a^2 - x^2\right)^{\frac{1}{2}} - \int \left(a^2 - x^2\right)^{\frac{1}{2}} \mathrm{d}x + a^2 \int \frac{\mathrm{d}x}{\left(a^2 - x^2\right)^{\frac{1}{2}}} \\
&= x\left(a^2 - x^2\right)^{\frac{1}{2}} - \int \left(a^2 - x^2\right)^{\frac{1}{2}} \mathrm{d}x + a^2 \sin^{-1}\left(\frac{x}{a}\right) \\
&= x\left(a^2 - x^2\right)^{\frac{1}{2}} - y(x) + a^2 \sin^{-1}\left(\frac{x}{a}\right)
\end{aligned}$$

In other words, one can have

$$y(x) = \left(\frac{x}{2}\right)\left(a^2 - x^2\right)^{\frac{1}{2}} + \left(\frac{a^2}{2}\right) \sin^{-1}\left(\frac{x}{a}\right) + c$$

(b) $f(x) = \left(a^2 + x^2\right)^{\frac{1}{2}}$

HINT: It is easy to verify that

$$\begin{aligned}
y(x) &= \int f(x)\,\mathrm{d}x \\
&= \int \left(a^2 + x^2\right)^{\frac{1}{2}} \mathrm{d}x \\
&= x\left(a^2 + x^2\right)^{\frac{1}{2}} - \int \left(\frac{x^2}{\left(a^2 + x^2\right)^{\frac{1}{2}}}\right) \mathrm{d}x \\
&= x\left(a^2 + x^2\right)^{\frac{1}{2}} - \int \left(a^2 + x^2\right)^{\frac{1}{2}} \mathrm{d}x + a^2 \int \frac{\mathrm{d}x}{\left(a^2 + x^2\right)^{\frac{1}{2}}} \\
&= x\left(a^2 + x^2\right)^{\frac{1}{2}} - \int \left(a^2 + x^2\right)^{\frac{1}{2}} \mathrm{d}x + a^2 \sinh^{-1}\left(\frac{x}{a}\right) \\
&= x\left(a^2 + x^2\right)^{\frac{1}{2}} - y(x) + a^2 \sinh^{-1}\left(\frac{x}{a}\right)
\end{aligned}$$

In other words, one can have

$$y(x) = \left(\frac{x}{2}\right)\left(a^2 + x^2\right)^{\frac{1}{2}} + \left(\frac{a^2}{2}\right) \sinh^{-1}\left(\frac{x}{a}\right) + c$$

(c) $f(x) = (x^2 - a^2)^{\frac{1}{2}}$

HINT: It is easy to verify that

$$\begin{aligned}
y(x) &= \int f(x)\,\mathrm{d}x \\
&= \int \left(x^2 - a^2\right)^{\frac{1}{2}} \mathrm{d}x \\
&= x\left(x^2 - a^2\right)^{\frac{1}{2}} - \int \left(\frac{x^2}{(x^2 - a^2)^{\frac{1}{2}}}\right) \mathrm{d}x \\
&= x\left(x^2 - a^2\right)^{\frac{1}{2}} - \int \left(x^2 - a^2\right)^{\frac{1}{2}} \mathrm{d}x + a^2 \int \frac{\mathrm{d}x}{(x^2 - a^2)^{\frac{1}{2}}} \\
&= x\left(x^2 - a^2\right)^{\frac{1}{2}} - \int \left(x^2 - a^2\right)^{\frac{1}{2}} \mathrm{d}x + a^2 \cosh^{-1}\left(\frac{x}{a}\right) \\
&= x\left(x^2 - a^2\right)^{\frac{1}{2}} - y(x) + a^2 \cosh^{-1}\left(\frac{x}{a}\right)
\end{aligned}$$

In other words, one can have

$$y(x) = \left(\frac{x}{2}\right)\left(x^2 - a^2\right)^{\frac{1}{2}} + \left(\frac{a^2}{2}\right)\cosh^{-1}\left(\frac{x}{a}\right) + c$$

(d) $f(x) = \dfrac{1}{(a^2 - x^2)^{\frac{1}{2}}}$

HINT: It is easy to verify that

$$\begin{aligned}
y(x) &= \int f(x)\,\mathrm{d}x \\
&= \int \frac{\mathrm{d}x}{(a^2 - x^2)^{\frac{1}{2}}} \\
&= \sin^{-1}\left(\frac{x}{a}\right) + c
\end{aligned}$$

(e) $f(x) = \dfrac{1}{(x^2 - a^2)^{\frac{1}{2}}}$

HINT: It is easy to verify that

$$\begin{aligned}
y(x) &= \int f(x)\,\mathrm{d}x \\
&= \int \frac{\mathrm{d}x}{(x^2 - a^2)^{\frac{1}{2}}} \\
&= \cosh^{-1}\left(\frac{x}{a}\right) + c
\end{aligned}$$

(f) $f(x) = \dfrac{1}{(a^2 + x^2)^{\frac{1}{2}}}$

HINT: Let $x = a \tan\theta$. It is easy to verify that

$$\begin{aligned} y(x) &= \int f(x)\,\mathrm{d}x \\ &= \int \frac{\mathrm{d}x}{(a^2 + x^2)^{\frac{1}{2}}} \\ &= \int \sec\theta\,\mathrm{d}\theta \\ &= \ln(\sec\theta + \tan\theta) + k \\ &= \ln\left(x + (a^2 + x^2)^{\frac{1}{2}}\right) + c \end{aligned}$$

HINT: It is easy to verify that

$$\begin{aligned} y(x) &= \int f(x)\,\mathrm{d}x \\ &= \int \frac{\mathrm{d}x}{(a^2 + x^2)^{\frac{1}{2}}} \\ &= \sinh^{-1}\left(\frac{x}{a}\right) + c \end{aligned}$$

(g) $f(x) = \dfrac{1}{a^2 - x^2}$

HINT: It is easy to verify that

$$\begin{aligned} y(x) &= \int f(x)\,\mathrm{d}x \\ &= \int \frac{\mathrm{d}x}{a^2 - x^2} \\ &= \left(\frac{1}{2a}\right)\int\left(\frac{1}{a-x} + \frac{1}{a+x}\right)\mathrm{d}x \\ &= \left(\frac{1}{2a}\right)(\ln(x+a) - \ln(a-x) + k) \\ &= \left(\frac{1}{2a}\right)\ln\left(\frac{a+x}{a-x}\right) + c \\ &= \ln\left(\frac{a+x}{a-x}\right)^{\frac{1}{2a}} + c \end{aligned}$$

(h) $f(x) = \dfrac{1}{x^2 - a^2}$

HINT: It is easy to verify that

$$y(x) = \int f(x)\,\mathrm{d}x$$

$$
\begin{aligned}
&= \int \frac{\mathrm{d}x}{x^2 - a^2} \\
&= \left(\frac{1}{2a}\right) \int \left(\frac{1}{x-a} - \frac{1}{x+a}\right) \mathrm{d}x \\
&= \left(\frac{1}{2a}\right) (\ln(x-a) - \ln(x+a) + k) \\
&= \left(\frac{1}{2a}\right) \ln\left(\frac{x-a}{x+a}\right) + c \\
&= \ln\left(\frac{x-a}{x+a}\right)^{\frac{1}{2a}} + c
\end{aligned}
$$

(i) $f(x) = \dfrac{1}{x^2 + a^2}$

HINT: It is easy to verify that

$$
\begin{aligned}
y(x) &= \int f(x)\,\mathrm{d}x \\
&= \int \frac{\mathrm{d}x}{x^2 + a^2} \\
&= \tan^{-1}\left(\frac{x}{a}\right) + c
\end{aligned}
$$

(j) $f(x) = \dfrac{x\,e^x}{(1+x)^2}$

HINT: It is easy to verify that

$$
\begin{aligned}
y(x) &= \int f(x)\,\mathrm{d}x \\
&= \int \frac{x\,e^x}{(1+x)^2}\,\mathrm{d}x \\
&= \int \left(\frac{e^x}{1+x}\right) \mathrm{d}x - \int \left(\frac{e^x}{(1+x)^2}\right) \mathrm{d}x \\
&= \frac{e^x}{1+x} + \int \left(\frac{e^x}{(1+x)^2}\right) \mathrm{d}x - \int \left(\frac{e^x}{(1+x)^2}\right) \mathrm{d}x \\
&= \frac{e^x}{1+x} + c
\end{aligned}
$$

(k) $f(x) = \dfrac{e^x[1 + x\ln(x)]}{x}$

HINT: It is easy to verify that

$$
\begin{aligned}
y(x) &= \int f(x)\,\mathrm{d}x \\
&= \int \left(\frac{e^x(1 + x\ln x)}{x}\right) \mathrm{d}x
\end{aligned}
$$

$$
\begin{aligned}
&= \int \left(\frac{e^x}{x}\right) \mathrm{d}x + \int e^x \ln x \, \mathrm{d}x \\
&= e^x \ln x - \int e^x \ln x \, \mathrm{d}x + \int e^x \ln x \, \mathrm{d}x \\
&= e^x \ln x + c
\end{aligned}
$$

(l) $f(x) = \dfrac{1}{\ln x} - \dfrac{1}{\ln^2 x}$

HINT: It is easy to verify that

$$
\begin{aligned}
y(x) &= \int f(x) \, \mathrm{d}x \\
&= \int \left(\frac{1}{\ln x} - \frac{1}{\ln^2 x}\right) \mathrm{d}x \\
&= \int \frac{\mathrm{d}x}{\ln x} - \int \frac{\mathrm{d}x}{\ln^2 x} \\
&= \frac{x}{\ln x} + \int \frac{\mathrm{d}x}{\ln^2 x} - \int \frac{\mathrm{d}x}{\ln^2 x} \\
&= \frac{x}{\ln x} + c
\end{aligned}
$$

2. Solve the differential equation $y' = f(x)$, where

(a) $f(x) = \sin(\ln x)$
HINT: It is easy to verify that

$$
\begin{aligned}
y(x) &= \int f(x) \, \mathrm{d}x \\
&= \int \sin(\ln x) \, \mathrm{d}x \\
&= x \sin(\ln x) - \int \cos(\ln x) \, \mathrm{d}x \\
&= x \sin(\ln x) - x \cos(\ln x) - \int \sin(\ln x) \, \mathrm{d}x
\end{aligned}
$$

In other words, one can have

$$
y(x) = \frac{x\,(\sin(\ln x) - \cos(\ln x))}{2} + c
$$

(b) If $f(x) = \dfrac{1}{\sin x + \cos x}$

HINT: It is easy to verify that

$$
\begin{aligned}
y(x) &= \int f(x) \, \mathrm{d}x \\
&= \int \frac{\mathrm{d}x}{\sin x + \cos x}
\end{aligned}
$$

$$
\begin{aligned}
&= \frac{1}{\sqrt{2}} \int \frac{\mathrm{d}x}{\frac{\sin x}{\sqrt{2}} + \frac{\cos x}{\sqrt{2}}} \\
&= \frac{1}{\sqrt{2}} \int \frac{\mathrm{d}x}{\sin\left(\frac{\pi}{4} + x\right)} \\
&= \frac{1}{\sqrt{2}} \int \operatorname{cosec}\left(\frac{\pi}{4} + x\right) \mathrm{d}x \\
&= \mp \left(\frac{1}{\sqrt{2}}\right) \ln\left(\operatorname{cosec}\left(\frac{\pi}{4} + x\right) \pm \cot\left(\frac{\pi}{4} + x\right)\right)
\end{aligned}
$$

(c) If $f(x) = \dfrac{1}{\sin x + \cos x}$

HINT: It is easy to verify that

$$
\begin{aligned}
y(x) &= \int f(x)\,\mathrm{d}x \\
&= \int \frac{\mathrm{d}x}{\sin x + \cos x} \\
&= \frac{1}{\sqrt{2}} \int \frac{\mathrm{d}x}{\frac{\sin x}{\sqrt{2}} + \frac{\cos x}{\sqrt{2}}} \\
&= \frac{1}{\sqrt{2}} \int \frac{\mathrm{d}x}{\cos\left(\frac{\pi}{4} - x\right)} \\
&= \frac{1}{\sqrt{2}} \int \sec\left(\frac{\pi}{4} - x\right) \mathrm{d}x \\
&= -\left(\frac{1}{\sqrt{2}}\right) \ln\left(\sec\left(\frac{\pi}{4} - x\right) + \tan\left(\frac{\pi}{4} - x\right)\right)
\end{aligned}
$$

(d) $f(x) = \sec x$

HINT: It is easy to verify that

$$
\begin{aligned}
y(x) &= \int f(x)\,\mathrm{d}x \\
&= \int \sec x\,\mathrm{d}x \\
&= \ln(\sec x + \tan x) + c
\end{aligned}
$$

(e) $f(x) = \operatorname{cosec} x$

HINT: It is easy to verify that

$$\begin{aligned} y(x) &= \int f(x)\,\mathrm{d}x \\ &= \int \operatorname{cosec} x\,\mathrm{d}x \\ &= \mp \ln(\operatorname{cosec} x \pm \cot x) + c \end{aligned}$$

(f) $f(x) = \sec^3 x$

HINT: It is easy to verify that

$$\begin{aligned} y(x) &= \int f(x)\,\mathrm{d}x \\ &= \int \sec^3 x\,\mathrm{d}x \\ &= \frac{\sec x \tan x + \ln(\sec x + \tan x)}{2} + c \end{aligned}$$

(g) $f(x) = \operatorname{cosec}^3 x$

HINT: It is easy to verify that

$$\begin{aligned} y(x) &= \int f(x)\,\mathrm{d}x \\ &= \int \operatorname{cosec}^3 x\,\mathrm{d}x \\ &= \frac{\mp \ln(\operatorname{cosec} x \pm \cot x) - \operatorname{cosec} x \cot x}{2} + c \end{aligned}$$

(h) $f(x) = \tan^m(x)\sec^{2n}(x)$

HINT: It is easy to verify that

$$\begin{aligned} y(x) &= \int f(x)\,\mathrm{d}x \\ &= \int \tan^m(x)\sec^{2n}(x)\,\mathrm{d}x \\ &= \int \tan^m(x)\sec^{2(n-1)}(x)\sec^2 x\,\mathrm{d}x \\ &= \int \tan^m(x)\,\left(1+\tan^2(x)\right)^{n-1}\sec^2 x\,\mathrm{d}x \end{aligned}$$

$$
\begin{aligned}
&= \int \tan^m(x)\, \left(1+\tan^2(x)\right)^{n-1}\, \mathrm{d}(\tan x) \\
&= \int \left(\sum_{i=0}^{n-1} \binom{n-1}{i} \tan^{m+2i}(x) \right) \mathrm{d}(\tan x) \\
&= \sum_{i=0}^{n-1} \binom{n-1}{i} \int \tan^{m+2i}(x) \mathrm{d}(\tan x) \\
&= \sum_{i=0}^{n-1} \binom{n-1}{i} \left(\frac{\tan^{m+2i+1}(x)}{m+2i+1} \right) + c
\end{aligned}
$$

(i) $f(x) = \tan^{2m+1}(x)\sec^n(x)$

HINT: It is easy to verify that

$$
\begin{aligned}
y(x) &= \int f(x)\,\mathrm{d}x \\
&= \int \tan^{2m+1}(x)\sec^n(x)\,\mathrm{d}x \\
&= \int \tan^{2m}(x)\sec^{n-1}(x)\sec x\tan x\,\mathrm{d}x \\
&= \int \sec^{n-1}(x)\, \left(\sec^2(x)-1\right)^m \sec x \tan x\,\mathrm{d}x \\
&= \int \sec^{n-1}(x)\, \left(\sec^2(x)-1\right)^m \mathrm{d}(\sec x) \\
&= \int \left(\sum_{i=0}^{m} (-1)^{m-i} \binom{m}{i} \sec^{n-1+2i}(x) \right) \mathrm{d}(\sec x) \\
&= \sum_{i=0}^{m} \binom{m}{i} (-1)^{m-i} \int \sec^{n-1+2i}(x) \mathrm{d}(\sec x) \\
&= \sum_{i=0}^{m} (-1)^{m-i} \binom{m}{i} \left(\frac{\sec^{n-1+2i+1}(x)}{n-1+2i+1} \right) + c
\end{aligned}
$$

(j) $f(x) = x^n \ln(x)$ where $n \geq 0$.

HINT: It is easy to verify that

$$
\begin{aligned}
y(x) &= \int f(x)\,\mathrm{d}x \\
&= \int x^n \ln(x)\,\mathrm{d}x \\
&= \int \ln(x)\,\mathrm{d}\left(\frac{x^{n+1}}{n+1}\right) \\
&= \left(\frac{x^{n+1}}{n+1}\right)\ln(x) - \frac{1}{n+1}\int x^n\,\mathrm{d}x \\
&= \left(\frac{x^{n+1}}{n+1}\right)\ln(x) - \frac{x^{n+1}}{(n+1)^2} + c
\end{aligned}
$$

(k) $f(x) = \ln^n(x)$ where $n \geq 1$.

HINT: It is easy to verify that

$$\begin{aligned} y(x) &= \int f(x)\,\mathrm{d}x \\ &= \int \ln^n(x)\,\mathrm{d}x \\ &= \int \ln^n(x)\,\mathrm{d}(x) \\ &= x\ln^n(x) - n\int \ln^{n-1}(x)\,\mathrm{d}x \\ &\vdots \\ &= x\left(\sum_{i=0}^{n}(-1)^i\left(\prod_{j=0}^{i-1}(n-j)\right)\ln^{n-i}(x)\right) + c \end{aligned}$$

(l) $f(x) = x^m \ln^n(x)$ where $n \geq 1$ and $m \geq 0$.

HINT: It is easy to verify that

$$\begin{aligned} y(x) &= \int f(x)\,\mathrm{d}x \\ &= \int x^m \ln^n(x)\,\mathrm{d}x \\ &= \int \ln^n(x)\,\mathrm{d}\left(\frac{x^{m+1}}{m+1}\right) \\ &= \left(\frac{x^{m+1}}{m+1}\right)\ln^n(x) - \frac{n}{m+1}\int x^m \ln^{n-1}(x)\,\mathrm{d}x \\ &\vdots \\ &= \frac{x^{m+1}}{m+1}\left(\sum_{i=0}^{n}(-1)^i\left(\frac{\prod_{j=0}^{i-1}(n-j)}{(m+1)^i}\right)\ln^{n-i}(x)\right) + c \end{aligned}$$

(m) $f(x) = x\sin x$

HINT: It is easy to verify that

$$\begin{aligned} y(x) &= \int f(x)\,\mathrm{d}x \\ &= \int x\sin x\,\mathrm{d}x \\ &= -x\cos x + \sin x + c \end{aligned}$$

(n) $f(x) = x\cos x$
HINT: It is easy to verify that

$$\begin{aligned} y(x) &= \int f(x)\,\mathrm{d}x \\ &= \int x\cos x\,\mathrm{d}x \\ &= x\sin x + \cos x + c \end{aligned}$$

(o) $f(x) = e^x \cos x$
HINT: It is easy to verify that

$$\begin{aligned} y(x) &= \int f(x)\,\mathrm{d}x \\ &= \int e^x \cos x\,\mathrm{d}x \\ &= e^x \sin x + e^x \cos x - \int e^x \cos x\,\mathrm{d}x \\ &= e^x \sin x + e^x \cos x - y(x) \end{aligned}$$

In other words, one can have

$$y(x) = \left(\frac{1}{2}\right)(e^x \sin x + e^x \cos x)$$

(p) $f(x) = e^x \sin x$
HINT: It is easy to verify that

$$\begin{aligned} y(x) &= \int f(x)\,\mathrm{d}x \\ &= \int e^x \sin x\,\mathrm{d}x \\ &= -e^x \cos x + e^x \sin x - \int e^x \sin x\,\mathrm{d}x \\ &= -e^x \cos x + e^x \sin x - y(x) \end{aligned}$$

In other words, one can have

$$y(x) = \left(\frac{1}{2}\right)(e^x \sin x - e^x \cos x)$$

(q) $f(x) = (\tan x + \cot x)^2$

HINT: It is easy to verify that

$$\begin{aligned} y(x) &= \int f(x)\,\mathrm{d}x \\ &= \int (\tan x + \cot x)^2\,\mathrm{d}x \\ &= \int \left(\tan^2 x + \cot^2 x + 2\right)\mathrm{d}x \\ &= \int \left(\sec^2 x + \operatorname{cosec}^2 x\right)\mathrm{d}x \\ &= \tan x - \cot x + c \end{aligned}$$

(r) $f(x) = \sec^2 x \operatorname{cosec}^2 x$

HINT: It is easy to verify that

$$\begin{aligned} y(x) &= \int f(x)\,\mathrm{d}x \\ &= \int \sec^2 x \operatorname{cosec}^2 x\,\mathrm{d}x \\ &= \int \left(\frac{1}{\cos^2 x \sin^2 x}\right)\mathrm{d}x \\ &= \int \left(\frac{\sin^2 x + \cos^2 x}{\cos^2 x \sin^2 x}\right)\mathrm{d}x \\ &= \int \left(\frac{1}{\cos^2 x} + \frac{1}{\sin^2 x}\right)\mathrm{d}x \\ &= \int \left(\sec^2 x + \operatorname{cosec}^2 x\right)\mathrm{d}x \\ &= \tan x - \cot x + c \end{aligned}$$

(s) $f(x) = a^x\, e^{bx}$

HINT: It is easy to verify that

$$\begin{aligned} y(x) &= \int f(x)\,\mathrm{d}x \\ &= \int a^x\, e^{bx}\,\mathrm{d}x \\ &= \int e^{\ln(a^x)}\, e^{bx}\,\mathrm{d}x \\ &= \int e^{x\ln(a)}\, e^{bx}\,\mathrm{d}x \end{aligned}$$

$$= \int e^{(\ln a+b)x}\,dx$$

$$= \frac{e^{(\ln a+b)x}}{\ln a+b} + c$$

3. Solve the differential equation $(1+y^2)dx + (x - e^{-\tan^{-1}(y)})dy = 0$.

   HINT: It is easy to verify that the given differential equation can be changed in the form $x' + \frac{x}{1+y^2} = \frac{e^{-\tan^{-1}(y)}}{1+y^2}$. In other words, $x = e^{-\tan^{-1}(y)}\left(\tan^{-1}(y) + c\right)$.

4. Solve the differential equation $y' - \cos(x)\,y = x\,e^{\sin x}$.

   HINT: It is easy to verify that $\mu(x) = e^{-\sin x}$ is the integrating factor. Hence $y(x) = \left[\left(\frac{x^2}{2}\right) + c\right] e^{\sin x}$.

5. Solve the differential equation $y' - y\cot x = -2\cot x$.

   HINT: It is easy to verify that $\mu(x) = \operatorname{cosec} x$ is the integrating factor. Hence $y(x) = 2 + c\sin x$.

6. Solve the differential equation $y' + y\tan x = 2\tan x$.

   HINT: It is easy to verify that $\mu(x) = \sec x$ is the integrating factor. Hence $y(x) = 2 + c\cos x$.

7. Solve the differential equation $\cos^2(x)\,y' + y = 2$.

   HINT: It is easy to verify that $\mu(x) = e^{\tan x}$ is the integrating factor. Hence $y(x) = 1 + c\,e^{-\tan x}$.

8. Solve the differential equation $y' + 2\tan(x)\,y = \sin x$.

   HINT: It is easy to verify that $\mu(x) = \frac{1}{\cos^2 x}$ is the integrating factor. Hence $y(x) = c\cos^2(x) + \cos(x)$.

9. Solve the differential equation $y' + \tan(x)\,y = \sin x$.

   HINT: It is easy to verify that $\mu(x) = \frac{1}{\cos x}$ is the integrating factor. Hence $y(x) = c\cos x - \cos x\ln(\cos x)$.

10. Solve the differential equation $\sin(x)\,y' + \cos(x)\,y = 1$.

    HINT: It is easy to verify that $\mu(x) = \sin x$ is the integrating factor. Hence $y(x) = x\operatorname{cosec} x + c\operatorname{cosec} x$.

11. Solve the differential equation $\cos(x)\,y' + \sin(x)\,y = 1$.

    HINT: It is easy to verify that $\mu(x) = \frac{1}{\cos x}$ is the integrating factor. Hence $y(x) = \sin x + c\cos x$.

12. Solve the differential equation $y' + \cos(x)\,y = \cos x$.

    HINT: It is easy to verify that $\mu(x) = e^{\sin x}$ is the integrating factor. Hence $y(x) = 1 + c\,e^{-\sin x}$.

13. Solve the differential equation $\cos^2(x)\,y' + y = \tan x$.

    HINT: It is easy to verify that $\mu(x) = e^{\tan x}$ is the integrating factor. Hence $y(x) = \tan x - 1 + c\,e^{-\tan x}$.

14. Solve the differential equation $y' + \sec(x)\, y = \tan x$.

    HINT: It is easy to verify that $\mu(x) = \sec x + \tan x$ is the integrating factor. Hence $y(x) = \frac{1}{\sec x + \tan x}(\sec x + \tan x - x + c)$.

15. Solve the differential equation $\cos(x)\, xy' + (\sin x\,(x) + \cos x)\, y = \tan x$.

    HINT: It is easy to verify that $\mu(x) = \frac{x}{\cos x}$ is the integrating factor. Hence $y(x) = \frac{\sec x \sin^2 x}{2x} + c\left(\frac{\cos x}{x}\right)$.

16. Solve the differential equation $y' + \cot(x)\, y = 2x \operatorname{cosec} x$.

    HINT: It is easy to verify that $\mu(x) = \sin x$ is the integrating factor. Hence $y(x) = x^2 \operatorname{cosec} x + c \operatorname{cosec} x$.

17. Solve the differential equation $y' + y \cot x = e^{\cos x}$.

    HINT: It is easy to verify that $\mu(x) = \sin(x)$ is the integrating factor. Hence $y(x) = c \operatorname{cosec} x - \operatorname{cosec}(x)\, e^{\cos x}$.

18. Solve the differential equation $xy' + y(1 + x \cot x) = x$.

    HINT: It is easy to verify that $\mu(x) = x \sin x$ is the integrating factor. Hence $y(x) = \frac{1}{x \sin x}(c + \sin x - x \cos x)$.

19. Solve the differential equation $y' + y = e^{-x}$.

    HINT: It is easy to verify that $\mu(x) = e^x$ is the integrating factor. Hence $y(x) = e^{-x}(c + x)$.

20. Solve the differential equation $x\, y' - y = -2x$.

    HINT: It is easy to verify that $\mu(x) = \frac{1}{x}$ is the integrating factor. Hence $y(x) = cx - 2x \ln(x)$.

21. Solve the differential equation $x\, y' + 2y = 2$.

    HINT: It is easy to verify that $\mu(x) = x^2$ is the integrating factor. Hence $y(x) = 1 + \frac{c}{x^2}$.

22. Solve the differential equation $y' + \frac{3y}{x} = \frac{1}{x^4}$.

    HINT: It is easy to verify that $\mu(x) = x^3$ is the integrating factor. Hence $y(x) = \frac{1}{x^3}[\ln(x) + c]$.

23. Solve the differential equation $y' + 2xy = 2x^3$.

    HINT: It is easy to verify that $\mu(x) = e^{x^2}$ is the integrating factor. Hence $y(x) = x^2 - 1 + c\,e^{-x^2}$.

2.3 First Degree Nonlinear Equations

In the previous sections, we have considered only linear first order differential equations. In this section, we deal with nonlinear first order and first degree differential equations. The most general type of first order and first degree nonlinear differential equation can be represented by $p(x, y)y' = q(x, y)$. In other words, $p(x, y)\, \mathrm{d}y = q(x, y)\, \mathrm{d}x$. In general, all the equations of the type $p(x, y)y' = q(x, y)$ may not have solutions. Again, the set of all equations of the type $p(x, y)y' = q(x, y)$ which have solutions has to satisfy a necessary and sufficient condition called the *exact condition*.

Theorem 2.3.1 A differential equation $p(x, y)\,dy + q(x, y)\,dx = 0$ is exact if and only if $\frac{\partial p}{\partial x} = \frac{\partial q}{\partial y}$.

Proof: Assume that the given differential equation is exact. Hence there exists a function $\phi(x, y) = c$, such that $d[\phi(x, y)] = p(x, y)\,dy + q(x, y)\,dx = 0$. In other words, $\left(\frac{\partial \phi}{\partial x}\right) dx + \left(\frac{\partial \phi}{\partial y}\right) dy = q(x, y)\,dx + p(x, y)\,dy = 0$. Hence by comparing the coefficients of both dx and dy, we have

$$\frac{\partial \phi}{\partial x} = q(x, y) \tag{2.9}$$

and

$$\frac{\partial \phi}{\partial y} = p(x, y) \tag{2.10}$$

Again, by differentiating Eq.(2.9) partially with respect to y and Eq.(2.10) with respect to x, we have

$$\frac{\partial q}{\partial y} = \frac{\partial p}{\partial x}$$

Assume that $\frac{\partial p}{\partial x} = \frac{\partial q}{\partial y}$. Let $V(x, y) = \int q(x, y)\,dx$. In other words, $\frac{\partial V}{\partial x} = q(x, y)$. Again, we have

$$\begin{aligned}\frac{\partial^2 V}{\partial y \partial x} &= \frac{\partial q}{\partial y}\\ &= \frac{\partial p}{\partial x}\end{aligned}$$

In other words, $p(x, y) = \frac{\partial V}{\partial y} + \psi(y)$. Again, it is easy to verify that

$$\begin{aligned}0 &= p(x, y)\,dy + q(x, y)\,dx\\ &= \left(\frac{\partial V}{\partial y} + \psi(y)\right) dy + \left(\frac{\partial V}{\partial x}\right) dx\\ &= \left(\frac{\partial V}{\partial y}\right) dy + \left(\frac{\partial V}{\partial x}\right) dx + \psi(y)dy\\ &= d(V(x, y) + \int \psi(y)\,dy)\end{aligned} \tag{2.11}$$

It is clear from Eq.(2.11) that $p(x, y)\,dy + q(x, y)\,dx = 0$ is exact.

Theorem 2.3.2 If $p(x, y)\,dy + q(x, y)\,dx = 0$ is any first order and first degree differential equation with $\frac{\partial p}{\partial x} = \frac{\partial q}{\partial y}$, then $\int q(x, y)\,dx + \int \left(p(x, y) - \frac{\partial}{\partial y}\left(\int q(x, y)\,dx\right)\right) dy = c$ is the corresponding solution.

Proof: Let $p(x, y) = \frac{\partial V}{\partial y} + \psi(y)$ and $V(x, y) = \int q(x, y)\, dx$. It is clear from Eq.(2.11) that

$$\begin{aligned} 0 &= d(V(x, y) + \int \psi(y)\, dy) \\ &= d\left(\int q(x, y)\, dx + \int \psi(y)\, dy\right) \end{aligned} \tag{2.12}$$

One can easily show that $\int q(x, y)\, dx + \int \psi(y)\, dy = c$ is the solution of the given equation where

$$\begin{aligned} \psi(y) &= p(x, y) - \frac{\partial V}{\partial y} \\ &= p(x, y) - \frac{\partial}{\partial y}\left(\int q(x, y)\, dx\right) \end{aligned} \tag{2.13}$$

Again, the corresponding solution in terms of $p(x, y)$ and $q(x, y)$ can be given by replacing the expression for $\psi(y)$ given in Eq.(2.13) in Eq.(2.12). Hence one can have

$$\begin{aligned} c &= \int q(x, y)\, dx + \int \left(p(x, y) - \frac{\partial}{\partial y}\left(\int q(x, y)\, dx\right)\right) dy \\ &= \int p(x, y)\, dy + \int \left(q(x, y) - \frac{\partial}{\partial x}\left(\int p(x, y)\, dy\right)\right) dx \end{aligned} \tag{2.14}$$

EXAMPLE 2.27 Solve the differential equation $(y \sin x + \sin y + y)\, dx + (x - \cos x + x \cos y)\, dy = 0$.

Solution: Clearly, $q(x, y) = y \sin(x) + \sin(y) + y$ and $p(x, y) = \sin(x) + x \cos(y) + x$. Again, we have

$$\frac{\partial q}{\partial y} = \sin x + \cos y + 1$$

and

$$\frac{\partial p}{\partial x} = \sin x + \cos y + 1$$

Hence the given differential equation is exact according to the condition $\frac{\partial q}{\partial y} = \frac{\partial p}{\partial x}$. The corresponding solution according to Eq.(2.14) is given by

$$\begin{aligned} c &= \int q(x, y)\, dx + \int \left(p(x, y) - \frac{\partial}{\partial y}\left(\int q(x, y)\, dx\right)\right) dy \\ &= -y \cos x + x \sin y + xy + \int \left(p(x, y) - \frac{\partial}{\partial y}\left(\int q(x, y)\, dx\right)\right) dy \\ &= x \sin y - y \cos x + xy \end{aligned}$$

EXAMPLE 2.28 Solve the differential equation $e^y\,\mathrm{d}x + (xe^y + 2y)\,\mathrm{d}y = 0$.

Solution: Clearly, $q(x,\, y) = e^y$ and $p(x,\, y) = xe^y + 2y$. Again, we have

$$\frac{\partial q}{\partial y} = e^y$$

and

$$\frac{\partial p}{\partial x} = e^y$$

Hence the given differential equation is exact according to the condition $\frac{\partial q}{\partial y} = \frac{\partial p}{\partial x}$. The corresponding solution according to Eq.(2.14) is given by

$$\begin{aligned} c &= \int q(x,\, y)\,\mathrm{d}x + \int \left(p(x,\, y) - \frac{\partial}{\partial y}\left(\int q(x,\, y)\,\mathrm{d}x\right)\right)\,\mathrm{d}y \\ &= xe^y + \int (xe^y + 2y - xe^y)\,\mathrm{d}y \\ &= xe^y + y^2 \end{aligned}$$

EXAMPLE 2.29 Solve the differential equation $xe^{x^2+y^2}\,\mathrm{d}x + y\left(e^{x^2+y^2} + 1\right)\,\mathrm{d}y = 0$.

Solution: Clearly, $q(x,\, y) = xe^{x^2+y^2}$ and $p(x,\, y) = y\left(e^{x^2+y^2} + 1\right)$. Again, we have

$$\frac{\partial q}{\partial y} = 2xye^{x^2+y^2}$$

and

$$\frac{\partial p}{\partial x} = 2xye^{x^2+y^2}$$

Hence the given differential equation is exact according to the condition $\frac{\partial q}{\partial y} = \frac{\partial p}{\partial x}$. The corresponding solution according to Eq.(2.14) is given by

$$\begin{aligned} c &= \int q(x,\, y)\,\mathrm{d}x + \int \left(p(x,\, y) - \frac{\partial}{\partial y}\left(\int q(x,\, y)\,\mathrm{d}x\right)\right)\,\mathrm{d}y \\ &= \left(\frac{1}{2}\right) e^{x^2+y^2} + \int \left(y\left(e^{x^2+y^2} + 1\right) - ye^{x^2+y^2}\right)\mathrm{d}y \\ &= \left(\frac{1}{2}\right) e^{x^2+y^2} + \left(\frac{1}{2}\right) y^2 \\ &= \left(\frac{1}{2}\right)\left(e^{x^2+y^2} + y^2\right) \end{aligned}$$

Exercises 2.3

1. Solve the differential equation $(\cos x + y\sin x)\mathrm{d}x - \cos x\,\mathrm{d}y = 0$.
   HINT: According to Eq.(2.14), $\sin x - y\cos x = c$.
2. Solve the differential equation $(\cos x + \sin x)\mathrm{d}x + \cos y\,\mathrm{d}y = 0$.
   HINT: According to Eq.(2.14), $\sin x - \cos x + \sin y = c$.
3. Solve the differential equation $(2x + e^x \sin y)\,\mathrm{d}x + e^x \cos y\,\mathrm{d}y = 0$.
   HINT: According to Eq.(2.14), $x^2 + e^x \sin y = c$.
4. Solve the differential equation $(2x + \sin y)\,\mathrm{d}x + x\cos y\,\mathrm{d}y = 0$.
   HINT: According to Eq.(2.14), $x^2 + x\sin(y) = c$.
5. Solve the differential equation $2x\sin 3y\,\mathrm{d}x + 3x^2\cos 3y\,\mathrm{d}y = 0$.
   HINT: According to Eq.(2.14), $x^2 \sin 3y = c$.
6. Solve the differential equation $(y^2 - 2xy)\,\mathrm{d}x + (2xy - x^2)\,\mathrm{d}y = 0$.
   HINT: According to Eq.(2.14), $xy^2 - x^2y = c$.
7. Solve the differential equation $(3x^2y + 8xy^2)\,\mathrm{d}x + (x^3 + 8x^2y + 12y^2)\,\mathrm{d}y = 0$.
   HINT: According to Eq.(2.14), $x^3y + 4x^2y^2 + 4y^3 = c$.
8. Solve the differential equation $(2x + e^y)\mathrm{d}x + xe^y\mathrm{d}y = 0$.
   HINT: According to Eq.(2.14), $x^2 + x\,e^y = c$.
9. Solve the differential equation $2x\ln y\,\mathrm{d}x + y^{-1}x^2\,\mathrm{d}y = 0$.
   HINT: According to Eq.(2.14), $x^2\ln y = c$.
10. Solve the differential equation $3xe^{3y}\,\mathrm{d}y + e^{3y}\,\mathrm{d}x = 0$.
    HINT: According to Eq.(2.14), $3x\,e^{3y} = c$.

2.4 First Degree Nonexact Differential Equation

In the previous section, we studied differential equations of the type $p(x, y)\mathrm{d}y + q(x, y)\mathrm{d}x = 0$, for which $\frac{\partial p}{\partial x} = \frac{\partial q}{\partial y}$. In this section, we have studied differential equations of the type $p(x, y)\mathrm{d}y + q(x, y)\mathrm{d}x = 0$ for which $\frac{\partial p}{\partial x} \neq \frac{\partial q}{\partial y}$. Assume that there exists a function $\mu(x, y)$, such that the modified differential equation $\mu(x, y)p(x, y)\mathrm{d}y + \mu(x, y)q(x, y)\mathrm{d}x = 0$ is exact. Let $m(x, y) = \mu(x, y)p(x, y)$ and $n(x, y) = \mu(x, y)q(x, y)$. According to Theorem 2.3.1, we have

$$\begin{aligned}
\mu\left(\frac{\partial p}{\partial x}\right) + p\left(\frac{\partial \mu}{\partial x}\right) &= \frac{\partial(\mu p)}{\partial x} \\
&= \frac{\partial(\mu q)}{\partial y} \\
&= \mu\left(\frac{\partial q}{\partial y}\right) + q\left(\frac{\partial \mu}{\partial y}\right)
\end{aligned}$$

In other words, we have

$$\frac{1}{\mu}\left(p\left(\frac{\partial\mu}{\partial x}\right)-q\left(\frac{\partial\mu}{\partial y}\right)\right)=\frac{\partial q}{\partial y}-\frac{\partial p}{\partial x} \tag{2.15}$$

Again, Eq.(2.15) reduces to

$$\left(\frac{\partial\mu}{\mu\,\partial x}\right)=\frac{\frac{\partial q}{\partial y}-\frac{\partial p}{\partial x}}{p} \tag{2.16}$$

when $\frac{\partial\mu}{\partial y}=0$, and

$$\left(\frac{\partial\mu}{\mu\,\partial y}\right)=\frac{\frac{\partial p}{\partial x}-\frac{\partial q}{\partial y}}{q} \tag{2.17}$$

when $\frac{\partial\mu}{\partial x}=0$. Depending upon the right hand side of Eqs.(2.16) and (2.17), one can decide what could the *integrating factor* $\mu(x, y)$.

Theorem 2.4.1 If $p(x, y)\mathrm{d}y+q(x, y)\mathrm{d}x=0$ is any differential equation with $\frac{\partial p}{\partial x}\neq\frac{\partial q}{\partial y}$ and $\frac{\frac{\partial q}{\partial y}-\frac{\partial p}{\partial x}}{p}=f(x)$, then $\mu(x)=e^{\int f(x)\,\mathrm{d}x}$.

Proof: According to the given conditions, Eq.(2.16) reduces to

$$\begin{aligned}\frac{\partial\mu}{\mu\partial x}&=\frac{\frac{\partial q}{\partial y}-\frac{\partial p}{\partial x}}{p}\\&=f(x)\end{aligned}$$

In other words, we have

$$\frac{\mathrm{d}\mu}{\mu}=f(x)\,\mathrm{d}x \tag{2.18}$$

The conclusion follows by integrating Eq.(2.18).

EXAMPLE 2.30 Show that the differential equation $y'+g(x)y=f(x)$ is not exact, and hence find the corresponding integrating factor.

Solution: It is easy to verify that the given differential equation can be transformed into $p(x, y)=1$ and $q(x, y)=g(x)y-f(x)$. Clearly, $\frac{\partial q}{\partial y}-\frac{\partial p}{\partial x}=g(x)\neq 0$. Again, we have

$$\frac{\frac{\partial q}{\partial y}-\frac{\partial p}{\partial x}}{p}=g(x)$$

According to Theorem 2.4.1, we have

$$\mu(x)=e^{\int g(x)\,\mathrm{d}x}$$

which is the same as given in Eq.(2.7).

EXAMPLE 2.31 Solve the differential equation $(x - x^2y)\mathrm{d}y = y\mathrm{d}x$.

Solution: Clearly, $p(x, y) = x^2y - x$ and $q(x, y) = y$. Again, we have

$$\begin{aligned}\frac{\frac{\partial q}{\partial y} - \frac{\partial p}{\partial x}}{p} &= -\frac{2}{x}\\ &= f(x)\end{aligned}$$

According to Theorem 2.4.1, we have

$$\begin{aligned}\mu(x) &= e^{\int f(x)\,\mathrm{d}x}\\ &= e^{-\int\left(\frac{2}{x}\right)\mathrm{d}x}\\ &= \frac{1}{x^2}\end{aligned}$$

Hence $m(x, y) = y - \frac{1}{x}$ and $n(x, y) = \frac{y}{x^2}$. According to Eq.(2.14), we have

$$\begin{aligned}c &= \int n(x, y)\,\mathrm{d}x + \int \left(m(x, y) - \frac{\partial}{\partial y}\left(\int n(x, y)\,\mathrm{d}x\right)\right)\mathrm{d}y\\ &= -\left(\frac{y}{x}\right) + \int\left(y - \frac{1}{x} + \frac{1}{x}\right)\mathrm{d}y\\ &= -\left(\frac{y}{x}\right) + \frac{y^2}{2}\\ &= \frac{xy^2 - 2y}{2x}\end{aligned}$$

EXAMPLE 2.32 Solve the differential equation $(x^2 + y^2)\mathrm{d}x = 2xy\mathrm{d}y$.

Solution: Clearly, $p(x, y) = -2xy$ and $q(x, y) = x^2 + y^2$. Again, we have

$$\begin{aligned}\frac{\frac{\partial q}{\partial y} - \frac{\partial p}{\partial x}}{p} &= -\frac{2}{x}\\ &= f(x)\end{aligned}$$

According to Theorem 2.4.1, we have

$$\begin{aligned}\mu(x) &= e^{\int f(x)\,\mathrm{d}x}\\ &= e^{-\int\left(\frac{2}{x}\right)\mathrm{d}x}\\ &= \frac{1}{x^2}\end{aligned}$$

Hence $m(x, y) = -2\left(\frac{y}{x}\right)$ and $n(x, y) = 1 + \frac{y^2}{x^2}$. According to Eq.(2.14), we have

$$\begin{aligned} c &= \int n(x, y)\,\mathrm{d}x + \int \left(m(x, y) - \frac{\partial}{\partial y}\left(\int n(x, y)\,\mathrm{d}x\right)\right) \mathrm{d}y \\ &= x - \frac{y^2}{x} + \int \left(-\frac{2y}{x} + \frac{2y}{x}\right) \mathrm{d}y \\ &= x - \frac{y^2}{x} \\ &= \frac{x^2 - y^2}{x} \end{aligned}$$

EXAMPLE 2.33 Solve the differential equation $y\mathrm{d}x = x\mathrm{d}y$.

Solution: Clearly, $p(x, y) = -x$ and $q(x, y) = y$. Again, we have

$$\begin{aligned} \frac{\frac{\partial q}{\partial y} - \frac{\partial p}{\partial x}}{p} &= -\frac{2}{x} \\ &= f(x) \end{aligned}$$

According to Theorem 2.4.1, we have

$$\begin{aligned} \mu(x) &= e^{\int f(x)\,\mathrm{d}x} \\ &= e^{\int -\frac{2}{x}\mathrm{d}x} \\ &= \frac{1}{x^2} \end{aligned}$$

Hence $m(x, y) = -\left(\frac{1}{x}\right)$ and $n(x, y) = \frac{y}{x^2}$. According to Eq.(2.14), we have

$$\begin{aligned} c &= \int n(x, y)\,\mathrm{d}x + \int \left(m(x, y) - \frac{\partial}{\partial y}\left(\int n(x, y)\,\mathrm{d}x\right)\right) \mathrm{d}y \\ &= -\frac{y}{x} + \int \left(-\frac{1}{x} + \frac{1}{x}\right) \mathrm{d}y \\ &= -\frac{y}{x} \end{aligned}$$

Theorem 2.4.2 If $p(x, y)\mathrm{d}y + q(x, y)\mathrm{d}x = 0$ is any differential equation with $\frac{\partial p}{\partial x} \neq \frac{\partial q}{\partial y}$ and $\frac{\frac{\partial p}{\partial x} - \frac{\partial q}{\partial y}}{q} = g(y)$, then $\mu(y) = e^{\int g(y)\,\mathrm{d}y}$.

Proof: According to the given conditions, Eq.(2.17) reduces to

$$\begin{aligned} \frac{\partial \mu}{\mu \partial y} &= \frac{\frac{\partial p}{\partial x} - \frac{\partial q}{\partial y}}{q} \\ &= g(y) \end{aligned}$$

In other words, we have

$$\frac{\mathrm{d}\mu}{\mu} = h(y)\,\mathrm{d}y \tag{2.19}$$

The conclusion follows by integrating Eq.(2.19).

EXAMPLE 2.34 Solve the differential equation $(3x^2y^4 + 2xy)\mathrm{d}x + (2x^3y^3 - x^2)\mathrm{d}y = 0$.
Solution: Clearly, $p(x, y) = 2x^3y^3 - x^2$ and $q(x, y) = 3x^2y^4 + 2xy$. Again, we have

$$\begin{aligned}\frac{\frac{\partial p}{\partial x} - \frac{\partial q}{\partial y}}{q} &= -\frac{2}{y}\\ &= g(y)\end{aligned}$$

According to Theorem 2.4.2, we have

$$\begin{aligned}\mu(y) &= e^{\int g(y)\,\mathrm{d}y}\\ &= e^{-\int\left(\frac{2}{y}\right)\mathrm{d}x}\\ &= \frac{1}{y^2}\end{aligned}$$

Hence $m(x, y) = 2x^3y - \frac{x^2}{y^2}$ and $n(x, y) = 3x^2y^2 + 2\left(\frac{x}{y}\right)$. According to Eq.(2.14), we have

$$\begin{aligned}c &= \int n(x, y)\,\mathrm{d}x + \int\left(m(x, y) - \frac{\partial}{\partial y}\left(\int n(x, y)\,\mathrm{d}x\right)\right)\mathrm{d}y\\ &= x^3y^2 + \frac{x^2}{y}\\ &= \frac{x^3y^3 + x^2}{y}\end{aligned}$$

EXAMPLE 2.35 Solve the differential equation $y(axy + e^x)\mathrm{d}x = e^x\mathrm{d}y$.
Solution: Clearly, $p(x, y) = -e^x$ and $q(x, y) = y(axy + e^x)$. Again, we have

$$\begin{aligned}\frac{\frac{\partial p}{\partial x} - \frac{\partial q}{\partial y}}{q} &= -\frac{2}{y}\\ &= g(y)\end{aligned}$$

According to Theorem 2.4.2, we have

$$\begin{aligned}\mu(y) &= e^{\int g(y)\,\mathrm{d}y}\\ &= e^{-\int\left(\frac{2}{y}\right)\mathrm{d}x}\\ &= \frac{1}{y^2}\end{aligned}$$

Hence $m(x, y) = -\left(\frac{e^x}{y^2}\right)$ and $n(x, y) = ax + \left(\frac{e^x}{y}\right)$. According to Eq.(2.14), we have

$$\begin{aligned}c &= \int n(x, y)\,\mathrm{d}x + \int\left(m(x, y) - \frac{\partial}{\partial y}\left(\int n(x, y)\,\mathrm{d}x\right)\right)\mathrm{d}y\\ &= \frac{ax^2}{2} + \frac{e^x}{y}\\ &= \frac{ax^2y + 2e^x}{2y}\end{aligned}$$

EXAMPLE 2.36 Solve the differential equation $2xy\mathrm{d}x + (y^2 - x^2)\mathrm{d}y = 0$.
Solution: Clearly, $p(x, y) = y^2 - x^2$ and $q(x, y) = 2xy$. Again, we have

$$\begin{aligned}\frac{\frac{\partial p}{\partial x} - \frac{\partial q}{\partial y}}{q} &= -\frac{2}{y} \\ &= g(y)\end{aligned}$$

According to Theorem 2.4.2, we have

$$\begin{aligned}\mu(y) &= e^{\int g(y)\,\mathrm{d}y} \\ &= e^{-\int \frac{2}{y}\,\mathrm{d}x} \\ &= \frac{1}{y^2}\end{aligned}$$

Hence $m(x, y) = 1 - \left(\frac{x^2}{y^2}\right)$ and $n(x, y) = 2\left(\frac{x}{y}\right)$. According to Eq.(2.14), we have

$$\begin{aligned}c &= \int n(x, y)\,\mathrm{d}x + \int \left(m(x, y) - \frac{\partial}{\partial y}\left(\int n(x, y)\,\mathrm{d}x\right)\right)\mathrm{d}y \\ &= \frac{x^2}{y} + y \\ &= \frac{x^2 + y^2}{y}\end{aligned}$$

Clearly, the solution is the set of all circles whose centre is on the y-axis and tangent to the x-axis.

EXAMPLE 2.37 Solve the differential equation $y\mathrm{d}x = x\mathrm{d}y$.
Solution: Clearly, $p(x, y) = -x$ and $q(x, y) = y$. Again, we have

$$\begin{aligned}\frac{\frac{\partial p}{\partial x} - \frac{\partial q}{\partial y}}{q} &= -\frac{2}{y} \\ &= g(y)\end{aligned}$$

According to Theorem 2.4.2, we have

$$\begin{aligned}\mu(x) &= e^{\int g(y)\,\mathrm{d}y} \\ &= e^{\int -\frac{2}{y}\,\mathrm{d}y} \\ &= \frac{1}{y^2}\end{aligned}$$

Hence $m(x, y) = -\left(\frac{x}{y^2}\right)$ and $n(x, y) = \frac{1}{y}$. According to Eq.(2.14), we have

$$\begin{aligned}c &= \int n(x, y)\,\mathrm{d}x + \int \left(m(x, y) - \frac{\partial}{\partial y}\left(\int n(x, y)\,\mathrm{d}x\right)\right)\mathrm{d}y \\ &= \frac{x}{y} + \int \left(-\frac{x}{y^2} + \frac{x}{y^2}\right)\mathrm{d}y \\ &= \frac{x}{y}\end{aligned}$$

Theorem 2.4.3 If $p(x, y)\mathrm{d}y + q(x, y)\mathrm{d}x = 0$ is any differential equation with $\frac{\partial p}{\partial x} \neq \frac{\partial q}{\partial y}$ and $\frac{\frac{\partial q}{\partial y} - \frac{\partial p}{\partial x}}{py-qx} = h(u)$, where $u = xy$, then $\mu(x, y) = e^{\int h(u)\,\mathrm{d}u}$.

Proof: Let $u = xy$. Accordingly, Eq.(2.15) reduces to

$$\begin{aligned}
\left(\frac{\partial q}{\partial y} - \frac{\partial p}{\partial x}\right) &= \left(\frac{1}{\mu}\right)\left(p\left(\frac{\partial \mu}{\partial x}\right) - q\left(\frac{\partial \mu}{\partial y}\right)\right) \\
&= \left(\frac{1}{\mu}\right)\left(py\left(\frac{\mathrm{d}\mu}{\mathrm{d}u}\right) - qx\left(\frac{\mathrm{d}\mu}{\mathrm{d}u}\right)\right) \\
&= \left(\frac{1}{\mu}\right)\left(\frac{\mathrm{d}\mu}{\mathrm{d}u}\right)(py - qx)
\end{aligned}$$

In other words, we have

$$\begin{aligned}
\left(\frac{1}{\mu}\right)\left(\frac{\mathrm{d}\mu}{\mathrm{d}u}\right) &= \frac{\frac{\partial q}{\partial y} - \frac{\partial p}{\partial x}}{py - qx} \\
&= h(u)
\end{aligned}$$

which is equivalent to

$$\frac{\mathrm{d}\mu}{\mu} = h(u)\,\mathrm{d}u \tag{2.20}$$

The conclusion follows by integrating Eq.(2.20).

EXAMPLE 2.38 Solve the differential equation $(x^2y - x)\mathrm{d}y - y\mathrm{d}x = 0$.

Solution: Clearly, $p(x, y) = x^2y - x$ and $q(x, y) = -y$. Again, we have

$$\begin{aligned}
\frac{\frac{\partial q}{\partial y} - \frac{\partial p}{\partial x}}{py - qx} &= -\frac{2xy}{x^2y^2} \\
&= -\frac{2}{xy} \\
&= -\frac{2}{u} \\
&= h(u)
\end{aligned}$$

According to Theorem 2.4.3, we have

$$\begin{aligned}
\mu(x, y) &= e^{\int h(u)\,\mathrm{d}u} \\
&= \frac{1}{u^2} \\
&= \frac{1}{(xy)^2}
\end{aligned}$$

Clearly, $m(x, y) = \frac{x^2y-x}{(xy)^2}$ and $n(x, y) = -\left(\frac{y}{(xy)^2}\right)$. According to Eq.(2.14), we have

$$\begin{aligned} c &= \int n(x, y)\,\mathrm{d}x + \int \left(m(x, y) - \frac{\partial}{\partial y}\left(\int n(x, y)\,\mathrm{d}x\right)\right)\mathrm{d}y \\ &= \frac{1}{xy} + \int \left(\frac{1}{y}\right)\mathrm{d}y \\ &= \frac{1}{xy} + \ln y \end{aligned}$$

EXAMPLE 2.39 Solve the differential equation $(2xy^2 - y)\mathrm{d}x + x\mathrm{d}y = 0$.

Solution: Clearly, $p(x, y) = x$ and $q(x, y) = 2xy^2 - y$. Again, we have

$$\begin{aligned} \frac{\frac{\partial q}{\partial y} - \frac{\partial p}{\partial x}}{py - qx} &= \frac{4xy - 2}{2xy - 2x^2y^2} \\ &= \frac{2xy - 1}{xy - x^2y^2} \\ &= \frac{2u - 1}{u - u^2} \\ &= h(u) \end{aligned}$$

According to Theorem 2.4.3, we have

$$\begin{aligned} \mu(x, y) &= e^{\int h(u)\,\mathrm{d}u} \\ &= \frac{1}{u - u^2} \\ &= \frac{1}{xy - x^2y^2} \end{aligned}$$

Clearly, $m(x, y) = \frac{x}{xy-x^2y^2}$ and $n(x, y) = \frac{2xy^2-y}{xy-x^2y^2}$. According to Eq.(2.14), we have

$$\begin{aligned} c &= \int n(x, y)\,\mathrm{d}x + \int \left(m(x, y) - \frac{\partial}{\partial y}\left(\int n(x, y)\,\mathrm{d}x\right)\right)\mathrm{d}y \\ &= -\ln(xy - x^2y^2) + \int \left(\frac{1}{(1 - xy)y} - \frac{\partial}{\partial y}\left(-\ln(xy - x^2y^2)\right)\right)\mathrm{d}y \\ &= -\ln(xy - x^2y^2) + \int \left(\frac{1}{(1 - xy)y} + \frac{\partial}{\partial y}\left(\ln(xy - x^2y^2)\right)\right)\mathrm{d}y \\ &= \ln\left(\frac{1}{xy - x^2y^2}\right) + 2\int \left(\frac{1}{y}\right)\mathrm{d}y \\ &= \ln\left(\frac{1}{xy - x^2y^2}\right) + \ln y^2 \\ &= \ln\left(\frac{y}{1 - xy}\right) - \ln x \end{aligned}$$

EXAMPLE 2.40 Solve the differential equation $y\mathrm{d}x = x\mathrm{d}y$.

Solution: Clearly, $p(x, y) = -x$ and $q(x, y) = y$. Again, we have

$$\begin{aligned}\frac{\frac{\partial q}{\partial y} - \frac{\partial p}{\partial x}}{py - qx} &= -\frac{1}{xy} \\ &= h(xy)\end{aligned}$$

According to Theorem 2.4.3, we have

$$\begin{aligned}\mu(x) &= e^{\int h(xy)\,\mathrm{d}(xy)} \\ &= e^{\int -\left(\frac{1}{xy}\right)\mathrm{d}(xy)} \\ &= \frac{1}{xy}\end{aligned}$$

Hence $m(x, y) = -\left(\frac{1}{y}\right)$ and $n(x, y) = \frac{1}{x}$. According to Eq.(2.14), we have

$$\begin{aligned}c &= \int n(x, y)\,\mathrm{d}x + \int \left(m(x, y) - \frac{\partial}{\partial y}\left(\int n(x, y)\,\mathrm{d}x\right)\right)\mathrm{d}y \\ &= \ln(x) + \int \left(-\frac{1}{y}\right)\mathrm{d}y \\ &= \ln x - \ln y \\ &= \ln \frac{x}{y}\end{aligned}$$

Theorem 2.4.4 If $p(x, y)\mathrm{d}y + q(x, y)\mathrm{d}x = 0$ is any differential equation with $\frac{\partial p}{\partial x} \neq \frac{\partial q}{\partial y}$ and $\frac{\frac{\partial q}{\partial y} - \frac{\partial p}{\partial x}}{p-q} = h(v)$, where $v = x + y$, then $\mu(x, y) = e^{\int h(v)\,\mathrm{d}v}$.

Proof: Let $v = x + y$. Accordingly, Eq.(2.15) reduces to

$$\begin{aligned}\left(\frac{\partial q}{\partial y} - \frac{\partial p}{\partial x}\right) &= \frac{1}{\mu}\left(p\left(\frac{\partial \mu}{\partial x}\right) - q\left(\frac{\partial \mu}{\partial y}\right)\right) \\ &= \left(\frac{1}{\mu}\right)\left(p\left(\frac{\mathrm{d}\mu}{\mathrm{d}v}\right) - q\left(\frac{\mathrm{d}\mu}{\mathrm{d}v}\right)\right) \\ &= \left(\frac{1}{\mu}\right)\left(\frac{\mathrm{d}\mu}{\mathrm{d}v}\right)(p - q)\end{aligned}$$

In other words, we have

$$\begin{aligned}\left(\frac{1}{\mu}\right)\left(\frac{\mathrm{d}\mu}{\mathrm{d}v}\right) &= \frac{\frac{\partial q}{\partial y} - \frac{\partial p}{\partial x}}{p - q} \\ &= h(v)\end{aligned}$$

which is equivalent to

$$\frac{\mathrm{d}\mu}{\mu} = h(v)\,\mathrm{d}v \tag{2.21}$$

The conclusion follows by integrating Eq.(2.21).

EXAMPLE 2.41 Solve the differential equation $(x + xy)\mathrm{d}x + (xy + y)\mathrm{d}y = 0$.

Solution: Clearly, $p(x, y) = xy + y$ and $q(x, y) = x + xy$. Again, we have

$$\begin{aligned}\frac{\frac{\partial q}{\partial y} - \frac{\partial p}{\partial x}}{p - q} &= -1\\ &= h(x + y)\\ &= h(v)\end{aligned}$$

According to Theorem 2.4.4, we have

$$\begin{aligned}\mu(x, y) &= e^{\int h(v)\,\mathrm{d}v}\\ &= e^{-v}\\ &= e^{-(x+y)}\end{aligned}$$

Clearly, $m(x, y) = \frac{xy+y}{e^{x+y}}$ and $n(x, y) = \frac{x+xy}{e^{x+y}}$. According to Eq.(2.14), we have

$$\begin{aligned}c &= \int n(x, y)\,\mathrm{d}x + \int \left(m(x, y) - \frac{\partial}{\partial y}\left(\int n(x, y)\,\mathrm{d}x \right)\right) \mathrm{d}y\\ &= -\frac{(1 + y)(1 + x)}{e^{x+y}} + \int \left(\frac{(x + 1)y}{e^{x+y}} - \frac{y(1 + x)}{e^{x+y}} \right) \mathrm{d}y\\ &= -\frac{(1 + y)(1 + x)}{e^{x+y}}\end{aligned}$$

EXAMPLE 2.42 Solve the differential equation $(xy - y)\mathrm{d}x + (xy - x)\mathrm{d}y = 0$.

Solution: Clearly, $p(x, y) = xy - x$ and $q(x, y) = xy - y$. Again, we have

$$\begin{aligned}\frac{\frac{\partial q}{\partial y} - \frac{\partial p}{\partial x}}{p - q} &= -1\\ &= h(x + y)\\ &= h(v)\end{aligned}$$

According to Theorem 2.4.4, we have

$$\begin{aligned}\mu(x, y) &= e^{\int h(v)\,\mathrm{d}v}\\ &= e^{-v}\\ &= e^{-(x+y)}\end{aligned}$$

Clearly, $m(x, y) = \frac{xy-x}{e^{x+y}}$ and $n(x, y) = \frac{xy-y}{e^{x+y}}$. According to Eq.(2.14), we have

$$\begin{aligned}
c &= \int n(x, y)\,\mathrm{d}x + \int \left(m(x, y) - \frac{\partial}{\partial y}\left(\int n(x, y)\,\mathrm{d}x \right) \right) \mathrm{d}y \\
&= -\frac{xy}{e^{x+y}} + \int \left(\frac{(y-1)x}{e^{x+y}} + \frac{x(1-y)}{e^{x+y}} \right) \mathrm{d}y \\
&= -\frac{xy}{e^{x+y}}
\end{aligned}$$

Theorem 2.4.5 If $p(x, y)\mathrm{d}y + q(x, y)\mathrm{d}x = 0$ is any differential equation with $\frac{\partial p}{\partial x} \neq \frac{\partial q}{\partial y}$ and $\frac{\frac{\partial q}{\partial y} - \frac{\partial p}{\partial x}}{p+q} = h(w)$ where $w = x - y$, then $\mu(x, y) = e^{\int h(w)\,\mathrm{d}w}$.

Proof: Let $w = x - y$. Accordingly, Eq.(2.15) reduces to

$$\begin{aligned}
\left(\frac{\partial q}{\partial y} - \frac{\partial p}{\partial x} \right) &= \left(\frac{1}{\mu} \right) \left(p \left(\frac{\partial \mu}{\partial x} \right) - q \left(\frac{\partial \mu}{\partial y} \right) \right) \\
&= \left(\frac{1}{\mu} \right) \left(p \left(\frac{\mathrm{d}\mu}{\mathrm{d}w} \right) + q \left(\frac{\mathrm{d}\mu}{\mathrm{d}w} \right) \right) \\
&= \left(\frac{1}{\mu} \right) \left(\frac{\mathrm{d}\mu}{\mathrm{d}w} \right) (p+q)
\end{aligned}$$

In other words, we have

$$\begin{aligned}
\left(\frac{1}{\mu} \right) \left(\frac{\mathrm{d}\mu}{\mathrm{d}w} \right) &= \frac{\frac{\partial q}{\partial y} - \frac{\partial p}{\partial x}}{p+q} \\
&= h(w)
\end{aligned}$$

which is equivalent to

$$\frac{\mathrm{d}\mu}{\mu} = h(w)\,\mathrm{d}w \tag{2.22}$$

The conclusion follows by integrating Eq.(2.22).

EXAMPLE 2.43 Solve the differential equation $(x + xy)\mathrm{d}x + (y - xy)\mathrm{d}y = 0$.

Solution: Clearly, $p(x, y) = y - xy$ and $q(x, y) = x + xy$. Again, we have

$$\begin{aligned}
\frac{\frac{\partial q}{\partial y} - \frac{\partial p}{\partial x}}{p+q} &= 1 \\
&= h(x-y) \\
&= h(w)
\end{aligned}$$

According to Theorem 2.4.5, we have

$$\begin{aligned}
\mu(x, y) &= e^{\int h(w)\,\mathrm{d}w} \\
&= e^{w} \\
&= e^{x-y}
\end{aligned}$$

Clearly, $m(x, y) = (y - xy)e^{x-y}$ and $n(x, y) = (x + xy)e^{x-y}$. According to Eq.(2.14), we have

$$\begin{aligned} c &= \int n(x, y)\,\mathrm{d}x + \int \left(m(x, y) - \frac{\partial}{\partial y}\left(\int n(x, y)\,\mathrm{d}x \right) \right) \mathrm{d}y \\ &= (1+y)(x-1)e^{x-y} + \int \left((1-x)ye^{x-y} - y(1-x)e^{x-y} \right) \mathrm{d}y \\ &= (1+y)(x-1)e^{x-y} \end{aligned}$$

EXAMPLE 2.44 Solve the differential equation $(xy - y)\mathrm{d}x - (xy + x)\mathrm{d}y = 0$.

Solution: Clearly, $p(x, y) = -(xy + x)$ and $q(x, y) = xy - y$. Again, we have

$$\begin{aligned} \frac{\frac{\partial q}{\partial y} - \frac{\partial p}{\partial x}}{p+q} &= -1 \\ &= h(x-y) \\ &= h(w) \end{aligned}$$

According to Theorem 2.4.5, we have

$$\begin{aligned} \mu(x, y) &= e^{\int h(w)\,\mathrm{d}w} \\ &= e^{-w} \\ &= e^{-(x-y)} \\ &= e^{y-x} \end{aligned}$$

Clearly, $m(x, y) = -(xy + x)e^{y-x}$ and $n(x, y) = (xy - y)e^{y-x}$. According to Eq.(2.14), we have

$$\begin{aligned} c &= \int n(x, y)\,\mathrm{d}x + \int \left(m(x, y) - \frac{\partial}{\partial y}\left(\int n(x, y)\,\mathrm{d}x \right) \right) \mathrm{d}y \\ &= -xye^{y-x} + \int \left(-(y+1)xe^{y-x} + x(1+y)e^{y-x} \right) \mathrm{d}y \\ &= -xye^{y-x} \end{aligned}$$

Theorem 2.4.6 If $p(x, y)\mathrm{d}y + q(x, y)\mathrm{d}x = 0$ is any differential equation in which both $p(x, y)$ and $q(x, y)$ are homogeneous function in x and y with $\frac{\partial p}{\partial x} \neq \frac{\partial q}{\partial y}$ and $qx + py \neq 0$, then $\mu(x, y) = \frac{1}{qx+py}$.

Proof: It is easy to verify that

$$\begin{aligned} q\mathrm{d}x + p\mathrm{d}y &= \frac{1}{2}\left((qx+py)\left(\frac{\mathrm{d}x}{x} + \frac{\mathrm{d}y}{y} \right) + (qx - py)\left(\frac{\mathrm{d}x}{x} - \frac{\mathrm{d}y}{y} \right) \right) \\ &= \frac{1}{2}\left((qx+py)\mathrm{d}(\ln(xy) + (px - qy))\mathrm{d}\left(\ln\left(\frac{x}{y} \right) \right) \right) \end{aligned}$$

In other words, we have

$$\begin{aligned}
\frac{q\mathrm{d}x + p\mathrm{d}y}{qx + py} &= \frac{1}{2}\left(\mathrm{d}(\ln(xy)) + \left(\frac{qx - py}{qx + py}\right)\mathrm{d}\left(\ln\left(\frac{x}{y}\right)\right)\right) \\
&= \frac{1}{2}\left(\mathrm{d}(\ln(xy)) + f\left(\frac{x}{y}\right)\mathrm{d}\left(\ln\left(\frac{x}{y}\right)\right)\right) \\
&= \frac{1}{2}\left(\mathrm{d}(\ln(xy)) + f\left(e^{\ln\left(\frac{x}{y}\right)}\right)\mathrm{d}\left(\ln\left(\frac{x}{y}\right)\right)\right) \\
&= \frac{1}{2}\left(\mathrm{d}(\ln(xy))\right) + \left(\frac{1}{2}\right) f\left(e^{\ln\left(\frac{x}{y}\right)}\right)\mathrm{d}\left(\ln\left(\frac{x}{y}\right)\right) \qquad (2.23)
\end{aligned}$$

Clearly, the right hand side of Eq.(2.23) is exact. Hence the integrating factor is $\mu(x, y) = \frac{1}{qx + py}$.

EXAMPLE 2.45 Solve the differential equation $(x^2y - 2xy^2)\mathrm{d}x + (3x^2y - x^3)\mathrm{d}y = 0$.

Solution: Clearly, $p(x, y) = 3x^2y - x^3$ and $q(x, y) = x^2y - 2xy^2$. Again, $qx + py = x^2y^2 \neq 0$. According to Theorem 2.4.6, the integrating factor is $\mu(x, y) = \frac{1}{x^2y^2}$. Hence $m(x, y) = \frac{3x^2y - x^3}{x^2y^2}$ and $n(x, y) = \frac{x^2y - 2xy^2}{x^2y^2}$. According to Eq.(2.14), we have

$$\begin{aligned}
c &= \int n(x, y)\,\mathrm{d}x + \int\left(m(x, y) - \frac{\partial}{\partial y}\left(\int n(x, y)\,\mathrm{d}x\right)\right)\mathrm{d}y \\
&= \frac{x}{y} - 2\ln x + 3\int\left(\frac{1}{y}\right)\mathrm{d}y \\
&= \frac{x}{y} - \ln(x^2) + 3\ln y \\
&= \frac{x}{y} - \ln(x^2) + \ln(y^3) \\
&= \ln\left(\frac{y^3}{x^2}\right) + \frac{x}{y}
\end{aligned}$$

EXAMPLE 2.46 Solve the differential equation $(x^2 - y^2)\mathrm{d}y = xy\mathrm{d}x$.

Solution: Clearly, $p(x, y) = x^2 - y^2$ and $q(x, y) = -xy$. Again, $qx + py = -y^3 \neq 0$. According to theorem 2.4.6, the integrating factor is $\mu(x, y) = -\left(\frac{1}{y^3}\right)$. Hence $m(x, y) = \frac{y^2 - x^2}{y^3}$ and $n(x, y) = \frac{x}{y^2}$. According to Eq.(2.14), we have

$$\begin{aligned}
c &= \int n(x, y)\,\mathrm{d}x + \int\left(m(x, y) - \frac{\partial}{\partial y}\left(\int n(x, y)\,\mathrm{d}x\right)\right)\mathrm{d}y \\
&= \frac{x^2}{2y^2} + \int\left(\frac{1}{y}\right)\mathrm{d}y \\
&= \frac{x^2}{2y^2} + \ln y
\end{aligned}$$

Theorem 2.4.7 If $p(x, y)\mathrm{d}y + q(x, y)\mathrm{d}x = 0$ is any differential equation in the form $f(xy)x\mathrm{d}y + g(xy)y\mathrm{d}x = 0$ with $\frac{\partial p}{\partial x} \neq \frac{\partial q}{\partial y}$ and $qx - py \neq 0$, then $\mu(x, y) = \frac{1}{qx - py}$.

Proof: It is easy to verify that

$$
\begin{aligned}
q\mathrm{d}x + p\mathrm{d}y &= \frac{1}{2}\left((qx+py)\left(\frac{\mathrm{d}x}{x}+\frac{\mathrm{d}y}{y}\right)+(qx-py)\left(\frac{\mathrm{d}x}{x}-\frac{\mathrm{d}y}{y}\right)\right)\\
&= \frac{1}{2}\left((qx+py)\mathrm{d}(\ln(xy))+(qx-py)\mathrm{d}\left(\ln\left(\frac{x}{y}\right)\right)\right)
\end{aligned}
$$

In other words, we have

$$
\begin{aligned}
\frac{q\mathrm{d}x + p\mathrm{d}y}{qx-py} &= \frac{1}{2}\left(\left(\frac{qx+py}{qx-py}\right)\mathrm{d}(\ln(xy))+\mathrm{d}\left(\ln\left(\frac{x}{y}\right)\right)\right)\\
&= \frac{1}{2}\left(\left(\frac{g(xy)xy+f(xy)xy}{g(xy)xy-f(xy)xy}\right)\mathrm{d}(\ln(xy))+\mathrm{d}\left(\ln\left(\frac{x}{y}\right)\right)\right)\\
&= \frac{1}{2}\left(\left(\frac{g(xy)+f(xy)}{g(xy)-f(xy)}\right)\mathrm{d}(\ln(xy))+\mathrm{d}\left(\ln\left(\frac{x}{y}\right)\right)\right)\\
&= \frac{1}{2}\left(\left(h(xy)\mathrm{d}(\ln(xy))+\mathrm{d}\left(\ln\left(\frac{x}{y}\right)\right)\right)\right)\\
&= \frac{1}{2}\left(h(e^{\ln(xy)})\mathrm{d}(\ln(xy))+\mathrm{d}\left(\ln\left(\frac{x}{y}\right)\right)\right)\\
&= \frac{1}{2}\,h(e^{\ln(xy)})\mathrm{d}(\ln(xy))+\frac{1}{2}\left(\mathrm{d}\left(\ln\left(\frac{x}{y}\right)\right)\right) \qquad (2.24)
\end{aligned}
$$

Clearly, the right hand side of Eq.(2.24) is exact. Hence the integrating factor is $\mu(x, y) = \frac{1}{qx-py}$.

EXAMPLE 2.47 Solve the differential equation $(x^2y^2+xy+1)y\mathrm{d}x+(x^2y^2-xy+1)x\mathrm{d}y = 0$.
Solution: Clearly, $p(x, y) = (x^2y^2-xy+1)x$ and $q(x, y) = (x^2y^2+xy+1)y$. Again, $qx-py = 2x^2y^2 \neq 0$. According to Theorem 2.4.7, $\mu(x, y) = \frac{1}{2x^2y^2}$. Hence $m(x, y) = \frac{x^2y^2-xy+1}{2xy^2}$ and $n(x, y) = \frac{x^2y^2+xy+1}{2x^2y}$. According to Eq.(2.14), we have

$$
\begin{aligned}
c &= \int n(x, y)\,\mathrm{d}x + \int\left(m(x, y) - \int\left(\frac{\partial n}{\partial y}\right)\mathrm{d}x\right)\mathrm{d}y\\
&= xy + \ln x - \frac{1}{xy} - \int\left(\frac{1}{y}\right)\mathrm{d}y\\
&= xy + \ln x - \frac{1}{xy} - \ln y\\
&= xy + \ln\left(\frac{x}{y}\right) - \frac{1}{xy}
\end{aligned}
$$

EXAMPLE 2.48 Solve the differential equation $(2x^2y^2+xy)y\mathrm{d}x+(xy-x^2y^2)x\mathrm{d}y = 0$.
Solution: Clearly, $p(x, y) = (xy-x^2y^2)x$ and $q(x, y) = (2x^2y^2+xy)y$. Again, $qx-py = 3x^3y^3 \neq 0$. According to Theorem 2.4.7, $\mu(x, y) = \frac{1}{3x^3y^3}$. Hence $m(x, y) = \frac{xy-x^2y^2}{3x^2y^3}$ and $n(x, y) = \frac{2x^2y^2+xy}{3x^3y^2}$. According to Eq.(2.14), we have

$$\begin{aligned}c &= \int n(x,\, y)\,\mathrm{d}x + \int \left(m(x,\, y) - \frac{\partial}{\partial y}\left(\int n(x,\, y)\,\mathrm{d}x \right) \right) \mathrm{d}y \\ &= 2\ln x - \frac{1}{xy} - \int \left(\frac{1}{y} \right) \mathrm{d}y \\ &= \ln(x^2) - \frac{1}{xy} - \ln y \\ &= \ln\left(\frac{x^2}{y} \right) - \frac{1}{xy}\end{aligned}$$

Exercises 2.4

1. If $qx + py = 0$, then the differential equation $p(x,\, y)\mathrm{d}y + q(x,\, y)\mathrm{d}x = 0$ has solution.

   HINT: It is easy to verify that $p = -\frac{qx}{y}$. Hence the given differential equation reduces to $\frac{q}{y}(y\,\mathrm{d}x - x\,\mathrm{d}y) = 0$. In other words, $y\,\mathrm{d}x - x\,\mathrm{d}y = 0$ which has solution.

2. If $qx - py = 0$, then the differential equation $p(x,\, y)\mathrm{d}y + q(x,\, y)\mathrm{d}x = 0$ has solution.

   HINT: It is easy to verify that $p = \frac{qx}{y}$. Hence the given differential equation reduces to $\frac{q}{y}(y\,\mathrm{d}x + x\,\mathrm{d}y) = 0$. In other words, $y\,\mathrm{d}x + x\,\mathrm{d}y = 0$ which has solution.

3. If $px + qy = 0$, then the differential equation $p(x,\, y)\mathrm{d}y + q(x,\, y)\mathrm{d}x = 0$ has solution.

   HINT: It is easy to verify that $p = -\frac{qy}{x}$. Hence the given differential equation reduces to $\frac{q}{x}(x\,\mathrm{d}x - y\,\mathrm{d}y) = 0$. In other words, $x\,\mathrm{d}x - y\,\mathrm{d}y = 0$ which has solution.

4. If $px - qy = 0$, then the differential equation $p(x,\, y)\mathrm{d}y + q(x,\, y)\mathrm{d}x = 0$ has solution.

   HINT: It is easy to verify that $p = \frac{qy}{x}$. Hence the given differential equation reduces to $\frac{q}{x}(x\,\mathrm{d}x + y\,\mathrm{d}y) = 0$. In other words, $x\,\mathrm{d}x + y\,\mathrm{d}y = 0$ which has solution.

5. Find the integrating factor $\mu(x,\, y)$ for the differential equation $(3x^2 - y^2)\mathrm{d}y = 2xy\mathrm{d}x$.

   HINT: According to Theorem 2.4.2, $\mu(x,\, y) = e^{\int -\left(\frac{4}{y}\right)\mathrm{d}y} = \frac{1}{y^4}$.

6. Find the integrating factor $\mu(x,\, y)$ for the differential equation $(x^2 - xy)\mathrm{d}y = (1 - xy)\mathrm{d}x$.

   HINT: According to Theorem 2.4.1, $\mu(x,\, y) = e^{\int -\left(\frac{1}{x}\right)\mathrm{d}x} = \frac{1}{x}$.

7. Find the integrating factor $\mu(x,\, y)$ for the differential equation $(x + 3x^3y^4)\mathrm{d}y = -y\mathrm{d}x$.

   HINT: According to Theorem 2.4.3, $\mu(x,\, y) = e^{\int -\left(\frac{3}{xy}\right)\mathrm{d}(xy)} = \frac{1}{(xy)^3}$.

8. Find the integrating factor $\mu(x,\, y)$ for the differential equation $(2x^2y^3 - x)\mathrm{d}y = y\mathrm{d}x$.

   HINT: According to Theorem 2.4.3, $\mu(x,\, y) = e^{\int -\left(\frac{2}{xy}\right)\mathrm{d}(xy)} = \frac{1}{(xy)^2}$.

9. Find the integrating factor $\mu(x,\, y)$ for the differential equation $2xy\mathrm{d}y + (x + 3y^2)\mathrm{d}x = 0$.

   HINT: According to Theorem 2.4.1, $\mu(x,\, y) = e^{\int \left(\frac{2}{x}\right)\mathrm{d}x} = x^2$.

10. Find the integrating factor $\mu(x, y)$ for the differential equation $(ye^y - 2x)\mathrm{d}y = y\mathrm{d}x$.

    HINT: According to Theorem 2.4.2, $\mu(x, y) = e^{\int \left(\frac{1}{y}\right) \mathrm{d}y} = y$.

11. Find the integrating factor $\mu(x, y)$ for the differential equation $(y^2 + xy + 1)\mathrm{d}y + (x^2 + xy + 1)\mathrm{d}x = 0$.

    HINT: According to Theorem 2.4.4, $\mu(x, y) = e^{\int -\left(\frac{1}{x+y}\right) \mathrm{d}(x+y)} = \frac{1}{x+y}$.

12. Find the integrating factor $\mu(x, y)$ for the differential equation $e^{-x}\mathrm{d}y + e^y\mathrm{d}x = 0$.

    HINT: According to Theorem 2.4.5, $\mu(x, y) = e^{\int \mathrm{d}(x-y)} = e^{x-y}$.

13. Find the integrating factor $\mu(x, y)$ for the differential equation $e^x\mathrm{d}y + e^{-y}\mathrm{d}x = 0$.

    HINT: According to Theorem 2.4.5, $\mu(x, y) = e^{\int -\mathrm{d}(x-y)} = e^{y-x}$.

14. Find the integrating factor $\mu(x, y)$ for the differential equation $\cos(x + y)\,\mathrm{d}x - \sin(x + y)\,\mathrm{d}y = 0$.

    HINT: According to Theorem 2.4.5, $\mu(x, y) = e^{\int \mathrm{d}(x-y)} = e^{x-y}$.

15. Find the integrating factor $\mu(x, y)$ for the differential equation $\sin(x + y)\,\mathrm{d}x - \cos(x + y)\,\mathrm{d}y = 0$.

    HINT: According to Theorem 2.4.5, $\mu(x, y) = e^{\int \mathrm{d}(y-x)} = e^{y-x}$.

16. Find the integrating factor $\mu(x, y)$ for the differential equation $3y^2\mathrm{d}y + (x^3 + xy^3)\mathrm{d}x = 0$.

    HINT: According to Theorem 2.4.1, $\mu(x, y) = e^{\int x\,\mathrm{d}x} = e^{\frac{x^2}{2}}$.

17. Find the integrating factor $\mu(x, y)$ for the differential equation $e^x\,\mathrm{d}x + (e^x\cot(y) + 2y\sec(y))\mathrm{d}y = 0$.

    HINT: According to Theorem 2.4.2, $\mu(x, y) = e^{\int \cot y\,\mathrm{d}y} = \sin y$.

18. Find the integrating factor $\mu(x, y)$ for the differential equation $x\cos(y)\mathrm{d}y + (x+2)\sin y\mathrm{d}x$.

    HINT: According to Theorem 2.4.1, $\mu(x, y) = e^{\int \left(1+\frac{1}{x}\right) \mathrm{d}x} = x\,e^x$.

19. Find the integrating factor $\mu(x, y)$ for the differential equation $(x + y)\mathrm{d}y = (2xy - y\ln y)\mathrm{d}x$.

    HINT: According to Theorem 2.4.2, $\mu(x, y) = e^{\int -\left(\frac{1}{y}\right) \mathrm{d}y} = \frac{1}{y}$.

2.5 Method of Regrouping

There are certain first order nonlinear differential equation, which are non-exact but difficult for finding integrating factor can be solved by regrouping the terms. There is no general procedure for this method.

EXAMPLE 2.49 Solve the differential equation $y(x^3 - y)\,\mathrm{d}x - x(x^3 + y)\,\mathrm{d}y = 0$.

Solution: One can easily verify that

$$\begin{aligned}
y(x^3 - y)\,\mathrm{d}x - x(x^3 + y)\,\mathrm{d}y = 0 \quad &\Leftrightarrow \quad yx^3\,\mathrm{d}x - x^4\,\mathrm{d}y - xy\,\mathrm{d}y - y^2\,\mathrm{d}x = 0\\
&\Leftrightarrow \quad y\,\mathrm{d}x - x\,\mathrm{d}y - \frac{y^2\,\mathrm{d}x - xy\,\mathrm{d}y}{x^3} = 0\\
&\Leftrightarrow \quad \frac{y\,\mathrm{d}x - x\,\mathrm{d}y}{y^2} - \frac{y^2\,\mathrm{d}x - xy\,\mathrm{d}y}{y^2x^3} = 0\\
&\Leftrightarrow \quad \mathrm{d}\left(\frac{x}{y}\right) - \frac{\mathrm{d}(xy)}{yx^3} = 0\\
&\Leftrightarrow \quad \left(\frac{x}{y}\right)\mathrm{d}\left(\frac{x}{y}\right) - \frac{\mathrm{d}(xy)}{y^2x^2} = 0\\
&\Leftrightarrow \quad \left(\frac{x}{y}\right)\mathrm{d}\left(\frac{x}{y}\right) - \frac{\mathrm{d}(xy)}{(xy)^2} = 0\\
&\Leftrightarrow \quad \int\left(\frac{x}{y}\right)\mathrm{d}\left(\frac{x}{y}\right) - \int\frac{\mathrm{d}(xy)}{(xy)^2} = c
\end{aligned}$$

In other words, the required solution can be given by $x^3 + 2y = 2c\,xy^2$.

EXAMPLE 2.50 Solve the differential equation $y(x^2y^2 - 1)\,\mathrm{d}x + x(x^2y^2 + 1)\,\mathrm{d}y = 0$.

Solution: One can easily verify that

$$\begin{aligned}
y(x^2y^2 - 1)\,\mathrm{d}x + x(x^2y^2 + 1)\,\mathrm{d}y = 0 \quad &\Leftrightarrow \quad x\,\mathrm{d}y - y\,\mathrm{d}x + x^2y^3\,\mathrm{d}x + x^3y^2\,\mathrm{d}y = 0\\
&\Leftrightarrow \quad \frac{x\mathrm{d}y - y\,\mathrm{d}x}{x^2} + y^3\,\mathrm{d}x + xy^2\,\mathrm{d}y = 0
\end{aligned}$$

$$\begin{aligned}
&\Leftrightarrow \quad \mathrm{d}\left(\frac{y}{x}\right) + y^2\,(y\,\mathrm{d}x + x\,\mathrm{d}y) = 0\\
&\Leftrightarrow \quad \mathrm{d}\left(\frac{y}{x}\right) + y^2\,\mathrm{d}(xy) = 0\\
&\Leftrightarrow \quad \left(\frac{x}{y}\right)\mathrm{d}\left(\frac{y}{x}\right) + xy\,\mathrm{d}(xy) = 0\\
&\Leftrightarrow \quad \left(\frac{y}{x}\right)^{-1}\mathrm{d}\left(\frac{y}{x}\right) + xy\,\mathrm{d}(xy) = 0\\
&\Leftrightarrow \quad \int\left(\frac{y}{x}\right)^{-1}\mathrm{d}\left(\frac{y}{x}\right) + \int xy\,\mathrm{d}(xy) = c
\end{aligned}$$

In other words, the required solution can be given by $\ln\left(\frac{y}{x}\right) + \frac{(xy)^2}{2} = c$.

EXAMPLE 2.51 Solve the differential equation $(x + y)\,\mathrm{d}x + (x - y)\,\mathrm{d}y = 0$.

Solution: One can easily verify that

$$\begin{aligned}
(x + y)\,\mathrm{d}x + (x - y)\,\mathrm{d}y = 0 \quad &\Leftrightarrow \quad x\,\mathrm{d}x - y\,\mathrm{d}y + y\,\mathrm{d}x + x\,\mathrm{d}y = 0\\
&\Leftrightarrow \quad x\,\mathrm{d}x - y\,\mathrm{d}y + \mathrm{d}(xy) = 0\\
&\Leftrightarrow \quad \int x\,\mathrm{d}x - \int y\,\mathrm{d}y + \int \mathrm{d}(xy) = c
\end{aligned}$$

In other words, the required solution can be given by $\frac{x^2}{2} - \frac{y^2}{2} + xy = c$

EXAMPLE 2.52 Solve the differential equation $(x + y)\,\mathrm{d}y + (x - y)\,\mathrm{d}x = 0$.

Solution: One can easily verify that

$$\begin{aligned}
(x + y)\,\mathrm{d}y + (x - y)\,\mathrm{d}x = 0 \quad &\Leftrightarrow \quad x\,\mathrm{d}x + y\,\mathrm{d}y + x\,\mathrm{d}y - y\,\mathrm{d}x = 0 \\
&\Leftrightarrow \quad \frac{x\,\mathrm{d}x + y\,\mathrm{d}y}{x^2 + y^2} + \frac{x\,\mathrm{d}y - y\,\mathrm{d}x}{x^2 + y^2} = 0 \\
&\Leftrightarrow \quad \frac{1}{2}\left(\frac{\mathrm{d}(x^2 + y^2)}{x^2 + y^2}\right) + \frac{x\,\mathrm{d}y - y\,\mathrm{d}x}{x^2 + y^2} = 0 \\
&\Leftrightarrow \quad \frac{1}{2}\left(\frac{\mathrm{d}(x^2 + y^2)}{x^2 + y^2}\right) + \frac{\frac{x\,\mathrm{d}y - y\,\mathrm{d}x}{x^2}}{\frac{x^2+y^2}{x^2}} = 0 \\
&\Leftrightarrow \quad \frac{1}{2}\left(\frac{\mathrm{d}(x^2 + y^2)}{x^2 + y^2}\right) + \frac{\mathrm{d}\left(\frac{y}{x}\right)}{1 + \left(\frac{y}{x}\right)^2} = 0 \\
&\Leftrightarrow \quad \frac{1}{2}\int\left(\frac{\mathrm{d}(x^2 + y^2)}{x^2 + y^2}\right) + \int\frac{\mathrm{d}\left(\frac{y}{x}\right)}{1 + \left(\frac{y}{x}\right)^2} = c
\end{aligned}$$

In other words, the required solution can be given by $\ln\left(\sqrt{x^2 + y^2}\right) + \tan^{-1}\left(\frac{y}{x}\right) = c$

EXAMPLE 2.53 Solve the differential equation $\left(x^{n+1}y^n + ay\right)\mathrm{d}y + \left(x^n y^{n+1} + ax\right)\mathrm{d}x = 0$.

Solution: One can easily verify that

$$\begin{aligned}
\left(x^{n+1}y^n + ay\right)\,\mathrm{d}y + \left(x^n y^{n+1} + ax\right)\,\mathrm{d}x = 0 \quad &\Leftrightarrow \quad a(y\,\mathrm{d}y + x\,\mathrm{d}x) + x^n y^n(x\,\mathrm{d}y + y\,\mathrm{d}x) = 0 \\
&\Leftrightarrow \quad a(y\,\mathrm{d}y + x\,\mathrm{d}x) + x^n y^n\,\mathrm{d}(xy) = 0 \\
&\Leftrightarrow \quad a\int(y\,\mathrm{d}y + x\,\mathrm{d}x) + \int x^n y^n\,\mathrm{d}(xy) = c
\end{aligned}$$

In other words, the required solution can be given by $\frac{x^2}{2} + \frac{y^2}{2} + \frac{(xy)^{n+1}}{n+1} = c$

EXAMPLE 2.54 Solve the differential equation $\left(x^{n+1}y^n + ay\right)\mathrm{d}x + \left(x^n y^{n+1} + ax\right)\mathrm{d}y = 0$.

Solution: One can easily verify that

$$\begin{aligned}
\left(x^{n+1}y^n + ay\right)\,\mathrm{d}x + \left(x^n y^{n+1} + ax\right)\,\mathrm{d}y = 0 \quad &\Leftrightarrow \quad a(y\,\mathrm{d}x + x\,\mathrm{d}y) + x^n y^n(x\,\mathrm{d}x + y\,\mathrm{d}y) = 0 \\
&\Leftrightarrow \quad a\,\mathrm{d}(xy) + x^n y^n(x\,\mathrm{d}x + y\,\mathrm{d}y) = 0 \\
&\Leftrightarrow \quad \frac{a\,\mathrm{d}(xy)}{x^n y^n} + x\,\mathrm{d}x + y\,\mathrm{d}y = 0 \\
&\Leftrightarrow \quad a\int\frac{\mathrm{d}(xy)}{x^n y^n} + \int x\,\mathrm{d}x + \int y\,\mathrm{d}y = c
\end{aligned}$$

In other words, the required solution can be given by $\frac{x^2}{2} + \frac{y^2}{2} - \frac{a}{(n-1)\,(xy)^{n-1}} = c$

EXAMPLE 2.55 Solve the differential equation $y(x^3\, e^{xy} - y)\, \mathrm{d}x + x(y + x^3\, e^{xy})\, \mathrm{d}y = 0$.

Solution: One can easily verify that

$$\begin{aligned}
y(x^3\, e^{xy} - y)\, \mathrm{d}x + x(xy + x^3\, e^{xy})\, \mathrm{d}y = 0 \quad &\Leftrightarrow \quad yx^3\, e^{xy}\, \mathrm{d}x + x^4\, e^{xy}\, \mathrm{d}y + xy\, \mathrm{d}y - y^2\, \mathrm{d}x = 0 \\
&\Leftrightarrow \quad x^3(y\, e^{xy}\, \mathrm{d}x + x\, e^{xy}\, \mathrm{d}y) + xy\, \mathrm{d}y - y^2\, \mathrm{d}x = 0 \\
&\Leftrightarrow \quad y\, e^{xy}\, \mathrm{d}x + x\, e^{xy}\, \mathrm{d}y + \frac{xy\, \mathrm{d}y - y^2\, \mathrm{d}x}{x^3} = 0 \\
&\Leftrightarrow \quad \mathrm{d}\,(e^{xy}) + \frac{xy\, \mathrm{d}y - y^2\, \mathrm{d}x}{x^3} = 0 \\
&\Leftrightarrow \quad \mathrm{d}\,(e^{xy}) + \left(\frac{y}{x}\right)\left(\frac{x\, \mathrm{d}y - y\, \mathrm{d}x}{x^2}\right) = 0 \\
&\Leftrightarrow \quad \mathrm{d}\,(e^{xy}) + \left(\frac{y}{x}\right) \mathrm{d}\left(\frac{y}{x}\right) = 0 \\
&\Leftrightarrow \quad \int \mathrm{d}\,(e^{xy}) + \int \left(\frac{y}{x}\right) \mathrm{d}\left(\frac{y}{x}\right) = c
\end{aligned}$$

In other words, the required solution can be given by $e^{xy} + \frac{1}{2}\left(\frac{y}{x}\right)^2 = c$

EXAMPLE 2.56 Solve the differential equation $(x^3y^3 + 1)\, \mathrm{d}x + x^4y^2\, \mathrm{d}y = 0$.

Solution: One can easily verify that

$$\begin{aligned}
(x^3y^3 + 1)\, \mathrm{d}x + x^4y^2\, \mathrm{d}y = 0 \quad &\Leftrightarrow \quad x^3y^3\, \mathrm{d}x + x^4y^2\, \mathrm{d}y + \mathrm{d}x = 0 \\
&\Leftrightarrow \quad x^3y^2(y\, \mathrm{d}x + x\, \mathrm{d}y) + \mathrm{d}x = 0 \\
&\Leftrightarrow \quad (xy)^2\, \mathrm{d}(xy) + \frac{\mathrm{d}x}{x} = 0 \\
&\Leftrightarrow \quad \int (xy)^2\, \mathrm{d}(xy) + \int \frac{\mathrm{d}x}{x} = c
\end{aligned}$$

In other words, the required solution can be given by $\frac{(xy)^3}{3} + \ln(x) = c$

EXAMPLE 2.57 Solve the differential equation $y(y^3 - x)\, \mathrm{d}x + x(y^3 + x)\, \mathrm{d}y = 0$.

Solution: One can easily verify that

$$\begin{aligned}
y(y^3 - x)\, \mathrm{d}x + x(y^3 + x)\, \mathrm{d}y = 0 \quad &\Leftrightarrow \quad y^4\, \mathrm{d}x + xy^3\, \mathrm{d}y + x^2\, \mathrm{d}y - xy\, \mathrm{d}x = 0 \\
&\Leftrightarrow \quad y^3(y\, \mathrm{d}x + x\, \mathrm{d}y) + x(x\, \mathrm{d}y - y\, \mathrm{d}x) = 0 \\
&\Leftrightarrow \quad (y\, \mathrm{d}x + x\, \mathrm{d}y) + \frac{x(x\, \mathrm{d}y - y\, \mathrm{d}x)}{y^3} = 0 \\
&\Leftrightarrow \quad \mathrm{d}(xy) + \frac{x^3}{y^3}\frac{(x\, \mathrm{d}y - y\, \mathrm{d}x)}{x^2} = 0
\end{aligned}$$

$$\Leftrightarrow \quad \mathrm{d}(xy) + \left(\frac{x}{y}\right)^3 \mathrm{d}\left(\frac{y}{x}\right) = 0$$

$$\Leftrightarrow \quad \int \mathrm{d}(xy) + \int \left(\frac{y}{x}\right)^{-3} \mathrm{d}\left(\frac{y}{x}\right) = c$$

In other words, the required solution can be given by $xy - \frac{1}{2}\left(\frac{x}{y}\right)^2 = c$

EXAMPLE 2.58 Solve the differential equation $y(1 - x^3y)\,\mathrm{d}x - x(x^3y + 1)\,\mathrm{d}y = 0$.

Solution: One can easily verify that

$$\begin{aligned}
y(1 - x^3y)\,\mathrm{d}x - x(x^3y + 1)\,\mathrm{d}y = 0 \quad &\Leftrightarrow \quad y\,\mathrm{d}x - x\,\mathrm{d}y - (x^3y^2\,\mathrm{d}x + x^4y\,\mathrm{d}y) = 0 \\
&\Leftrightarrow \quad \frac{y\,\mathrm{d}x - x\,\mathrm{d}y}{y^2} - \frac{x^3y(y\,\mathrm{d}x + x\,\mathrm{d}y)}{y^2} = 0 \\
&\Leftrightarrow \quad \mathrm{d}\left(\frac{x}{y}\right) - \frac{x^3}{y}\,\mathrm{d}(xy) = 0 \\
&\Leftrightarrow \quad \left(\frac{y}{x}\right)^2 \mathrm{d}\left(\frac{x}{y}\right) - xy\,\mathrm{d}(xy) = 0 \\
&\Leftrightarrow \quad \left(\frac{x}{y}\right)^{-2} \mathrm{d}\left(\frac{x}{y}\right) - xy\,\mathrm{d}(xy) = 0 \\
&\Leftrightarrow \quad \int \left(\frac{x}{y}\right)^{-2} \mathrm{d}\left(\frac{x}{y}\right) - \int xy\,\mathrm{d}(xy) = c
\end{aligned}$$

In other words, the required solution can be given by $\frac{y}{x} + \frac{(xy)^2}{2} = c$

Exercises 2.5

1. Solve the following nonlinear differential equations by regrouping.

   (a) $(x^3 + xy^2 + y)\,\mathrm{d}x + (y^3 + x^2y + x)\,\mathrm{d}y = 0$.
   HINT: It is easy to verify that

   $$(x^3 + xy^2 + y)\,\mathrm{d}x + (y^3 + x^2y + x)\,\mathrm{d}y \quad \Leftrightarrow \quad \mathrm{d}(xy) + \mathrm{d}\left(\frac{x^4}{4} + \frac{y^4}{4}\right) + \frac{1}{2}\mathrm{d}(x^2y^2)$$

   (b) $y\,\mathrm{d}x + (x + x^3y^2)\,\mathrm{d}y = 0$.
   HINT: It is easy to verify that

   $$\begin{aligned}
   y\,\mathrm{d}x + (x + x^3y^2)\,\mathrm{d}y = 0 \quad &\Leftrightarrow \quad \mathrm{d}(xy) + x^3y^2\,\mathrm{d}y = 0 \\
   &\Leftrightarrow \quad \frac{\mathrm{d}(xy)}{x^3y^3} + \frac{\mathrm{d}y}{y} = 0
   \end{aligned}$$

(c) $(4x^3y^3 - 2xy)\,\mathrm{d}x + (3x^4y^2 - x^2)\,\mathrm{d}y = 0$.
HINT: It is easy to verify that

$$(4x^3y^3 - 2xy)\,\mathrm{d}x + (3x^4y^2 - x^2)\,\mathrm{d}y = 0 \quad \Leftrightarrow \quad \mathrm{d}(x^4y^3) - \mathrm{d}(x^2y) = 0$$

(d) $3x^2y\,\mathrm{d}x + (y^4 - x^3)\,\mathrm{d}y = 0$.
HINT: It is easy to verify that

$$\begin{aligned} 3x^2y\,\mathrm{d}x + (y^4 - x^3)\,\mathrm{d}y = 0 \quad &\Leftrightarrow \quad 3x^2y\,\mathrm{d}x - x^3\,\mathrm{d}y + y^4\,\mathrm{d}y = 0 \\ &\Leftrightarrow \quad \frac{3x^2y\,\mathrm{d}x - x^3\,\mathrm{d}y}{y^2} + y^2\,\mathrm{d}y = 0 \\ &\Leftrightarrow \quad \mathrm{d}\left(\frac{x^3}{y}\right) + \mathrm{d}\left(\frac{y^3}{3}\right) = 0 \end{aligned}$$

(e) $y\,\mathrm{d}x - x\,\mathrm{d}y - xy^3\,\mathrm{d}y = 0$.
HINT: It is easy to verify that

$$\begin{aligned} y\,\mathrm{d}x - x\,\mathrm{d}y - xy^3\,\mathrm{d}y = 0 \quad &\Leftrightarrow \quad \frac{y\,\mathrm{d}x - x\,\mathrm{d}y}{y^2} - xy\,\mathrm{d}y = 0 \\ &\Leftrightarrow \quad \mathrm{d}\left(\frac{x}{y}\right) - xy\,\mathrm{d}y = 0 \\ &\Leftrightarrow \quad \left(\frac{y}{x}\right)\mathrm{d}\left(\frac{x}{y}\right) - y^2\,\mathrm{d}y = 0 \\ &\Leftrightarrow \quad \left(\frac{x}{y}\right)^{-1}\mathrm{d}\left(\frac{x}{y}\right) - \mathrm{d}\left(\frac{y^3}{3}\right) = 0 \end{aligned}$$

(f) $x\,\mathrm{d}y - (x^5 + x^3y^2 + y)\,\mathrm{d}x = 0$.
HINT: It is easy to verify that

$$\begin{aligned} x\,\mathrm{d}y - (x^5 + x^3y^2 + y)\,\mathrm{d}x = 0 \quad &\Leftrightarrow \quad x\,\mathrm{d}y - y\,\mathrm{d}x - (x^5 + x^3y^2)\,\mathrm{d}x = 0 \\ &\Leftrightarrow \quad \frac{x\,\mathrm{d}y - y\,\mathrm{d}x}{x^2} - (x^3 + xy^2)\,\mathrm{d}x = 0 \\ &\Leftrightarrow \quad \mathrm{d}\left(\frac{y}{x}\right) - x(x^2 + y^2)\,\mathrm{d}x = 0 \\ &\Leftrightarrow \quad \mathrm{d}\left(\frac{y}{x}\right) - x^3\left(1 + \frac{y^2}{x^2}\right)\mathrm{d}x = 0 \\ &\Leftrightarrow \quad \frac{\mathrm{d}\left(\frac{y}{x}\right)}{1 + \left(\frac{y}{x}\right)^2} - x^3\,\mathrm{d}x = 0 \\ &\Leftrightarrow \quad \frac{\mathrm{d}\left(\frac{y}{x}\right)}{1 + \left(\frac{y}{x}\right)^2} - \mathrm{d}\left(\frac{x^4}{4}\right)x = 0 \end{aligned}$$

(g) $x\,\mathrm{d}y - (y + x^2 + 9y^2)\,\mathrm{d}x = 0$.
HINT: It is easy to verify that

$$\begin{aligned} x\,\mathrm{d}y - (y + x^2 + 9y^2)\,\mathrm{d}x = 0 \quad &\Leftrightarrow \quad x\,\mathrm{d}y - y\,\mathrm{d}x - x^2\left(1 + \frac{9y^2}{x^2}\right)\mathrm{d}y = 0 \\ &\Leftrightarrow \quad \frac{x\,\mathrm{d}y - y\,\mathrm{d}x}{x^2} - \left(1 + \left(\frac{3y}{x}\right)^2\right)\mathrm{d}y = 0 \\ &\Leftrightarrow \quad \mathrm{d}\left(\frac{3y}{x}\right) - 3\left(1 + \left(\frac{3y}{x}\right)^2\right)\mathrm{d}y = 0 \\ &\Leftrightarrow \quad \frac{\mathrm{d}\left(\frac{3y}{x}\right)}{1 + \left(\frac{3y}{x}\right)^2} - 3\,\mathrm{d}y = 0 \end{aligned}$$

(h) $y(2xy + e^x)\,\mathrm{d}x - e^x\,\mathrm{d}y = 0$.
HINT: It is easy to verify that

$$\begin{aligned} y(2xy + e^x)\,\mathrm{d}x - e^x\,\mathrm{d}y = 0 \quad &\Leftrightarrow \quad 2xy^2\,\mathrm{d}x + y\,e^x\,\mathrm{d}x - e^x\,\mathrm{d}y = 0 \\ &\Leftrightarrow \quad 2x\,\mathrm{d}x + \frac{y\,e^x\,\mathrm{d}x - e^x\,\mathrm{d}y}{y^2} = 0 \\ &\Leftrightarrow \quad 2x\,\mathrm{d}x + e^x\left(\frac{y\,\mathrm{d}x - \mathrm{d}y}{y^2}\right) = 0 \\ &\Leftrightarrow \quad 2x\,\mathrm{d}x + \mathrm{d}\left(\frac{e^x}{y}\right) = 0 \end{aligned}$$

(i) $(y^2\,e^{xy^2} + 4x^3)\,\mathrm{d}x + (2xy\,e^{xy^2} - 3y^2)\,\mathrm{d}y = 0$.
HINT: It is easy to verify that

$$(y^2\,e^{xy^2} + 4x^3)\,\mathrm{d}x + (2xy\,e^{xy^2} - 3y^2)\,\mathrm{d}y = 0 \quad \Leftrightarrow \quad \mathrm{d}\left(e^{xy^2}\right) + \mathrm{d}(x^4 - y^3) = 0$$

(j) $y(1 - x^2y^4)\,\mathrm{d}x - (x(1 + x^2y^4)\mathrm{d}y = 0$.
HINT: It is easy to verify that

$$\begin{aligned} y(1 - x^2y^4)\,\mathrm{d}x - x(1 + x^2y^4)\mathrm{d}y = 0 \quad &\Leftrightarrow \quad y\,\mathrm{d}x - x^2y^5\,\mathrm{d}x - x^3y^4\,\mathrm{d}y - x\,\mathrm{d}y = 0 \\ &\Leftrightarrow \quad y\,\mathrm{d}x - x\,\mathrm{d}y - x^2y^5\,\mathrm{d}x - x^3y^4\,\mathrm{d}y = 0 \\ &\Leftrightarrow \quad \frac{y\,\mathrm{d}x - x\,\mathrm{d}y}{y^2} - x^2y^3\,\mathrm{d}x - x^3y^2\,\mathrm{d}y = 0 \\ &\Leftrightarrow \quad \mathrm{d}\left(\frac{x}{y}\right) - x^2y^2\,(y\,\mathrm{d}x + x\,\mathrm{d}y) = 0 \\ &\Leftrightarrow \quad \mathrm{d}\left(\frac{x}{y}\right) - (xy)^2\,\mathrm{d}(xy) = 0 \end{aligned}$$

(k) $(2x^3y^4 - y)\,dx + (x^4y^3 + x)dy = 0$.

HINT: It is easy to verify that

$$\begin{aligned}(2x^3y^4 - y)dx + (x^4y^3 + x)dy = 0 \;&\Leftrightarrow\; 2x^3y^4\,dx + x^4y^3\,dy + y\,dx - x\,dy = 0\\ &\Leftrightarrow\; 2x^3y^2\,dx + x^4y\,dy + \frac{y\,dx - x\,dy}{y^2} = 0\\ &\Leftrightarrow\; x^2y(2xy\,dx + x^2\,dy) + d\left(\frac{x}{y}\right) = 0\\ &\Leftrightarrow\; x^2y\,d(x^2y) + d\left(\frac{x}{y}\right) = 0\end{aligned}$$

2.6 Equation Type $(ax + by + c)dx = (a'x + b'y + c')dy$

Any equation of the type $(ax + by + c)dx = (a'x + b'y + c')dy$ can be transformed into an equation of the type $(au + bz)du = (a'u + b'z)dz$ by some suitable transform $x = u + \alpha$ and $y = z + \beta$, where α and β are constants to be determined, provided $ab' \neq a'b$. If $ab' = a'b$, then one can replace $ax + by = v$ in order to make the modified differential equation applicable to variables separable form.

Replacing x and y in terms of u and z, we have

$$ax + by + c = au + bz + (a\alpha + b\beta + c)$$

and

$$a'x + b'y + c' = a'u + b'z + (a'\alpha + b'\beta + c')$$

If one assumes that

$$a\alpha + b\beta + c = 0 \tag{2.25}$$

and

$$a'\alpha + b'\beta + c' = 0 \tag{2.26}$$

then the given differential equation reduces to the required type

$$(au + bz)du = (a'u + b'z)dz \tag{2.27}$$

which can be solved by changing the dependent variable z using $z = uv$. Again, Eq.(2.27) in terms of the variable v becomes

$$\begin{aligned}\frac{du}{u} &= \left(\frac{a' + b'v}{a + (b - a')v - b'v^2}\right)dv\\ &= -\frac{1}{2}\left(\frac{2a' + 2b'v}{b'v^2 + (a' - b)v - a}\right)dv \end{aligned}\tag{2.28}$$

which is in variable separable form. Hence the solution of Eq.(2.28) can be obtained by the formula given in Eq.(2.2). Integration of the right hand side of Eq.(2.28) is simple when $a' = -b$, otherwise one has to apply partial fraction to the right hand side of Eq.(2.28). Again, unique solution of Eqs.(2.25) and (2.26) is possible if and only if $ab' \neq a'b$. Assume that $ab' \neq a'b$. Hence the solution of Eqs.(2.25) and (2.26) is

$$\alpha = \frac{b'c - bc'}{a'b - ab'} \tag{2.29}$$

and

$$\beta = \frac{ac' - a'c}{a'b - ab'} \tag{2.30}$$

Again, the general solution of the given differential equation can be obtained from the solution of Eq.(2.28) by replacing $u = x - \alpha$ and $y = z - \beta$ where α and β are given in Eqs.(2.29) and (2.30).

EXAMPLE 2.59 Find the general solution of the differential equation $-(x - 2y + 3)\mathrm{d}x = (2x - 4y + 5)\mathrm{d}y$.

Solution: Clearly, $a = -1$, $b = 2$, $c = -3$, $a' = 2$, $b' = -4$ and $c' = 5$. Again, $ab' = a'b$. Hence $x - 2y = v$. Accordingly, the given differential equation reduces to

$$\frac{\mathrm{d}v}{\mathrm{d}x} = \frac{4v + 11}{2v + 5}$$

In other words, we have

$$\begin{aligned}\mathrm{d}x &= \left(\frac{2v + 5}{4v + 11}\right)\mathrm{d}v \\ &= \left(\frac{1}{2} - \frac{1}{2}\left(\frac{1}{4v + 11}\right)\right)\mathrm{d}v \end{aligned} \tag{2.31}$$

Again, solution of Eq.(2.31) is

$$8x = 4v - \ln(4v + 11) + c$$

The general solution of the given differential equation is

$$4x + 8y + \ln(4x - 8y + 11) = c$$

EXAMPLE 2.60 Find the general solution of the differential equation $(3x - 2y + 2)\mathrm{d}x = (2x - 4y + 5)\mathrm{d}y$.

Solution: Clearly, $a = 2$, $b = -2$, $c = 2$, $a' = 2$, $b' = -4$ and $c' = 5$. Again, $ab' \neq a'b$. According to Eqs.(2.29) and (2.30), $\alpha = \frac{2}{3}$ and $\beta = 2$. Accordingly, Eq.(2.28) reduces to

$$\frac{\mathrm{d}u}{u} = \frac{1}{2}\left(\frac{8v - 4}{2 - 4v + 4v^2}\right)\mathrm{d}v \tag{2.32}$$

Again, the general solution of Eq.(2.32) is

$$\frac{u^2}{2 - 4v + 4v^2} = c$$

The general solution of the given differential equation is

$$\frac{\left(x-\frac{2}{3}\right)^4}{\left(x-\frac{2}{3}\right)^2-2\left(x-\frac{2}{3}\right)(y-2)+2(y-2)^2}=2c$$

EXAMPLE 2.61 Find the general solution of the differential equation $(2x+3y+2)\mathrm{d}x = (3x+2y+3)\mathrm{d}y$.

Solution: Clearly, $a = 2$, $b = 3$, $c = 2$, $a' = 3$, $b' = 2$ and $c' = 3$. Again, $ab' \neq a'b$ and $a' \neq -b$. According to Eqs.(2.29) and (2.30), $\alpha = -1$ and $\beta = 0$. Accordingly, Eq.(2.28) reduces to

$$\begin{aligned}\frac{\mathrm{d}u}{u} &= -\frac{1}{2}\left(\frac{2v+3}{v^2-1}\right)\mathrm{d}v \\ &= \frac{1}{4}\left(\frac{1}{v+1}\right)\mathrm{d}v - \frac{5}{4}\left(\frac{1}{v-1}\right)\mathrm{d}v \end{aligned} \tag{2.33}$$

Again, the general solution of Eq.(2.33) is

$$\frac{u^4(v-1)^5}{(v+1)} = c$$

The general solution of the given differential equation is

$$\frac{(y-x-1)^5}{x+y+1} = c$$

EXAMPLE 2.62 Find the general solution of the differential equation $(3x+2y+3)\mathrm{d}x = (2x+3y+2)\mathrm{d}y$.

Solution: Clearly, $a = 3$, $b = 2$, $c = 3$, $a' = 2$, $b' = 3$ and $c' = 2$. Again, $ab' \neq a'b$ and $a' \neq -b$. According to Eqs.(2.29) and (2.30), $\alpha = -1$ and $\beta = 0$. Accordingly, Eq.(2.28) reduces to

$$\begin{aligned}\frac{\mathrm{d}u}{u} &= -\frac{1}{3}\left(\frac{3v+2}{v^2-1}\right)\mathrm{d}v \\ &= -\frac{1}{6}\left(\frac{1}{v+1}\right)\mathrm{d}v - \frac{5}{6}\left(\frac{1}{v-1}\right)\mathrm{d}v \end{aligned} \tag{2.34}$$

Again, the general solution of Eq.(2.34) is

$$(v+1)(v-1)^5u^6 \quad = \quad c$$

The general solution of the given differential equation is

$$(y-x-1)^5(x+y+1) \quad = \quad c$$

Exercises 2.6

1. Solve the differential equation $(y + x)\mathrm{d}x + (x - y)\mathrm{d}y = 0$.

   HINT: One can verify that $xy + \frac{x^2 - y^2}{2} = c$.

2. Solve the differential equation $(x + 3)\mathrm{d}x + 9(y + 1)\mathrm{d}y = 0$.

   HINT: One can verify that $3x + y + \frac{x^2 + 9y^2}{2} = c$.

3. Solve the differential equation $(3x + 1)\mathrm{d}x + (y + 1)\mathrm{d}y = 0$.

   HINT: One can verify that $x + y + \frac{3x^2 + y^2}{2} = c$.

4. Solve the differential equation $(y - x)\,\mathrm{d}x + (x - y)\,\mathrm{d}y = 0$.

   HINT: One can verify that $xy - \frac{x^2 + y^2}{2} = c$.

5. Solve the differential equation $(2x + y + 1)\mathrm{d}x + (4x + 2y - 1)\mathrm{d}y = 0$.

   HINT: One can verify that $x + 2y + \ln(2x + y - 1) = c$.

6. Solve the differential equation $(y - x + 1)\mathrm{d}x + (x - y - 5)\mathrm{d}y = 0$.

   HINT: One can verify that $xy + x - 5y - \frac{x^2 + y^2}{2} = c$.

7. Solve the differential equation $(x + 2y - 1)\mathrm{d}x + (2x - y + 3)\mathrm{d}y = 0$.

   HINT: One can verify that $c = (y - 1)^2 - 4(y - 1)(x + 1) - (x - 1)^2$.

8. Solve the differential equation $(y - x - 3)\mathrm{d}x + (x + y + 1)\mathrm{d}y = 0$.

   HINT: One can verify that $c = (y - 1)^2 + 2(y - 1)(x + 2) - (x + 2)^2$.

9. Solve the differential equation $(y - x)\mathrm{d}x + (x - y + 1)\mathrm{d}y = 0$.

   HINT: One can verify that $xy + y - \frac{x^2 + y^2}{2} = c$.

2.7 Equation Reducible to Separable Form

Certain first order differential equations are not separable as such, but can be made separable by a simple change of variables. The change of variables depends upon the type of differential equation given. In particular, this is possible whenever $p(x, y)$ and $q(x, y)$ of the differential equation $p(x, y)\,dy + q(x, y)\,dx = 0$ are homogeneous in x and y.

EXAMPLE 2.63 Solve the differential equation $(xy' - y)\cos\frac{2y}{x} + 3x^4 = 0$.

Solution: Let $y = vx$. Clearly, we have $y' = v + xv'$. Hence the given differential equation reduces to

$$v'\cos 2v + 3x^2 = 0$$

which is of the variables separable from. Hence the corresponding general solution is

$$\sin 2v + 2x^3 = 2c$$

In other words, the general solution of the given equation is

$$\sin\left(\frac{2y}{x}\right) + 2x^3 = 2c$$

EXAMPLE 2.64 Solve the differential equation $\left(2yy' - \frac{y^2}{x}\right)\cos\frac{y^2}{x} + 3x^3 = 0$.
Solution: Let $y^2 = vx$. Clearly, we have $2yy' = v + xv'$. Hence the given differential equation reduces to

$$v'\cos v + 3x^2 = 0$$

which is of the variables separable form. Hence the corresponding general solution is

$$\sin v + x^3 = c$$

In other words, the general solution of the given equation is

$$\sin\left(\frac{y^2}{x}\right) + x^3 = c$$

EXAMPLE 2.65 Solve the differential equation $(x^2 - y^2)y' = 2xy$.
Solution: Let $y = vx$. Clearly, we have $y' = v + xv'$. Hence we have

$$\begin{aligned} v + xv' &= y' \\ &= \frac{2xy}{x^2 - y^2} \\ &= \frac{2v}{1 - v^2} \end{aligned}$$

In other words, we have

$$\begin{aligned} xv' &= \frac{2v}{1 - v^2} - v \\ &= \frac{v + v^3}{1 - v^2} \end{aligned} \tag{2.35}$$

Clearly, Eq.(2.35) is of the variables separable form. Again, the solution of Eq.(2.35) can be given by

$$\begin{aligned} \ln(x) &= \int \frac{dx}{x} \\ &= \int \left(\frac{1 - v^2}{v + v^3}\right) dv \\ &= \int \frac{dv}{v} - \int \left(\frac{2v}{1 + v^2}\right) dv \\ &= \ln v - \ln(1 + v^2) + \ln\left(\frac{1}{2c}\right) \\ &= \ln\left(\frac{v}{1 + v^2}\right) - \ln\left(\frac{1}{2c}\right) \end{aligned}$$

In other words, we have

$$\frac{1}{2c} = \frac{v}{x(1+v^2)}$$
$$= \frac{y}{x^2+y^2}$$

The solution of the given differential equation is the set of all circles which are tangential to the x-axis and given by $x^2 + y^2 = 2cy$.

Exercises 2.7

1. Solve the differential equation $y' = (x + e^y - 1)e^{-y}$.

   HINT: Let $v = e^y + x$. Clearly, the given differential equation reduces to $v' = v$. In other words, one can have $e^y + x = c\,e^x$.

2. Solve the differential equation $xy' = e^{-xy} - y$.

   HINT: Let $v = yx$. Clearly, the given differential equation reduces to $v' = e^{-v}$. In other words, one can have $xy = \ln(x + c)$.

3. Solve the differential equation $y' = (y - x)^2 + 1$.

   HINT: Let $v = y - x$. Clearly, the given differential equation reduces to $v' = v^2$. In other words, one can have $(y - x)(x + c) = -1$.

4. Solve the differential equation $y' = \tan(x + y) - 1$.

   HINT: Let $v = y + x$. Clearly, the given differential equation reduces to $v' = \tan(v)$. In other words, one can have $\sin(y + x) = c\,e^x$.

5. Solve the differential equation $2x^2yy' = \tan(x^2y^2) - 2xy^2$.

   HINT: Let $v = y^2x^2$. Clearly, the given differential equation reduces to $v' = \tan(v)$. In other words, one can have $\sin(y^2x^2) = c\,e^x$.

6. Solve the differential equation $y' = x(y - x)^2 + 1$.

   HINT: Let $v = y - x$. Clearly, the given differential equation reduces to $v' = xv^2$. In other words, one can have $(y - x)(x^2 + 2c) = -2$.

7. Solve the differential equation $2xyy' = x^2 - y^2$.

   HINT: Let $v = y^2x$. Clearly, the given differential equation reduces to $v' = x^2$. In other words, one can have $xy^2 = \frac{x^3}{3} + c$.

2.8 Equation Reducible to Linear Form

Any nonlinear differential equation of the form $g'(y)y' + g(y)p(x) = q(x)$ can be transferred into a linear differential equation of the form $z' + p(x)z = q(x)$ by transforming the dependent variable $z = g(y)$. Clearly, $z' + p(x)z = q(x)$ is a first order linear nonhomogeneous differential equation. According to Eq.(2.8), the solution of the differential equation $z' + p(x)z = q(x)$ is given by

$$z(x) = e^{-\int p(x)\,\mathrm{d}x}\left(\int e^{\int p(x)\,\mathrm{d}x} q(x)\,\mathrm{d}x + c\right) \tag{2.36}$$

EXAMPLE 2.66 Solve the differential equation $\cos(y)\, y' + 2\sin(y)x = x$.

Solution: Clearly, $g(y) = \sin y$, $p(x) = 2x$ and $q(x) = x$. According to Eq.(2.36), we have

$$\begin{aligned} z(x) &= e^{-\int p(x)\,\mathrm{d}x}\left(\int e^{\int p(x)\,\mathrm{d}x} q(x)\,\mathrm{d}x + c\right) \\ &= e^{-\int 2x\,\mathrm{d}x}\left(\int e^{\int 2x\,\mathrm{d}x} x\,\mathrm{d}x + c\right) \\ &= e^{-x^2}\left(\int e^{x^2}\, x\,\mathrm{d}x + c\right) \\ &= e^{-x^2}\left(\frac{e^{x^2}}{2} + c\right) \\ &= \frac{1}{2} + c\,e^{-x^2} \end{aligned}$$

In other words, the required solution is given by

$$y = \sin^{-1}\left(\frac{1}{2} + c\,e^{-x^2}\right)$$

EXAMPLE 2.67 Solve the first order non-linear equation $\cos(y)\, y' - 2x\sin(y) = -2x^3$.

Solution: Clearly, $p(x) = -2x$ and $f(x) = -2x^3$. Let $u = \sin y$. Accordingly, the given differential equation becomes

$$u' - 2xu = -2x^3 \tag{2.37}$$

Again, the solution of Eq.(2.37) according to the formula in Eq.(2.8) is given by

$$\begin{aligned} u(x) &= e^{-\int p(x)\,\mathrm{d}x}\left(e^{\int p(x)\,\mathrm{d}x} f(x)\,\mathrm{d}x + c\right) \\ &= e^{x^2}\left(\int -2x^3 e^{-x^2}\,\mathrm{d}x + c\right) \\ &= e^{x^2}\left(\int x^2\,\mathrm{d}\left(e^{-x^2}\right) + c\right) \\ &= e^{x^2}\left(x^2\,e^{-x^2} + e^{-x^2} + c\right) \\ &= x^2 + 1 + ce^{x^2} \end{aligned}$$

In other words, the solution of the given differential equation is

$$\begin{aligned} \sin y &= u(x) \\ &= 1 + x^2 + c\,e^{x^2} \end{aligned}$$

Exercises 2.8

1. Solve the differential equation $y' + x\sin(2y) = \cos^2(y)\,x^3$.

   HINT: Let $z = \tan(y)$. Clearly, the given differential equation can be looked in the form $z' + 2xz = x^3$. Hence the corresponding solution is $\tan(y) = \frac{x^2}{2} - \frac{1}{2} + c\,e^{-x^2}$.

2. Solve the differential equation $xy^2y' + y^3 = x\cos x$.

   HINT: Let $z = y^3$. Clearly, the given differential equation can be looked in the form $z' + \frac{3z}{x} = 3\cos x$. Hence the corresponding solution is $y^3 = \left(3 - \frac{18}{x^2}\right)\sin x + \left(\frac{9}{x} - \frac{18}{x^3}\right)\cos x + \frac{c}{x^3}$.

3. Solve the differential equation $\cot y\,\mathrm{cosec}(y)\,y' + \frac{\mathrm{cosec}\,y}{x} = \frac{1}{x^2}$.

   HINT: Let $z = -\mathrm{cosec}\,y$. Clearly, the given differential equation can be looked in the form $z' - \frac{z}{x} = \frac{1}{x^2}$. Hence the corresponding solution is $\mathrm{cosec}(y) = \frac{1}{2x} - cx$.

4. Solve the differential equation $\cos(y)\,y' + \sin y\cot x = -\cot(x)\,e^{\sin x}$.

   HINT: Let $z = \sin y$. Clearly, the given differential equation can be looked in the form $z' + \cot(x)\,z = -\cot(x)\,e^{\sin x}$. Hence the corresponding solution is $\sin y = c\ \mathrm{cosec}\,x - \mathrm{cosec}\,x\,e^{\sin x}$.

5. Solve the differential equation $\sec(y)\tan(y)\,y' - \sec y = -x$.

   HINT: Let $z = \sec y$. Clearly, the given differential equation can be looked in the form $z' - z = -x$. Hence the corresponding solution is $\sec(y) = x + 1 - c\,e^x$.

6. Solve the differential equation $x^3y' + 4x^2\tan y = e^x\sec y$.

   HINT: Let $z = \sin y$. Clearly, the given differential equation can be looked in the form $z' + \frac{4z}{x} = \frac{e^x}{x^3}$. Hence the corresponding solution is $\sin y = \frac{1}{x^4}(xe^x - e^x + c)$.

7. Solve the differential equation $\cot(y)\,y' + \cot x = \sin y\sin^2 x$.

   HINT: Let $z = \mathrm{cosec}\,y$. Clearly, the given differential equation can be looked in the form $z' - z\cot x = -\sin^2 x$. Hence the corresponding solution is $\mathrm{cosec}(y) = \sin x(\cos x + c)$.

8. Solve the differential equation $\tan(y)\,y' + \tan x = \cos y\cos^2 x$.

   HINT: Let $z = \sec y$. Clearly, the given differential equation can be looked in the form $z' + z\tan x = \cos^2 x$. Hence the corresponding solution is $\sec y = \cos x(\sin x + c)$.

9. Solve the differential equation $y' + x\sin 2y = x^3\cos^2 y$.

   HINT: Let $z = \tan y$. Clearly, the given differential equation can be looked in the form $z' + 2zx = x^3$. Hence the corresponding solution is $\tan y = \frac{x^2}{2} - \frac{1}{2} + c\,e^{-x^2}$.

10. Solve the differential equation $y' + x\sin 2y = x^3\cos^2 y$.

    HINT: Clearly, the given differential equation can be transferred to $\sec^2(y)\,y' + 2x\tan y = x^3$. Let $\tan y = z$. Hence the given differential equation can be looked in the form $z' + 2xz = x^3$. In other words, $\tan y = \frac{x^3-1}{2} + c\,e^{-x^2}$ is the corresponding solution.

11. Solve the differential equation $x^3\,y' + 4x^2 \tan y = e^x \sec y$.

    HINT: Clearly, the given differential equation can be transferred to $x^3 \cos(y)\,y' + 4x^2 \sin y = e^x$. Let $\sin y = z$. Hence the given differential equation can be looked in the form $z' + \frac{4z}{x} = \frac{e^x}{x^3}$. In other words, $\sin y = \frac{1}{x^4}\,(x\,e^x - e^x + c)$ is the corresponding solution.

12. Solve the differential equation $\sin(y)\,y' = \cos x(2\cos y - \sin^2 x)$.

    HINT: Let $z = -\cos y$. Hence the given differential equation can be looked in the form $z' + 2\cos(x)\,z = -\cos x \sin^2 x$. In other words, $\cos y = \frac{\sin^2 x - \sin x}{2} + \frac{1}{4} - c\,e^{-2\sin x}$ is the corresponding solution.

13. Solve the differential equation $\sin(y)\,y' = \sin x(2\cos y - \cos^2 x)$.

    HINT: Let $z = -\cos y$. Hence the given differential equation can be looked in the form $z' + 2\sin(x)\,z = -\sin x \cos^2 x$. In other words, $\cos(y) = \frac{\cos^2 x + \cos x}{2} + \frac{1}{4} - c\,e^{2\cos x}$ is the corresponding solution.

14. Solve the differential equation $\cos(y)\,y' = \sin x(2\sin y - \cos^2 x)$.

    HINT: Let $z = \sin y$. Hence the given differential equation can be looked in the form $z' - 2\sin(x)\,z = -\sin x \cos^2 x$. In other words, $\sin y = \frac{\cos^2 x - \cos x}{2} + \frac{1}{4} + c\,e^{-2\cos x}$ is the corresponding solution.

15. Solve the differential equation $\cos(y)\,y' = \cos x(2\sin y - \sin^2 x)$.

    HINT: Let $z = \sin(y)$. Hence the given differential equation can be looked in the form $z' - 2\cos(x)\,z = -\cos(x)\sin^2(x)$. In other words, $\sin y = \frac{\sin^2 x + \sin x}{2} + \frac{1}{4} + c\,e^{2\sin x}$ is the corresponding solution.

16. Show that the differential equation $y' + p(x)y \ln y = f(x)y$ can be transferred to $z' + p(x)z = f(x)$ by changing the dependent variable $z = \ln y$.

17. Solve the equation $y' + \left(\frac{y}{x}\right)\ln y = ye^x$.

    HINTS: Let $z = \ln(y)$. Hence the given differential equation can be looked in the form $z' + \frac{z}{x} = e^x$. In other words, one can have $\ln(y) = \frac{1}{x}(xe^x - e^x + c)$.

18. Solve the differential equation $xy' = 2x^2 y + y\ln(y)$.

    HINTS: Let $z = \ln(y)$. Hence the given differential equation can be looked in the form $z' - \frac{z}{x} = 2x$. In other words, one can have $\ln(y) = 2x^2 + cx$.

2.9 Bernoulli Equation

Any differential equation of the form $y' - p(x)y = f(x)y^n$, where $n \geq 1$, is called a *Bernoulli equation.* One can rewrite the given equation in the form

$$\frac{y'}{y^n} - \frac{p(x)}{y^{n-1}} = f(x) \tag{2.38}$$

Let $u = -\frac{1}{y^{n-1}}$. Clearly, one can have $u' = (n-1)\left(\frac{y'}{y^n}\right)$. In terms of u and x, Eq.(2.38) becomes

$$u' + (n-1)p(x)u = (n-1)f(x)$$

In other words, we have

$$u' + \overline{p}(x)u = \overline{f}(x) \tag{2.39}$$

where $\overline{p}(x) = (n-1)p(x)$ and $\overline{f}(x) = (n-1)f(x)$. Clearly, Eq.(2.39) is a first order linear nonhomogeneous equation which can be solved using the formula given in Eq.(2.8).

EXAMPLE 2.68 Solve the first order nonlinear equation $y' - 2xy = 2xy^2$.

Solution: Clearly, $n = 2$, $p(x) = 2x$ and $f(x) = 2x$. According to Eq.(2.38), the given differential equation can be written in the form

$$\frac{y'}{y^2} - 2\left(\frac{x}{y}\right) = 2x \tag{2.40}$$

Let $u = -\frac{1}{y}$. Accordingly, Eq.(2.40) becomes

$$u' + 2xu = 2x \tag{2.41}$$

Again, the solution of Eq.(2.41) according to the formula in Eq.(2.8) is given by

$$\begin{aligned}
u(x) &= e^{-\int \overline{p}(x)\,\mathrm{d}x}\left(e^{\int \overline{p}(x)\,\mathrm{d}x}\overline{f}(x)\,\mathrm{d}x + c\right) \\
&= e^{-x^2}\left(\int 2xe^{x^2}\,\mathrm{d}x + c\right) \\
&= e^{-x^2}\left(e^{x^2} + c\right) \\
&= 1 + ce^{-x^2}
\end{aligned}$$

In other words, the solution of the given differential equation is

$$\begin{aligned}
y(x) &= -\frac{1}{u(x)} \\
&= -\frac{1}{1 + ce^{-x^2}} \\
&= \frac{1}{ke^{-x^2} - 1}
\end{aligned}$$

EXAMPLE 2.69 Solve the first order nonlinear equation $xy(1 + xy^2)y' = 1$.

Solution: One can rewrite the given differential equation by $x' - xy = x^2y^3$. Clearly, $n = 2$, $p(y) = y$ and $f(y) = y^3$. According to Eq.(2.38), the given differential equation can be written in the form

$$\frac{x'}{x^2} - \frac{y}{x} = y^3 \tag{2.42}$$

Let $u = -\frac{1}{x}$. Accordingly, Eq.(2.42) becomes

$$u' + yu = -y^3 \tag{2.43}$$

Again, the solution of Eq.(2.43) according to the formula in Eq.(2.8) is given by

$$\begin{aligned} u(y) &= e^{-\int \overline{p}(y)\,\mathrm{d}y}\left(e^{\int \overline{p}(y)\,\mathrm{d}y}\overline{f}(y)\,\mathrm{d}y + c\right) \\ &= e^{-\frac{y^2}{2}}\left(\int y^3 e^{\frac{y^2}{2}}\,\mathrm{d}y + c\right) \\ &= e^{-\frac{y^2}{2}}\left(y^2 e^{\frac{y^2}{2}} - 2e^{\frac{y^2}{2}} + c\right) \\ &= y^2 - 2 + ce^{-\frac{y^2}{2}} \end{aligned}$$

In other words, the solution of the given differential equation is

$$\begin{aligned} x(y) &= -\frac{1}{u(y)} \\ &= -\frac{1}{y^2 - 2 + ce^{-\frac{y^2}{2}}} \\ &= \frac{1}{ke^{-\frac{y^2}{2}} + 2 - y^2} \end{aligned}$$

EXAMPLE 2.70 Solve the first order nonlinear equation $y' - xy = x^3y^2$.

Solution: Clearly, $n = 2$, $p(x) = x$ and $f(x) = x^3$. According to Eq.(2.38), the given differential equation can be written in the form

$$\frac{y'}{y^2} - \frac{x}{y} = x^3 \tag{2.44}$$

Let $u = -\frac{1}{y}$. Accordingly, Eq.(2.44) becomes

$$u' + xu = x^3 \tag{2.45}$$

Again, the solution of Eq.(2.45) according to the formula in Eq.(2.8) is given by

$$\begin{aligned} u(x) &= e^{-\int \overline{p}(x)\,\mathrm{d}x}\left(e^{\int \overline{p}(x)\,\mathrm{d}x}\overline{f}(x)\,\mathrm{d}x + c\right) \\ &= e^{-\frac{x^2}{2}}\left(\int x^3 e^{\frac{x^2}{2}}\,\mathrm{d}x + c\right) \\ &= e^{-\frac{x^2}{2}}\left(x^2 e^{\frac{x^2}{2}} - 2e^{\frac{x^2}{2}} + c\right) \\ &= x^2 - 2 + ce^{-\frac{x^2}{2}} \end{aligned}$$

In other words, the solution of the given differential equation is

$$y(x) = -\frac{1}{u(x)}$$
$$= -\frac{1}{x^2 - 2 + ce^{-\left(\frac{x^2}{2}\right)}}$$
$$= \frac{1}{ke^{-\left(\frac{x^2}{2}\right)} + 2 - x^2}$$

EXAMPLE 2.71 Solve the first order nonlinear equation $xy' - y = xy^4$.

Solution: Clearly, $n = 4$, $p(x) = -\frac{1}{x}$ and $f(x) = 1$. According to Eq.(2.38), the given differential equation can be written in the form

$$\frac{y'}{y^4} - \frac{1}{xy^3} = 1 \tag{2.46}$$

Let $u = -\frac{1}{y^3}$. Accordingly, Eq.(2.46) becomes

$$u' + \left(\frac{3}{x}\right)u = 3 \tag{2.47}$$

Again, the solution of Eq.(2.47) according to the formula in Eq.(2.8) is given by

$$u(x) = e^{-\int \overline{p}(x)\,\mathrm{d}x}\left(e^{\int \overline{p}(x)\,\mathrm{d}x}\overline{f}(x)\,\mathrm{d}x + c\right)$$
$$= e^{-3\ln x}\left(\int 3e^{3\ln x}\,\mathrm{d}x + c\right)$$
$$= x^{-3}\left(3\int x^3\,\mathrm{d}x + c\right)$$
$$= x^{-3}\left(\frac{3}{4}x^4 + c\right)$$
$$= \frac{3}{4}x + cx^{-3}$$

In other words, the solution of the given differential equation is

$$y(x) = -\left(\frac{1}{u(x)}\right)^{\frac{1}{3}}$$
$$= -\left(\frac{1}{\frac{3}{4}x + cx^{-3}}\right)^{\frac{1}{3}}$$

EXAMPLE 2.72 Solve the differential equation $2xy' + (x-1)y = y^{-1}\,x^2\,e^x$.

Solution: Clearly, $n = -1$, $p(x) = \frac{x-1}{x}$ and $f(x) = x\,e^x$. According to Eq.(2.38), the given differential equation can be written in the form

$$2yy' + \left(\frac{x-1}{x}\right)y^2 = x\,e^x \tag{2.48}$$

Let $u = y^2$. Accordingly, Eq.(2.48) becomes

$$u' + \left(\frac{x-1}{x}\right)u \quad = \quad x\,e^x \tag{2.49}$$

Again, the solution of Eq.(2.49) according to the formula in Eq.(2.8) is given by

$$\begin{aligned}
u(x) &= e^{-\int\left(\frac{x-1}{x}\right)\mathrm{d}x}\left(\int x\,e^x e^{-\int\left(\frac{x-1}{x}\right)\mathrm{d}x}\mathrm{d}x + c\right)\\
&= x\,e^{-x}\left(\int e^{2x}\,\mathrm{d}x + c\right)\\
&= x\,e^{-x}\left(\frac{e^{2x}}{2} + c\right)\\
&= \frac{x\,e^x}{2} + c\,x\,e^{-x}
\end{aligned}$$

In other words, the solution of the given differential equation is

$$y^2(x) = \frac{x\,e^x}{2} + c\,x\,e^{-x}$$

EXAMPLE 2.73 Solve the differential equation $2yy' - 2y^2x = x$.

Solution: Clearly, $p(x) = -2x$ and $f(x) = x$. Let $u(x) = y^2$. Accordingly, the given differential equation reduces to

$$u' - 2xu = x \tag{2.50}$$

Again, the solution of Eq.(2.50) according to the formula in Eq.(2.8) is given by

$$\begin{aligned}
u(x) &= e^{-\int p(x)\,\mathrm{d}x}\left(\int e^{\int p(x)\,\mathrm{d}x}q(x)\,\mathrm{d}x + c\right)\\
&= e^{\int 2x\,\mathrm{d}x}\left(\int e^{-\int 2x\,\mathrm{d}x}x\,\mathrm{d}x + c\right)\\
&= e^{x^2}\left(\int e^{-x^2}x\,\mathrm{d}x + c\right)\\
&= e^{x^2}\left(-\frac{e^{-x^2}}{2} + c\right)\\
&= -\frac{1}{2} + c\,e^{x^2}
\end{aligned}$$

In other words, the required solution is given by

$$y(x) = \left(-\frac{1}{2} + c\,e^{x^2}\right)^{\frac{1}{2}}$$

Exercises 2.9

1. Solve the equation $y' = \frac{y}{2x} + \frac{x^2}{2y}$.

   HINTS: Let $z = y^2$. Hence the given differential equation can be looked in the form $z' - \frac{z}{x} = x^2$. Hence the corresponding solution is $y^2 = \frac{x^3}{2} + cx$.

2. Solve the differential equation $3y^2y' + xy^3 = x$.

   HINT: Let $z = y^3$. Clearly, the given differential equation can be looked in the form $z' + zx = x$. Hence the corresponding solution is $y^3 = 1 + c\,e^{-\frac{x^2}{2}}$.

3. Solve the differential equation $yy' = \frac{e^{-x^3}}{x^2} - \frac{2y^2}{x}$.

   HINT: Let $z = y^2$. Clearly, the given differential equation can be looked in the form $z' + \frac{4z}{x} = \frac{2e^{-x^3}}{x^2}$. Hence the corresponding solution is $y^2 = \frac{c}{x^4} - \left(\frac{2}{3x^4}\right) e^{-x^3}$.

4. Solve the differential equation $2yxy' = 3y^2 - x^2$.

   HINT: Let $z = y^2$. Clearly, the given differential equation can be looked in the form $z' - \frac{3z}{x} = -x$. Hence the corresponding solution is $y^2 = \frac{c}{x^3} - \frac{x^2}{5}$.

5. Solve the equation $y' + \frac{y}{x} = y^3x^3$.

   HINT: Let $z = \frac{1}{y^2}$. Clearly, the given differential equation can be looked in the form $z' - \frac{2z}{x} = -2x^3$. Hence the corresponding solution is $y^2 = \frac{1}{cx^2 - x}$.

6. Solve the equation $xy' + y = x^3y^6$.

   HINT: Let $z = \frac{1}{y^5}$. Clearly, the given differential equation can be looked in the form $z' - \frac{5z}{x} = -5x^2$. Hence the corresponding solution is $y^5 = \frac{2}{5x^3 + 2cx^5}$.

7. Solve the equation $y' + xy = x^3y^3$.

   HINT: Let $z = \frac{1}{y^2}$. Clearly, the given differential equation can be looked in the form $z' - 2xz = -2x^3$. Hence the corresponding solution is $y^2 = \frac{1}{1 + x^2 + ce^{x^2}}$.

8. Solve the equation $y' - xy = x^3y^3$.

   HINT: Let $z = \frac{1}{y^2}$. Clearly, the given differential equation can be looked in the form $z' + 2xz = -2x^3$. Hence the corresponding solution is $y^2 = \frac{1}{1 - x^2 + ce^{-x^2}}$.

9. Solve the equation $y'(x^2y^3 - xy) = 1$.

   HINT: Let $z = \frac{1}{x}$. Clearly, the given differential equation can be looked in the form $z' - yz = -y^3$. Hence the corresponding solution is $x = \frac{1}{2 + y^2 + c\,e^{\frac{y^2}{2}}}$.

10. Solve the equation $xy' + y = x^4y^3$.

    HINT: Let $z = \frac{1}{y^2}$. Clearly, the given differential equation can be looked in the form $z' - \frac{2z}{x} = -2x^3$. Hence the corresponding solution is $y^2 = \frac{1}{cx^2 - x^4}$.

11. Solve the differential equation $(1 - x^2)y' - xy - axy^2 = 0$.

    HINT: Let $z = -\frac{1}{y}$. Clearly, the given differential equation can be looked in the form $z' + \frac{xz}{1-x^2} = \frac{ax}{1-x^2}$. Hence the corresponding solution is $y^2 = \frac{2}{2c\sqrt{1-x^2}-a}$.

12. Solve the differential equation $x^2y' = xy + y^2$.

    HINT: Let $z = -\frac{1}{y}$. Clearly, the given differential equation can be looked in the form $z' + \frac{z}{x} = \frac{1}{x^2}$. Hence the corresponding solution is $y = -\frac{x}{\ln(x)+c}$.

13. Solve the differential equation $y' = x^3y^2 + xy$.

    HINT: Let $z = -\frac{1}{y}$. Clearly, the given differential equation can be looked in the form $z' + x\,z = x^3$. Hence the corresponding solution is $y = -\frac{1}{x^2 - 2 + c\,e^{-\frac{x^2}{2}}}$

14. Solve the differential equation $xy' + y = xy^5$.

    HINT: Let $z = \frac{1}{y^4}$. Clearly, the given differential equation can be looked in the form $z' - \frac{4z}{x} = -4$. Hence the corresponding solution is $y^4 = \frac{3}{4x+3cx^4}$.

15. Solve the differential equation $y' + y\tan x = y^3$.

    HINT: Let $z = \frac{1}{y^2}$. Clearly, the given differential equation can be looked in the form $z' - 2\tan(x)\,z = -2$. Hence the corresponding solution is $y^2 = \frac{2\cos^2 x}{2c - \sin(2x) - 2x}$.

16. Solve the differential equation $xy' + y = \ln(x)\,y^2$.

    HINT: Let $z = \frac{1}{y}$. Clearly, the given differential equation can be looked in the form $z' - \frac{z}{x} = -\frac{\ln(x)}{x}$. In other words, one can have $y = \frac{1}{x\left(\frac{\ln(x)}{x} - \int \frac{dx}{x^2} + c\right)} = \frac{1}{1+\ln(x)+cx}$.

2.10 Reccati Equation

A differential equation of the form $y' + p(x)y + q(x)y^2 = r(x)$ is called a *Riccati equation.* Any differential equation of the Reccati type can be transformed into one of the Bernoulli type by a change of variable when a solution $y(x) = u(x)$ of the given Reccati equation is known. In other words, we have $y(x) = u(x)$, such that

$$\begin{aligned} r(x) &= y' + p(x)y + q(x)y^2 \\ &= u' + p(x)u + q(x)u^2 \end{aligned} \tag{2.51}$$

Let $y(x) = u(x) + v(x)$ be the general solution of the given equation, where $v(x)$ is unknown and to be determined. Accordingly, one can have

$$\begin{aligned} r(x) &= y' + p(x)y + q(x)y^2 \\ &= (u' + v') + p(x)(u + v) + q(x)(u + v)^2 \\ &= u' + p(x)u + q(x)u^2 + (v' + p(x)v + q(x)(v^2 + 2uv) \end{aligned} \tag{2.52}$$

Again, one can use the relation given in Eq.(2.51) in the relation given in Eq.(2.52). Hence we have

$$v' + p(x)v + q(x)(v^2 + 2uv) = 0 \tag{2.53}$$

In other words, Eq.(2.53) can be rewritten in the form

$$\begin{aligned} -q(x)v^2 &= v' + p(x)v + 2q(x)uv \\ &= v' + (p(x) + 2q(x)u(x))v \end{aligned} \tag{2.54}$$

Clearly, Eq.(2.54) is a Bernoulli equation which can be easily solved according to the principle discussed in Section 2.9.

EXAMPLE 2.74 Solve the equation $y' = y^2 - \frac{2}{x^2}$.
Solution: Clearly, $u(x) = \frac{1}{x}$ is one of the solutions of the given differential equation in which $p(x) = 0$, $q(x) = -1$ and $r(x) = -\frac{2}{x^2}$. Let $y(x) = u(x) + v(x)$. According to Eq.(2.54), we have

$$v^2 = v' - 2\left(\frac{v}{x}\right)$$

In other words, we have

$$v' = 2\left(\frac{v}{x}\right) + v^2 \tag{2.55}$$

which is a Bernoulli equation with $n = 2$. Let $z(x) = -\frac{1}{v}$. According to Eq.(2.39), Eq.(2.55) is transformed into

$$z' + 2\left(\frac{z}{x}\right) = 1 \tag{2.56}$$

The solution of Eq.(2.56) according to the formula given in Eq.(2.8) is given by

$$\begin{aligned} z(x) &= x^{-2}\left(\int x^2\, \mathrm{d}x + c\right) \\ &= \frac{x}{3} + \frac{c}{x^2} \end{aligned}$$

In other words, we have

$$\begin{aligned} y(x) &= u(x) + v(x) \\ &= \frac{1}{x} - \frac{1}{z(x)} \\ &= \frac{1}{x} + \frac{3x^2}{3c - x^3} \end{aligned}$$

as the required solution of the given differential equation.

EXAMPLE 2.75 Solve the equation $y' + xy = y^2 - \frac{2}{x^2} + 1$.

Solution: Clearly, $u(x) = \frac{1}{x}$ is one of the solution of the given differential equation in which $p(x) = x$, $q(x) = -1$ and $r(x) = 1 - \frac{2}{x^2}$. Let $y(x) = u(x) + v(x)$. According to Eq.(2.54), we have

$$v^2 = v' + \left(x - \frac{2}{x}\right) v$$

In other words, we have

$$v' = \left(\frac{2}{x} - x\right) v + v^2 \tag{2.57}$$

which is a Bernoulli equation. Let $z(x) = -\frac{1}{v}$. Using Eq.(2.39), Eq.(2.57) is transformed into

$$z' + \left(\frac{2}{x} - x\right) z = 1 \tag{2.58}$$

The solution of Eq.(2.58) according to the formula given in Eq.(2.8) is given by

$$\begin{aligned} z(x) &= x^{-2} e^{\frac{x^2}{2}} \left(\int x^2 e^{-\frac{x^2}{2}} \, dx + c \right) \\ &= x^{-2} e^{\frac{x^2}{2}} \left(\int e^{-\frac{x^2}{2}} \, dx - x e^{-\frac{x^2}{2}} + c \right) \end{aligned}$$

In other words, we have

$$\begin{aligned} y(x) &= u(x) + v(x) \\ &= \frac{1}{x} - \frac{1}{z(x)} \\ &= \frac{1}{x} + \frac{x^2}{e^{\frac{x^2}{2}} \left(x e^{\frac{x^2}{2}} - \int e^{-\frac{x^2}{2}} \, dx - c \right)} \end{aligned}$$

as the required solution of the given differential equation.

Exercises 2.10

1. Solve the differential equation $2x^2 y' = 2xy + (x-1)y^2 - x^2(x-1)$.
2. Solve the differential equation $(1 - x^2)y' + xy - axy^2 = 1 - ax^3$.
3. Solve the differential equation $x^2 y' = xy + y^2 - x^2$.
4. Solve the differential equation $x^2 y' - xy + y^2 = x^2$.
5. Solve the differential equation $x^2 y' + xy - y^2 = x^2$.
6. Solve the differential equation $2x^2 y' - y^2 = x^2$.
7. Show that the Reccati equation $y' + p(x)y + q(x)y^2 = r(x)$ can be transformed into the higher order differential equation $u'' + \left(p(x) - \frac{q'(x)}{q(x)}\right) u' - r(x)q(x)u = 0$ by taking $y(x) = \frac{u'(x)}{q(x)u(x)}$.

8. Show that the Reccati equation $y' + p(x)y + q(x)y^2 = r(x)$ can be transformed into the higher order differential equation $u'' - cu' - ku = 0$ if and only if $q(x) = e^{\int p(x)\,dx + cx}$ and $r(x) = \frac{k}{q(x)}$.
9. If $y(x) = y_1(x)$ is a particular solution of the Reccati equation $y' + p(x)y + q(x)y^2 = r(x)$, then $y(x) = y_1(x) + u(x)$ is a general solution of the Reccati equation $y' + p(x)y + q(x)y^2 = r(x)$, where $u' + (p(x) + 2q(x)y_1(x))u = -q(x)u^2$.

2.11 Higher Degree Equations

So far, we have discussed differential equations with only first order and first degree, in other words, differential equations containing only the y' term. In the present section, we shall consider the set of all first order but higher degree differential equations satisfying one of the following conditions:

1. Equations solvable for y'
2. Equations solvable for y
3. Equations solvable for x
4. Equations missing one variable
5. Equations homogeneous in x and y.

The general form of the first order and nth degree equation can be expressed as

$$(y')^n + P_1(x, y)(y')^{n-1} + \cdots + P_{n-1}(x, y)y' + P_n(x, y) = 0 \tag{2.59}$$

The objective of this section is to reduce the given differential equation into two variables by some suitable substitution, and separate the variables, so that each of the separated parts is in exact form.

2.11.1 Equations Solvable for y'

If Eq.(2.59) is solved for y', then one can rewrite Eq.(2.59) in the form

$$0 = \prod_{i=1}^{n} (y' - f_i(x, y))$$

Again, each $y' - f_i(x, y) = 0$ for $1 \leq i \leq n$ is a first order and first degree differential equation whose solution can be obtained by the formula given in Eq.(2.14). Let $g_i(y, x, c) = 0$ be the solution of the equation $y' - f_i(x, y) = 0$ for $1 \leq i \leq n$. Under this situation, the general solution of Eq.(2.59) is given by

$$\prod_{i=1}^{n} g_i(y, x, c) = 0 \tag{2.60}$$

EXAMPLE 2.76 Solve the differential equation $x^2(y')^2 + xyy' - 6(y')^2 = 0$.

Solution: The given equation can be rewritten as

$$(xy' + 3y)(xy' - 2y) = 0$$

Hence we have two homogeneous differential equations

$$xy' + 3y = 0 \tag{2.61}$$

and

$$xy' - 2y = 0 \tag{2.62}$$

Again, $yx^3 - c = 0$ is the solution of Eq.(2.61) and $y - cx^2 = 0$ is the solution of Eq.(2.62) by the formula given in Eq.(2.2). Hence the general solution of the given differential equation is

$$(yx^3 - c)(y - cx^2) = 0$$

EXAMPLE 2.77 Solve the differential equation $x^2(y')^2 - xyy' - 6(y')^2 = 0$.

Solution: The given equation can be rewritten as

$$(xy' - 3y)(xy' + 2y) = 0$$

Hence we have two homogeneous differential equations

$$xy' - 3y = 0 \tag{2.63}$$

and

$$xy' - 2y = 0 \tag{2.64}$$

Again, $y - cx^3 = 0$ is the solution of Eq.(2.63) and $yx^2 - c = 0$ is the solution of Eq.(2.64) by the formula given in Eq.(2.2). Hence the general solution of the given differential equation is

$$(y - cx^3)(yx^2 - c) = 0$$

Exercises 2.11.1

1. Solve the differential equation $(y')^3 + 2x(y')^2 - y^2(y')^2 - 2xy^2y' = 0$.
2. Solve the differential equation $xy(y') + y'(3x^2 - 2y^2) - 6xy = 0$.
3. Solve the differential equation $(y')^2 - 5y' + 6 = 0$.
4. Solve the differential equation $(y')^2 - 2xy' + x^2 - y^2 = 0$.
5. Solve the differential equation $y(y')^2 - (x + y)y' + x = 0$.
6. Solve the differential equation $x(y')^2 + (y - x)y' - y = 0$.

2.11.2 Equations Solvable for y

If Eq.(2.59) is solved for y, then one can rewrite Eq.(2.59) in the form

$$y = F(x, y') \tag{2.65}$$

Again, one can differentiate Eq.(2.65) with respect to x. Hence we have

$$y' = \frac{\partial F}{\partial x} + \left(\frac{\partial F}{\partial y'}\right) y'' \tag{2.66}$$

Again, Eq.(2.66) contains only two variables x and y'. Hence the equation can be solved by the formula given in Eq.(2.14) or Eq.(2.8). Again, the solution of Eq.(2.66) can be expressed in the form

$$H(y', x, c) = 0 \tag{2.67}$$

In this situation, the general solution of Eq.(2.59) is obtained by eliminating y' between Eqs.(2.65) and (2.67) if an arbitrary constant is present; otherwise it is called the *singular solution.*

EXAMPLE 2.78 Solve the differential equation $y = 2xy' - x(y')^2$.

Solution: Differentiating the given equation with respect to x, we have

$$y' = 2y' + 2x\left(\frac{dy'}{dx}\right) - (y')^2 - 2xy'\left(\frac{dy'}{dx}\right)$$

In other words, one can write

$$\left(2y' + 2x\left(\frac{dy'}{dx}\right)\right)(1 - y') = 0$$

Again, the solution of the equation

$$2y' + 2x\left(\frac{dy'}{dx}\right) = 0$$

is

$$(y')^2 = \frac{c}{x} \tag{2.68}$$

Eliminating y' between Eq.(2.68) and the given equation, the general solution of the given equation is given by

$$(y + c)^2 = 4cx$$

Again, eliminating y' between $y' = 1$ and the given equation, we have a singular solution of the differential equation which is given by

$$y = x$$

EXAMPLE 2.79 Solve the differential equation $x - yy' = a(y')^2$.

Solution: Clearly, the given equation can be solved for y. Hence we have

$$y = \frac{x - a(y')^2}{y'} \tag{2.69}$$

Differentiating Eq.(2.69) with respect to x, we have

$$y' = \frac{1}{y'} - \left(\frac{x}{(y')^2}\right)\frac{dy'}{dx} - a\left(\frac{dy'}{dx}\right)$$

In other words, we have

$$(a(y')^2 + x)\frac{dy'}{dx} = y'(1 - (y')^2) \tag{2.70}$$

Again, one can rewrite Eq.(2.70) in the form

$$\frac{dx}{dy'} - \left(\frac{1}{y'(1 - (y')^2)}\right)x = \frac{ay'}{1 - (y')^2} \tag{2.71}$$

According to the formula given in Eq.(2.8), the solution of Eq.(2.71) is given by

$$\begin{aligned} x &= \frac{y'}{(1 - (y')^2)^{\frac{1}{2}}}\left(\int\left(\frac{ay'}{1 - (y')^2}\right)\left(\frac{(1 - (y')^2)^{\frac{1}{2}}}{y'}\right)dy' + c\right) \\ &= \frac{y'}{(1 - (y')^2)^{\frac{1}{2}}}\left(\int\left(\frac{a}{(1 - (y')^2)^{\frac{1}{2}}}\right)dy' + c\right) \\ &= \frac{y'}{(1 - (y')^2)^{\frac{1}{2}}}(c + a\sin^{-1}(y')) \end{aligned} \tag{2.72}$$

Eliminating x between Eq.(2.72) and the given equation, one can get

$$y = -ay' + \left(\frac{y'}{(1 - (y')^2)^{\frac{1}{2}}}\right)(c + a\sin^{-1}(y')) \tag{2.73}$$

Clearly, Eqs.(2.71) and (2.72) together constitute the solution of the given differential equation.

Exercises 2.11.2

1. Solve the differential equation $y - 3x - \ln y' = 0$.
2. Solve the differential equation $y - 2xy' + (y')^2 = 0$
3. Solve the differential equation $y' - \tan\left(\frac{y-x}{a}\right) = 0$.
4. Solve the differential equation $y - 3xy' - 4(y')^3$.
5. Solve the differential equation $x(y')^2 - 2yy' + ax = 0$.
6. Solve the differential equation $y - y'\tan y' - \ln(\cos y') = 0$.

2.11.3 Equations Solvable for x

If Eq.(2.59) is solved for x, then one can rewrite Eq.(2.59) in the form

$$x = G(y, y') \tag{2.74}$$

Again, one can differentiate Eq.(2.74) with respect to y. Hence we have

$$\frac{1}{y'} = \frac{\partial G}{\partial y} + \left(\frac{\partial G}{\partial y'}\right)\left(\frac{\partial y'}{\partial y}\right) \tag{2.75}$$

Again, Eq.(2.75) contains only two variables y and y'. Hence the equation can be solved by the formula given in Eq.(2.14). Again, the solution of Eq.(2.75) can be expressed in the form

$$K(y', y, c) = 0 \tag{2.76}$$

In this situation, the general solution of Eq.(2.59) is obtained by eliminating y' between Eqs.(2.74) and (2.76) if the arbitrary constant is present; otherwise is called the *singular solution.*

EXAMPLE 2.80 Solve the differential equation $y(y')^2 - 2xy' + y = 0$.

Solution: Clearly, the given equation can be solved for x. Hence we have

$$2x = yy' + \frac{y}{y'} \tag{2.77}$$

Differentiating Eq. (2.77) with respect to y, we have

$$\frac{2}{y'} = y' + y\left(\frac{dy'}{dy}\right) + \frac{1}{y'} - \left(\frac{y}{(y')^2}\right)\left(\frac{dy'}{dy}\right)$$

In other words, we have

$$\left(y' + y\left(\frac{dy'}{dy}\right)\right)\left(1 - \frac{1}{(y')^2}\right) = 0$$

Let $y' + y\left(\frac{dy'}{dy}\right) = 0$. In other words,

$$\frac{dy'}{y'} + \frac{dy}{y} = 0 \tag{2.78}$$

Clearly, $y' = \frac{c}{y}$ is the solution of Eq.(2.78). The general solution of the given equation can be obtained by eliminating y' between the given equation and $y' = \frac{c}{y}$. Hence the general solution is given by

$$\frac{c^2}{y^2} - 2x\left(\frac{c}{y}\right) + y = 0$$

Let $1 - \frac{1}{(y')^2} = 0$. In other words, we have

$$y' = \pm 1$$

Clearly, replacing $y' = \pm 1$ in the given equation leads to a singular solution which is given by

$$\begin{aligned} y \pm 2x + y &= 2y \pm 2x \\ &= 0 \end{aligned}$$

EXAMPLE 2.81 Solve the differential equation $(y')^2 - 2xy' + y = 0$.

Solution: Clearly, the given equation can be solved for x. Hence we have

$$2x = y' + \frac{y}{y'} \tag{2.79}$$

Differentiating Eq.(2.79) with respect to y, we have

$$\frac{2}{y'} = \frac{dy'}{dy} + \frac{1}{y'} - \left(\frac{y}{(y')^2}\right)\left(\frac{dy'}{dy}\right)$$

In other words, we have

$$\left(1 - \frac{1}{(y')^2}\right)\left(\frac{dy'}{dy}\right) = \frac{1}{y'} \tag{2.80}$$

One can rewrite Eq.(2.80) by

$$\frac{dy}{dy'} - \frac{y}{y'} = y' \tag{2.81}$$

Clearly, the solution of Eq.(2.81) is

$$y = (y')^2 + cy' \tag{2.82}$$

The general solution of the given equation can be obtained by eliminating y' between the given equation and Eq.(2.82). Hence the general solution is given by

$$y - 2x\left(\frac{-c \pm (c^2+4)^{\frac{1}{4}}}{2}\right) + y = 2y - 2x\left(\frac{-c \pm (c^2+4)^{\frac{1}{4}}}{2}\right)$$
$$= 0$$

Exercises 2.11.3

1. Solve the differential equation $y = 2xy' + y^2(y')^3$.
2. Solve the differential equation $x = y + a \ln y'$.
3. Solve the differential equation $x = (y')^3 - y' + 2$.
4. Solve the differential equation $(y')^3 - 4xyy' + 8y^2 = 0$.
5. Solve the differential equation $xyy' + (y')^2 - y^2 \ln y = 0$.
6. Solve the differential equation $x(y')^2 - 2yy' + 4x = 0$.
7. Solve the differential equation $y(y')^2 + y + 2xy' = 0$.
8. Solve the differential equation $y = 3xy' + 6y^2\,(y')^2$.
9. Solve the differential equation $(y')^2 + 2yy' \cot x = y^2$.

2.11.4 Equations Missing One Variable

If Eq.(2.59) does not contain one of the variables, then Eq.(2.59) can be expressed either in the form

$$U(y',\, x) = 0 \tag{2.83}$$

or in the form

$$V(y',\, y) = 0 \tag{2.84}$$

In other words, Eq.(2.59) contains only two variables. If Eq.(2.59) is solved for y', then Eqs.(2.83) and (2.84) respectively can be rewritten as:

$$y' = f(x) \tag{2.85}$$

and

$$y' = g(y) \tag{2.86}$$

which are easily integrable. If Eq.(2.76) is not solvable for y', then Eqs.(2.83) and (2.84) respectively can be rewritten as

$$x = f(y') \tag{2.87}$$

and

$$y = g(y') \tag{2.88}$$

which are in the form discussed in Subsections 2.11.2 and 2.11.3.

EXAMPLE 2.82 Solve the differential equation $x^2 = (y')^2(a^2 - x^2)$.

Solution: Clearly, the variable y is missing in the given differential equation. One can rewrite the given equation in the form

$$x = \frac{ay'}{(1+(y')^2)^{\frac{1}{2}}} \tag{2.89}$$

If one differentiate Eq.(2.89) with respect to y', then we have

$$\begin{aligned} \mathrm{d}y &= y'\mathrm{d}x \\ &= \frac{ay'\mathrm{d}y'}{(1+(y')^2)^{\frac{3}{2}}} \end{aligned} \tag{2.90}$$

Again, the integration of Eq.(2.90) gives

$$y + c = -\frac{a}{(1+(y')^2)^{\frac{1}{2}}} \tag{2.91}$$

The singular solution of the given equation can be obtained by eliminating y' between Eq.(2.91) and the given equation. Hence the required solution is

$$x^2 + (y+c)^2 = a^2$$

EXAMPLE 2.83 Solve the differential equation $y^2 + (y')^2 = a^2$.

Solution: Clearly, the variable x is missing in the given differential equation. One can rewrite the given equation in the form

$$y = (a^2 - (y')^2)^{\frac{1}{2}} \tag{2.92}$$

If Eq.(2.92) is differentiated with respect to y', then we have

$$\begin{aligned} \mathrm{d}x &= \frac{\mathrm{d}y}{y'} \\ &= -\frac{\mathrm{d}y'}{(a^2 - (y')^2)^{\frac{1}{2}}} \end{aligned} \tag{2.93}$$

Again, integration of Eq.(2.93) gives

$$x + c = \cos^{-1}\frac{y'}{a} \tag{2.94}$$

The singular solution of the given differential equation can be obtained by eliminating y' between Eq.(2.94) and the given differential equation. Hence the solution is

$$(x + c)^2 = y^2$$

EXAMPLE 2.84 Solve the differential equation $x = (y')^3 - y' - 1$.

Solution: Differentiating the given equation, we have

$$\begin{aligned} dy &= y'\mathrm{d}x \\ &= y'\left(3(y')^2 - 1\right)\mathrm{d}y' \\ &= \left((3y')^3 - y'\right)\mathrm{d}y' \end{aligned}$$

In other words, we have

$$y = \frac{3}{4}(y')^4 - \frac{1}{2}(y')^2 + c$$

Clearly, $x = (y')^3 - y' - 1$ and $y = \frac{3}{4}(y')^4 - \frac{1}{2}(y')^2 + c$ constitutes the parametric solution of the given equation.

Exercises 2.11.4

1. Solve the differential equation $y = 2y' + 3(y')^2$.
2. Solve the differential equation $x^2 = a^2(1 + (y')^2)$.
3. Solve the differential equation $x^2(1 + (y')^2) = (y')^2$.
4. Solve the differential equation $y^2 = 1 + (y')^2$.
5. Solve the differential equation $(y')^2 - 2xy' + 1 = 0$.
6. Solve the differential equation $(y')^2 + y^2 - 4 = 0$.

2.11.5 Equations Homogeneous in x and y

If Eq.(2.59) is homogeneous in x and y, then one can express Eq.(2.59) in the form

$$Q(y', z) = 0 \tag{2.95}$$

where $zx = y$. Easily, one can solve Eq.(2.95) for either z or y'. If Eq.(2.95) is solvable for y', then we have

$$\begin{aligned} h(z) &= y' \\ &= \frac{\mathrm{d}y}{\mathrm{d}x} \\ &= z + x\left(\frac{\mathrm{d}z}{\mathrm{d}x}\right) \\ &= x + xz' \end{aligned} \tag{2.96}$$

Again, one can find the solution of Eq.(2.96) using the formula given in Eq.(2.14). On the other hand, if Eq.(2.95) is solvable for z, then we have $g(y') = z$. In other words, we have

$$\begin{aligned} xg(y') &= zx \\ &= y \end{aligned} \tag{2.97}$$

One can easily differentiate Eq.(2.97) with respect to x. Hence we have

$$y' = g(y') + xg'(y')\left(\frac{\mathrm{d}y'}{\mathrm{d}x}\right)$$

or

$$\frac{1}{x}\,\mathrm{d}x = \left(\frac{g'(y')}{y' - g(y')}\right)\mathrm{d}(y') \tag{2.98}$$

One can easily integrate Eq.(2.98) to have a relation between x and y'. Let $y' = \phi(x)$ be the solution of Eq.(2.98). One can obtain another type of solution of the given differential equation by substituting the expression $y' = \phi(x)$ in the given differential equation, which is called the *singular solution* because it does not contain any arbitrary constant.

EXAMPLE 2.85 Solve the differential equation $(y'x - y)^2 = (y')^2 - \left(\frac{2y}{x}\right)y' + 1$.

Solution: This differential equation is not homogeneous in x and y, but it can be solved by using the substitution $y = vx$ for the purpose of separating variables. It is easy to verify that

$$\begin{aligned} (y'x - y)^2 &= (y')^2 - \left(\frac{2y}{x}\right)y' + 1 \\ &= \left(y' - \frac{y}{x}\right)^2 - \frac{y^2}{x^2} + 1 \\ &= \frac{(y'x - y)^2}{x^2} + \frac{x^2 - y^2}{x^2} \end{aligned}$$

In other words, we have

$$y'x - y = \left(\frac{x^2 - y^2}{x^2 - 1}\right)^{\frac{1}{2}} \tag{2.99}$$

Replacing $y = vx$ in Eq.(2.99), one can have

$$xv' = \left(\frac{1 - v^2}{x^2 - 1}\right)^{\frac{1}{2}} \tag{2.100}$$

Clearly, Eq.(2.100) has only two variables x and v. Again, one can separate the variables in Eq.(2.100). Hence we have

$$\frac{\mathrm{d}v}{(1 - v^2)^{\frac{1}{2}}} = \frac{\mathrm{d}x}{x(x^2 - 1)^{\frac{1}{2}}} \tag{2.101}$$

Again, Eq.(2.101) is in exact differential form. Hence the integration of Eq.(2.101) gives

$$\sin^{-1} v = -\sin^{-1}\frac{1}{x} + c$$

The general solution of the given differential equation is given by

$$\sin^{-1}\frac{y}{x} + \sin^{-1}\frac{1}{x} = c$$

EXAMPLE 2.86 Solve the differential equation $(y')^2x^2 - 2xyy' + 2y^2 - x^2 = 0$.

Solution: Clearly, the given differential equation is homogeneous in x and y. Let $y = xv$. Hence $y' = v + xv'$. In other words, the given differential equation reduces to

$$\begin{aligned} v + xv' &= \frac{vx \pm x(1 - v^2)^{\frac{1}{2}}}{x} \\ &= v \pm (1 - v^2)^{\frac{1}{2}} \end{aligned}$$

which is in variables separable form. Hence we have

$$\frac{\mathrm{d}v}{(1 - v^2)^{\frac{1}{2}}} = \pm\frac{\mathrm{d}x}{x} \tag{2.102}$$

Again, integration of Eq.(2.102) gives

$$\begin{aligned} \sin^{-1} v &= \pm \ln x + c \\ &= \pm \ln kx \end{aligned}$$

Hence the general solution of the given equation is

$$\sin^{-1}\frac{y}{x} = \pm \ln kx$$

Exercises 2.11.5

1. Solve the differential equation $xy^2((y')^2+2) - 2y'y^3 + x^3 = 0$.
2. Solve the differential equation $y^2 + xyy' - x^2(y')^2 = 0$.
3. Solve the differential equation $(y'x - y)^2 - (1+(y')^2)(x^2 - y^2) = 0$.
4. Solve the differential equation $(y')^2x^2 - 2xyy' + 2y^2 - x^2 = 0$
5. Solve the differential equation $2xyy' - y^2 + x^2 = 0$.
6. Solve the differential equation $xyy' - 2y^2 - 4x^2 = 0$.
7. Solve the differential equation $x^2y' = y^2 + xy + x^2$.
8. Solve the differential equation $x^2y' = y^2 + 5xy + 4x^2$.
9. Solve the differential equation $xy^3y' = y^4 + x^4$.
10. Solve the differential equation $y' = \frac{y}{x} - \operatorname{cosec}\frac{y}{x}$.
11. Solve the differential equation $y' = \frac{y}{x} + \sec\frac{y}{x}$.
12. Solve the differential equation $y' = \frac{e^{\frac{x}{y}}\left(\frac{x}{y}-1\right)}{1+e^{\frac{x}{y}}}$.
13. Solve the differential equation $y' = \frac{e^{\frac{x}{y}}}{1+\left(\frac{x}{y}\right)e^{\frac{x}{y}}}$.
14. Solve the differential equation $y' = \frac{y}{x} - \sin\frac{y}{x}\cos\frac{y}{x}$.
15. Solve the differential equation $y' = \frac{\left(\frac{x}{y}\right)^2\ln\left(\frac{x}{y}\right)-1}{\frac{x}{y}\ln\left(\frac{x}{y}\right)}$.

2.12 Higher Degree Equations of Special Type

In this section, we discuss two special types of higher degree differential equations. In particular, they are

1. Clairaut equation
2. Lagrange equation

2.12.1 Clairaut Equation

Any differential equation of the form $y = xy' + f(y')$ is called the *Clairaut equation.* One can differentiate the equation $y = xy' + f(y')$ with respect to x. In other words, we have

$$y' = xy'' + y' + f'(y')y'' \tag{2.103}$$

One can rewrite Eq.(2.103) in the form

$$(x + f'(y'))y'' = 0 \tag{2.104}$$

Equating each factor of Eq.(2.104) to zero, one can have

$$y'' = 0 \tag{2.105}$$

and

$$x + f'(y') = 0 \tag{2.106}$$

Again, the solution of Eq.(2.105) is $y' = c$. Hence putting $y' = c$ in the given differential equation, we have

$$y = xc + f(c) \tag{2.107}$$

which is the required solution and called the *integral solution.* One can eliminate y' between the given differential equation and Eq. (2.106) to have another solution, which is called the *singular solution* because it does not contain any arbitrary constant.

EXAMPLE 2.87 Solve the differential equation $y = xy' + \frac{a}{y'}$.

Solution: According to Eq.(2.107), the integral solution is given by $y = cx + \frac{a}{c}$. Again, the equivalent form of Eq. (2.106) is given by

$$x - \frac{a}{(y')^2} = 0 \tag{2.108}$$

Eliminating y' between the Eq.(2.108) and the given differential equation $y = xy' + \frac{a}{y'}$, the singular solution is given by $y^2 = 4ax$.

EXAMPLE 2.88 Solve the differential equation $(xy' - y)(x - yy') = 2y'$.

Solution: Let $x^2 = u$ and $y^2 = v$. Clearly, $\frac{\mathrm{d}u}{\mathrm{d}x} = 2x$ and $\frac{\mathrm{d}y}{\mathrm{d}v} = \frac{1}{2y}$. Again, one can easily verify that

$$\begin{aligned}\frac{\mathrm{d}y}{\mathrm{d}x} &= \left(\frac{\mathrm{d}y}{\mathrm{d}u}\right)\left(\frac{\mathrm{d}u}{\mathrm{d}x}\right)\\ &= \left(\frac{\mathrm{d}y}{\mathrm{d}u}\right)2x\\ &= \left(\frac{\mathrm{d}y}{\mathrm{d}v}\right)\left(\frac{\mathrm{d}v}{\mathrm{d}u}\right)2x\\ &= \left(\frac{x}{y}\right)\left(\frac{\mathrm{d}v}{\mathrm{d}u}\right)\end{aligned} \tag{2.109}$$

According to Eq.(2.109), the given differential equation reduces to

$$\begin{aligned}2\left(\frac{x}{y}\right)\left(\frac{\mathrm{d}v}{\mathrm{d}u}\right) &= 2y'\\ &= (xy' - y)(x - yy')\\ &= \left(x\left(\frac{x}{y}\right)\left(\frac{\mathrm{d}v}{\mathrm{d}u}\right) - y\right)\left(x - \left(\frac{x}{y}\right)\left(\frac{\mathrm{d}v}{\mathrm{d}u}\right)y\right)\\ &= \left(u\left(\frac{\mathrm{d}v}{\mathrm{d}u}\right) - v\right)\left(1 - \frac{\mathrm{d}v}{\mathrm{d}u}\right)\left(\frac{x}{y}\right)\end{aligned}$$

In other words, one can have

$$2\left(\frac{dv}{du}\right) = \left(u\left(\frac{dv}{du}\right) - v\right)\left(1 - \frac{dv}{du}\right) \tag{2.110}$$

Again, Eq.(2.110) can be rewritten in the form

$$v = u\left(\frac{dv}{du}\right) - \left(\frac{2\left(\frac{dv}{du}\right)}{1 - \frac{dv}{du}}\right) \tag{2.111}$$

which is a Clairaut equation. According to the formula given in Eq.(2.107), the solution of Eq.(2.111) is given by

$$v = cu - \frac{2c}{1-c}$$

In other words, the integral solution of the given equation is given by

$$y^2 = cx^2 - \frac{2c}{1-c}$$

Exercises 2.12.1

1. Solve the differential equation $y' = \ln(xy' - y)$.
2. Solve the differential equation $(y' - 1)e^{3x} + (y')^3 e^{2y} = 0$.
3. Solve the differential equation $y = 4xy' - 16y^3(y')^2$.
4. Solve the differential equation $y = xy' + (y')^3$.
5. Solve the differential equation $y' = \sin(y - xy')$.
6. Solve the differential equation $y = xy' + \ln y'$.
7. Solve the differential equation $y(y')^2 + y'x^3 = x^2 y$.
8. Solve the differential equation $y = xy' + (y')^2$.
9. Solve the differential equation $y = xy' + 2(y')^2$.
10. Solve the differential equation $y = 3xy' + 6(y')^2 y^2$.
11. Solve the differential equation $y = xy' - e^{y'}$.
12. Solve the differential equation $a^2 y' = (yy' + x)(xy' - y)$.
13. Solve the differential equation $(y' + 1)^2 = (x^2 y' + y^2)(xy' + y)$.
14. Solve the differential equation $y + y^2 + y'(2y - 2xy - x + 2) + x(y')^2(x - 2) = 0$.
15. Solve the differential equation $(x^2 + y^2)(1 + (y')^2) = 2(x + y)(x + yy') - (x + yy')^2$.

    HINT: Substitute $x^2 + y^2 = u$ and $x + y = v$.
16. Solve the differential equation $\sin y \cos^2 x = \cos^2(y(y')^2) + \sin x \cos y \cos yy'$.

    HINT: Substitute $\sin y = u$ and $\sin x = v$.
17. Solve the differential equation $y = xy' + (y' - (y')^2)$.
18. Solve the differential equation $y = xy' + \sin^{-1}(y')$.
19. Solve the differential equation $y = xy' + \frac{y'}{y'-1}$.

20. Solve the differential equation $x^2(y - xy') = y(y')^2$.

    HINT: Put $x^2 = u$ and $y^2 = v$. Accordingly, the given differential equation reduces to $v = uv' + (v')^2$, where $v' = \frac{dv}{du}$.

21. Solve the differential equation $y = 2xy' + y^2(y')^3$.

    HINT: Put $y^2 = v$. Accordingly, the given differential equation reduces to $v = xv' + \frac{v^3}{8}$.

22. Solve the differential equation $y + xy' = x^4(y')^2$.

    HINT: Put $xv = 1$. Accordingly, the given differential equation reduces to $y = uy' + (y')^2$ where $y' = \frac{dy}{du}$.

2.12.2 Lagrange Equation

Any differential equation of the form $y = xg(y') + f(y')$ is called a *Lagrange equation.* One can differentiate the equation $y = xg(y') + f(y')$ with respect to x. Hence we have

$$y' = g(y') + xg'(y')y'' + f'(y')y'' \tag{2.112}$$

Again, one can rewrite Eq.(2.112) in the form

$$\frac{dx}{dy'} = \left(\frac{g'(y')}{y' - g(y')}\right)x + \frac{f'(y')}{y' - g(y')} \tag{2.113}$$

Clearly, Eq.(2.113) is a first order linear nonhomogeneous differential equation which can be solved by the formula given in Eq.(2.8).

EXAMPLE 2.89 Solve the differential equation $y = 2xy' - (y')^3$.

Solution: Clearly, we have $g(y') = 2y'$ and $f(y') = -(y')^3$. According to Eq.(2.113), we have

$$\begin{aligned}\frac{dx}{dy'} &= \left(\frac{2}{y' - 2y'}\right)x - \frac{3(y')^2}{y' - 2y'} \\ &= -\frac{2x}{y'} + 3y' \end{aligned} \tag{2.114}$$

which is a first order linear nonhomogeneous differential equation. Hence the solution of Eq.(2.114), according to the formula given in Eq.(2.8), is given by

$$\begin{aligned} x &= (y')^{-2}\left(\int (y')^2 3y'\, dy' + c\right) \\ &= (y')^{-2}\left(\int 3(y')^3\, dy' + c\right) \\ &= \frac{3}{4}(y')^2 + c(y')^{-2} \end{aligned} \tag{2.115}$$

The general solution is given by eliminating y' between Eq.(2.115) and the given differential equation $y = 2xy' - (y')^3$.

EXAMPLE 2.90 Solve the differential equation $y = 2xy' - (y')^2$.

Solution: Clearly, we have $g(y') = 2y'$ and $f(y') = -(y')^2$. According to Eq.(2.113), we have

$$\begin{aligned}\frac{\mathrm{d}x}{\mathrm{d}y'} &= \left(\frac{2}{y' - 2y'}\right)x - \frac{2(y')}{y' - 2y'}\\ &= -\frac{2x}{y'} + 2 \qquad (2.116)\end{aligned}$$

which is a first order linear nonhomogeneous differential equation. Hence the solution of Eq.(2.116), according to the formula given in Eq.(2.8), is given by

$$\begin{aligned}x &= (y')^{-2}\left(\int 2(y')^2\,\mathrm{d}y' + c\right)\\ &= \frac{2}{3}y' + c(y')^{-2} \qquad (2.117)\end{aligned}$$

The general solution is given by eliminating y' between Eq.(2.117) and the given differential equation $y = 2xy' - (y')^2$.

Exercises 2.12.2

1. Solve the differential equation $yy' = x(y')^3 - 1$.
2. Solve the differential equation $y = xy' + \ln y'$.
3. Solve the differential equation $y = xy' + \sin y'$.
4. Solve the differential equation $y = x(y')^2 + (y')^2$.
5. Solve the differential equation $y = y(y')^2 + 2xy'$.
6. Solve the differential equation $y = x(1 + y') + (y')^2$.
7. Solve the differential equation $y = x + (y')^2 - \frac{2}{3}(y')^3$.
8. Solve the differential equation $y = 2xy' + (y')^2$.
9. Solve the differential equation $y = (y' + (y')^2)x + \frac{1}{y'}$.

2.13 Initial Value Problem

Every first order differential equation solution contains exactly one unknown parameter which appears by the process of integration. One can determine uniquely the unknown parameter when exactly one condition $y(x_0) = y_0$ is given.

EXAMPLE 2.91 Solve the initial value problem $y'\,x\ln(x) = y$ with $y(3) = \ln(81)$.

Solution: According to equation (2.2), the general solution is given by

$$\begin{aligned}y(x) &= e^{\int\left(\frac{1}{x\ln(x)}\right)\mathrm{d}x + \ln(c)}\\ &= e^{\ln(\ln(x)) + \ln(c)}\\ &= e^{\ln(c\ln(x))}\\ &= c\ln(x)\\ &= \ln(x^c)\end{aligned}$$

One can use the initial condition $y(3) = \ln(3^4)$. Hence we have $c = 4$ and the particular solution is $y(x) = \ln(x^4)$.

EXAMPLE 2.92 Solve the initial value problem $y' + x\,y = 0$ with $y(0) = 1$.
Solution: According to Eq.(2.2), the general solution is given by

$$y(x) = k\,e^{-\int x\,\mathrm{d}x}$$
$$= k\,e^{\left(-\frac{x^2}{2}\right)}$$

One can use the initial condition. Hence we have $k = 1$, and the particular solution is $y(x) = e^{\left(-\frac{x^2}{2}\right)}$.

EXAMPLE 2.93 Solve the initial value problem $y' + x\,y = x$ with $y(0) = 1$.
Solution: According to Eq.(2.4), the general solution is given by

$$y(x) = e^{\left(-\frac{x^2}{2}\right)}\left(\int x\,e^{\int x\,\mathrm{d}x}\,\mathrm{d}x + k\right)$$
$$= e^{\left(-\frac{x^2}{2}\right)}\left(\int x\,e^{\frac{x^2}{2}}\,\mathrm{d}x + k\right)$$
$$= e^{\left(-\frac{x^2}{2}\right)}\left(e^{\frac{x^2}{2}} + k\right)$$

One can use the initial condition. Hence we have $k = 0$, and the particular solution is $y(x) = 1$.

EXAMPLE 2.94 Solve the initial value problem $(x-1)y' + y = x - 1$ with $y(2) = \frac{1}{2}$.
Solution: According to Eq.(2.4), the general solution is given by

$$y(x) = \frac{1}{x-1}\left(\int (x-1)\,\mathrm{d}x + k\right)$$
$$= \frac{1}{x-1}\left(\frac{(x-1)^2}{2} + k\right)$$
$$= \frac{x-1}{2} + \frac{k}{x-1}$$

One can use the initial condition. Hence we have $k = 0$ and the particular solution is $y(x) = \frac{x-1}{2}$.

EXAMPLE 2.95 Solve the initial value problem $(x-1)y' + y = 1 - x$ with $y(2) = \frac{1}{2}$.
Solution: According to Eq.(2.4), the general solution is given by

$$y(x) = \frac{1}{x-1}\left(\int -(x-1)\,\mathrm{d}x + k\right)$$
$$= \frac{1}{x-1}\left(-\frac{(x-1)^2}{2} + k\right)$$
$$= -\frac{x-1}{2} + \frac{k}{x-1}$$

One can use the initial condition. Hence we have $k = 1$ and the particular solution is $y(x) = -\frac{x-1}{2} + \frac{1}{x-1}$.

EXAMPLE 2.96 Solve the initial value problem $y' = 4y^2$ with $y(0) = 1$.

Solution: Clearly, the given differential equation is of variables separable form. Hence the corresponding solution is

$$y(x) = -\left(\frac{1}{4x+k}\right)$$

One can use the initial condition. Hence we have $k = -1$ and the particular solution is $y(x) = \frac{1}{1-4x}$.

EXAMPLE 2.97 Solve the initial value problem $y' + y = 0$ with $y(0) = 1$.

Solution: According to Eq.(2.4), the general solution is given by

$$\begin{aligned} y(x) &= e^{-\int \mathrm{d}x} + k \\ &= e^{-x} + k \end{aligned}$$

One can use the initial condition. Hence we have $k = 0$ and the particular solution is $y(x) = e^{-x}$.

Exercises 2.13

1. Solve the initial value problem $y' + 2y = 2$ with $y(0) = 0$.
2. Solve the initial value problem $y' + 2y = 2$ with $y(0) = 2$.
3. Solve the initial value problem $y' + 2y = 2\,e^{-2x}$ with $y(0) = 2$.
4. Solve the initial value problem $y' + 2xy = e^{-x^2}$ with $y(1) = 0$.
5. Solve the initial value problem $y' - y = x$ with $y(0) = 1$.
6. Solve the initial value problem $y' - 2y = 4$ with $y(0) = 0$.
7. Solve the differential equation $xy' + y = xy^5$ with $y(1) = 1$.
8. Solve the differential equation $y' + 2xy = 2x^3$ with $y(0) = 0$.
9. Show that the solution of the differential equation $y' + p(x)\,y = f(x)$ with $y(x_0) = y_0$ can be represented by $y(x) = y(x_0)\,e^{-\int_{x_0}^{x} p(u)\,\mathrm{d}u} + e^{\int p(x)\,\mathrm{d}x}\left(\int_{x_0}^{x} f(u)e^{-\int p(u)\,\mathrm{d}u}\,\mathrm{d}u\right)$.

   HINT: Clearly, the corresponding integrating factor is $\mu(x) = e^{\int p(x)\,\mathrm{d}x}$. One can easily verify that the solution of the differential equation $y' + p(x)\,y = f(x)$ with initial condition $x = x_0$ can be given by

$$\begin{aligned} y(x)\mu(x) - y(x_0)\mu(x_0) &= \int_{x_0}^{x} \mathrm{d}\,(y(u)\,\mu(u)) \\ &= \int_{x_0}^{x} \mathrm{d}\left(y(u)\,e^{\int p(u)\,\mathrm{d}u}\right) \end{aligned}$$

$$= \int_{x_0}^{x} f(u)\,\mu(u)\,\mathrm{d}u$$

$$= \int_{x_0}^{x} f(u)\,e^{\int p(u)\,\mathrm{d}u}\,\mathrm{d}u$$

which is equivalent to

$$y(x) = \left(\frac{y(x_0)\mu(x_0)}{\mu(x)}\right) + \frac{1}{\mu(x)}\left(\int_{x_0}^{x} f(u)\,e^{\int p(u)\,\mathrm{d}u}\,\mathrm{d}u\right)$$

$$= y(x_0)\,e^{-\int_{x_0}^{x} p(u)\,\mathrm{d}u} + e^{-\int p(x)\,\mathrm{d}x}\left(\int_{x_0}^{x} f(u)\,e^{\int p(u)\,\mathrm{d}u}\,\mathrm{d}u\right)$$

Chapter 3

First Order Equation Applications

The differential equation of first order has a lot of applications in almost all branches of science and engineering. We shall formulate and solve a few problems in this connection for illustration.

3.1 Orthogonal Trajectories

Any curve which intersects every curve of a given family of curves at some constant angle θ is called a *θ-trajectory* of the given family of curves. If $\theta = 90°$, then the trajectory is called the *orthogonal trajectory* of the given family of curves. The first order differential equation has an important role in finding orthogonal trajectories of any given family of curves. There is a certain procedure for finding the orthogonal trajectories of family of curves, and it is as follows:

1. Find the differential equation associated with the given family of curves.
2. Replace y' by $-x'$ in the associated differential equation whenever x and y are space variables of the family of curves.
3. Replace r' by $-r^2\theta'$ in the associated differential equation whenever r and θ are space variables of the family of curves.
4. Solve the transformed differential equation for orthogonal trajectories.

EXAMPLE 3.1 Find the orthogonal trajectories of the family of curves $r = a(1 + \sin\theta)$.

Solution: Clearly, the corresponding differential equation is

$$r' = r\left(\frac{\cos\theta}{1+\sin\theta}\right) \tag{3.1}$$

Replacing r' by $-r^2\theta'$ in Eq.(3.1), we have

$$r\theta' = -\frac{\cos\theta}{1+\sin\theta} \tag{3.2}$$

Clearly, Eq.(3.2) is a first order differential equation which is of the variables separable form. In other words, we have

$$\begin{aligned}\frac{\mathrm{d}r}{r} &= -\frac{1+\sin\theta}{\cos\theta}\,\mathrm{d}\theta \\ &= -(\sec\theta+\tan\theta)\mathrm{d}\theta \end{aligned} \tag{3.3}$$

Clearly, integration of Eq.(3.3) gives

$$\begin{aligned} r &= c\left(\frac{\cos\theta}{\sec\theta+\tan\theta}\right) \\ &= c\left(\frac{\cos^2\theta}{1+\sin\theta}\right) \\ &= \frac{c}{1-\sin\theta} \end{aligned}$$

EXAMPLE 3.2 Find the orthogonal trajectory of the set of parabolas $xy = a$.
Solution: Clearly, the associated differential equation of the given set of parabolas is

$$y' = -\frac{y}{x} \tag{3.4}$$

Replacing y' by $-x'$ in Eq.(3.4), we have

$$x' = \frac{y}{x} \tag{3.5}$$

Clearly, Eq.(3.5) is of the variables separable form. In other words, we have

$$x\,\mathrm{d}x = y\,\mathrm{d}y \tag{3.6}$$

Integration of Eq.(3.6) gives

$$x^2 - y^2 = c$$

EXAMPLE 3.3 Find the orthogonal trajectories of the set of confocal conics $\frac{x^2}{a} + \frac{y^2}{a-\lambda} = 1$.
Solution: Clearly, the differentiation of $\frac{x^2}{a} + \frac{y^2}{a-\lambda} = 1$ gives

$$\frac{x}{a} + \left(\frac{y}{a-\lambda}\right) y' = 0 \tag{3.7}$$

Solving Eq.(3.7) for a, we have

$$a = \frac{\lambda x}{x+yy'} \tag{3.8}$$

In other words, we have

$$a - \lambda = -\frac{\lambda yy'}{x+yy'} \tag{3.9}$$

Substituting the expressions for a and $a - \lambda$ from Eqs.(3.8) and (3.9) in Eq.(3.7), we have

$$(x + yy')(xy' - y) - \lambda y' = 0 \tag{3.10}$$

Replacing y' by $-x'$ in Eq.(3.10), we have

$$(yx' - x)(xx' + y) - \alpha x' = 0 \tag{3.11}$$

where $\alpha = \lambda$. Clearly, the differential equations (3.10) and (3.11) are the same. Hence the given family of curve is a self-orthogonal trajectories.

Exercises 3.1

1. Find the orthogonal trajectories of $x + 2y = c$.
2. Find the orthogonal trajectories of $x^2 + 2y^2 = c$.
3. Find the orthogonal trajectories of $ay^2 = x^2$.
4. Find the orthogonal trajectories of $x^2 + y^2 = 2cy$.
5. Find the orthogonal trajectories of $x^2 + y^2 = 2cx$.
6. Find the orthogonal trajectories of $r^2 = c^2 \cos 2\theta$.
7. Find the orthogonal trajectories of $r = \frac{2c}{1+\cos\theta}$.

3.2 Growth Problem

This problem occurs in various fields like economic growth, growth of bacteria and growth of population. Malthus observed that the growth of humans, animals and bacteria is directly proportional to the population present. If $y(t)$ denotes the population of a given species at time t, then y', the rate of change of population of that species per unit time, is proportional to y. Mathematically, we have

$$y' = ky, \qquad k > 0 \tag{3.12}$$

Since y' is the rate of growth of y and Eq.(3.12) demonstrates that the rate of growth of y is proportional to y. The constant k is known as the *growth constant* of the substance. According to the formula given in Eq.(2.2), we have

$$\begin{aligned} y(t) &= ce^{\int p(t)\,dt} \\ &= ce^{\int k\,dt} \\ &= ce^{k\,(t-t_0)} \end{aligned} \tag{3.13}$$

It $y(t_0) = y_0$, then the unknown parameter c in Eq.(3.13) can be determined by

$$\begin{aligned} y_0 &= y(t_0) \\ &= ce^{t_0 - t_0} \\ &= c \end{aligned}$$

Substituting the value of c in Eq.(3.13), we have

$$y(t) = y_0 e^{k\,(t-t_0)} \tag{3.14}$$

The growth constant k is very much species population dependent. One can determine the growth constant k by using one more condition observed from the species population. It is convenient to express the rate of growth of a species population in terms of its *double-life*, which is the time required for double of a given population of species to grow.

EXAMPLE 3.4 A bacteria culture of population y is known to have a growth rate proportional to y itself. If the population triples between 6 P.M. and 7 P.M., then at what time will the population become 100 times what it was at 6 P.M.?

Solution: According to Eq.(3.14), we have

$$y(t) = y_0 e^{k\,(t-t_0)}$$

Clearly, $t_0 = 6$ and $y(6) = y_0$. Hence we have

$$y(t) = y_0 e^{k\,(t-6)}$$

Using the condition $y(7) = 3y_0$, we have

$$3 = e^k$$

In other words, $k = \ln 3$. Hence we have

$$y(t) = y_0 e^{\ln 3\,(t-6)}$$

Using the second condition, we have

$$e^{\ln 3(t-6)} = 100$$

In other words, the time required for making the population 100 times the population y_0 is $t = 6 + \left(\frac{\ln 100}{\ln 3}\right) = 10.19$ P.M.

EXAMPLE 3.5 If the growth of a culture of yeast is proportional to the amount $y(t)$ at time t and the amount is doubled in one day, then how much can be expected after one week?

Solution: According to Eq.(3.14), we have

$$y(t) = y_0 e^{k\,(t-t_0)}$$

Clearly, $t_0 = 0$. Hence we have

$$y(t) = y_0 e^{k\,t}$$

Using the half-life period condition, we have

$$2 = e^k$$

In other words, $k = \ln 2$. Hence we have

$$y(t) = y_0 e^{t \ln 2}$$

Population of yeast after one week is given by

$$y(7) = y_0 e^{7 \ln 2}$$

Exercises 3.2

1. If bacteria in a culture increases in number at a rate proportional to the number present and the original number of bacteria is doubled in 30 minutes, then find the time taken for the number to increase sixfold.
2. If bacteria in a nutrient solution increases at a rate proportional to the number present, the number doubles in 2 hours and 10^6 bacteria are at the end of 10 hours, then how many were there initially?
3. If a person deposits \$1 in a bank with interest of 5 per cent per year, then find the time required to double the invested amount when compounded continuously.
4. If y_0 bacteria are placed in a unlimited nutrient solution at time $t = 0$, $y(t)$ is the bacteria population of the colony at a later time t and the bacteria population at any moment is increasing at a rate proportional to the population at that moment, then find $y(t)$ as a function of t.
5. If in a culture of yeast the rate of growth is proportional to the population $y(t)$ present at time t and population doubles in 3 days, then how much can be expected after 1 week at the same rate of growth?
6. If the rate of growth is proportional to the amount of bacteria $y(t)$ at any time t and the amount of bacteria in a test tube doubles in 8 hours, then how long will it take to triple?

3.3 Decay Problem

This problem occurs in various fields like decay of radioactive elements in physics and chemistry. Rutherford showed that the radioactive decay of substance is directly proportional to the number of atoms of the substance present. If $y(t)$ denotes the number of atoms present at time t, then y', the number of atoms that disintegrate per unit time is proportional to y. Mathematically, we have

$$y' = ky, \qquad k < 0 \tag{3.15}$$

Since y' is the rate of growth of y, $-y'$ is its rate of decay, and Eq.(3.15) demonstrates that the rate of decay of y is proportional to y. The constant k is known as the *decay constant* of the substance. According to the formula given in Eq.(2.2), we have

$$\begin{aligned} y(t) &= ce^{\int p(t)\,dt} \\ &= ce^{\int k\,dt} \\ &= ce^{k\,(t-t_0)} \end{aligned} \tag{3.16}$$

It $y(t_0) = y_0$, then the unknown parameter c in Eq.(3.16) can be determined by

$$\begin{aligned} y_0 &= y(t_0) \\ &= ce^{t_0 - t_0} \\ &= c \end{aligned}$$

Substituting value of c in Eq.(3.16), we have

$$y(t) = y_0 e^{k\,(t-t_0)} \tag{3.17}$$

The decay constant k is very much radioactive substance dependent. One can determine the decay constant k by using one more condition observed from radioactive substances. It is convenient to express the rate of decay of a radioactive element in terms of its *half-life*, which is the time required for half of a given quantity of radioactive substance to decay.

EXAMPLE 3.6 If the rate of decay of radioactive material is proportional to the amount present and the half-life period of a sample of the material is 10 years, then how long does it take for 90% of the original amount of material to disintegrate?

Solution: According to Eq.(3.17), we have

$$y(t) = y_0 e^{k\,(t-t_0)}$$

Clearly, $t_0 = 0$. Hence we have

$$y(t) = y_0 e^{k\,t}$$

Using the half-life period condition, we have

$$\frac{1}{2} = e^{10k}$$

In other words, $k = -\frac{\ln 2}{10}$. Hence we have

$$y(t) = y_0 e^{-\left(\frac{\ln 2}{10}\right)t}$$

Using the second condition, we have

$$e^{-\left(\frac{\ln 2}{10}\right)t} = 1 - \frac{9}{10}$$

$$= \frac{1}{10}$$

In other words, the time required for disintegrating 90% of the original amount is $t = 10\left(\frac{\ln 10}{\ln 2}\right) = 33.22$ years.

EXAMPLE 3.7 If the evaporation rate of moisture from a sheet hung on a clothesline is proportional to the moisture content and one half of the moisture is lost in the first 20 minutes, then how long will it take to evaporate 95% of the moisture?

Solution: According to Eq.(3.17), we have

$$y(t) = y_0 e^{k\,(t-t_0)}$$

Clearly, $t_0 = 0$. Hence we have

$$y(t) = y_0 e^{k\,t}$$

Using the half-life period condition, we have

$$\frac{1}{2} = e^{20k}$$

In other words, $k = -\frac{\ln 2}{20}$. Hence we have

$$y(t) = y_0 e^{-\left(\frac{\ln 2}{20}\right)t}$$

Using the second condition, we have

$$\begin{aligned} e^{-\left(\frac{\ln 2}{20}\right)t} &= 1 - \frac{19}{20} \\ &= \frac{1}{20} \end{aligned}$$

In other words, the time required for disintegrating 95% of the original amount is $t = 20\left(\frac{\ln 20}{\ln 2}\right) = 86.44$ minutes.

Exercises 3.3

1. If a substance y decays exponentially, and only half of the quantity of the given substance remains after 2 years, then how long does it take for 5 pounds of the substance y to decay to 1 pound?
2. Many chemicals dissolve in water at a rate which is jointly proportional to the amount undissolved and the difference between the concentration of a saturated solution and the concentration of the actual solution. If a chemical of this kind is placed in a tank containing g gallons of water, y_0 amount is undissolved at time $t = 0$, y_1 amount is undissolved at time $t = t_1$ and s is the amount dissolved in the tank when the solution is saturated, find the amount y undissolved at time t.
3. If a wet sheet in a dryer loses its moisture content at a rate proportional to the total moisture content and it loses half of its moisture in 5 minutes, then how long will it take to lose 99% of its moisture?
4. If a wet sheet in a dryer loses its moisture content at a rate proportional to the total moisture content and it loses half of its moisture in 8 minutes, then when will it lose 95% of its moisture?

3.4 Mixing Problem

Many important problems in biology and engineering can be put into the mixing problem framework. A solution containing a fixed concentration of substance y flows into a tank containing another solution of substance y with certain concentration at a specified rate. The mixture is stirred together very rapidly, and then leaves the tank, again at a specified rate. The objective is to find the concentration of substance y in the tank at any time t. Let v be the number of gallons of solution kept in the tank with concentration c of the substance y. u gallons of another solution with concentration α of the substance y flows into the tank and u gallons of the mixture flows out at the rate w per minute. Let $s(t)$ be the amount of substance y in the tank at time t. The rate at which the substance y flows into the tank is αu, and the rate at which the substance y leaves the tank is $\left(\frac{w}{v+(u-w)t}\right) s(t)$. Since the rate of change of amount of the substance y in the tank at time t must equal the rate at which the substance enters the tank minus the rate at which it leaves the tank; therefore, we have

$$s' = \alpha u - \left(\frac{w}{v + (u-w)t}\right) s \tag{3.18}$$

Clearly, Eq.(3.18) is a first order constant coefficient nonhomogeneous equation. Let $p(t) = \frac{w}{v+(u-w)t}$ and $f(t) = \alpha u$. According to Eq.(2.8), the solution of Eq.(3.18) is given by

$$\begin{aligned}
s(t) &= e^{-\int p(t)\,\mathrm{d}t}\left(\int e^{\int p(t)\,\mathrm{d}t} f(t)\,\mathrm{d}t + k\right) \\
&= e^{-\int\left(\frac{w}{v+(u-w)t}\right)\mathrm{d}t}\left(\int \alpha u\, e^{\int\left(\frac{w}{v+(u-w)t}\right)\mathrm{d}t}\,\mathrm{d}t + k\right) \\
&= e^{-\left(\frac{w}{u-w}\right)\ln(v+(u-w)t)}\left(\alpha u \int e^{\left(\frac{w}{u-w}\right)\ln(v+(u-w)t)}\,\mathrm{d}t + k\right) \\
&= \left(\frac{1}{(v+(u-w)t)}\right)^{\frac{w}{u-w}}\left(\alpha(u-w)(v+(u-w)t)^{\left(\frac{u}{u-w}\right)} + k\right)
\end{aligned} \tag{3.19}$$

Again, the constant k, present in Eq.(3.19) can be determined using the initial condition $s(0) = cv$. Hence we have

$$\begin{aligned}
s(t) &= (\alpha(u-w)(v+(u-w)t)^{\frac{u}{u-w}} \\
&\quad + (c - \alpha(u-w))\, v^{\frac{u}{u-w}})\left(\frac{1}{(v+(u-w)t)}\right)^{\frac{w}{u-w}}
\end{aligned} \tag{3.20}$$

If $u = w$, then the differential equation given in Eq.(3.18) reduces to

$$s' = \alpha u - \left(\frac{u}{v}\right) s \tag{3.21}$$

According to the formula given in Eq.(2.8), the solution of Eq.(3.21) is given by

$$s(t) = \alpha v + (cv - \alpha v)e^{-\left(\frac{u}{v}\right)t} \tag{3.22}$$

EXAMPLE 3.8 A tank of 100 gallons capacity is initially full of pure water. Again, brine water containing $\frac{1}{4}$ pound of salt per gallon flows into the tank at the rate of 2 gallons per minute. The mixture flows out at the rate of 2 gallons per minute. If there is perfect mixing, then find the amount of salt in the tank after t minutes.

Solution: Clearly, $v = 100$ gallons, $c = 0$, $\alpha = 0.25$ and $u = 2 = w$. According to Eq.(3.22), the amount of salt $s(t)$ after t minutes is given by

$$\begin{aligned}
s(t) &= \alpha v + (cv - \alpha v)e^{-\left(\frac{u}{v}\right)t} \\
&= \frac{1}{2} - \left(\frac{1}{2}\, e^{-\frac{t}{50}}\right) \\
&= \frac{1}{2}\left(1 - e^{-\frac{t}{50}}\right)
\end{aligned}$$

EXAMPLE 3.9 A tank of 100 gallons capacity is initially full of pure water. Again, brine water containing 1 pound of salt per gallon flows into the tank at the rate of 1 gallon per minute, and at the same time pure water is allowed to run into the tank at the rate 1 gallon per minute. The mixture flows out at the rate of 2 gallons per minute. If there is perfect mixing, then find the amount of salt in the tank after t minutes.

Solution: Clearly, $v = 100$ gallons, $c = 0$, $\alpha = 0.5$ and $u = 2 = w$. According to Eq.(3.22), the amount of salt $s(t)$ after t minutes is given by

$$
\begin{aligned}
s(t) &= \alpha v + (cv - \alpha v)e^{-\left(\frac{u}{v}\right)t} \\
&= 1 - e^{-\frac{t}{50}}
\end{aligned}
$$

EXAMPLE 3.10 A tank contains 100 gallons of fresh water. If two gallons of brine each containing 1 pound of dissolved salt run into the tank per minute, and the mixture kept uniform by stirring runs out at the rate of 1 gallon per minute, then find the amount of salt present when the tank contains 150 gallons of brine.

Solution: Clearly, $v = 100$ gallons, $c = 0$, $\alpha = 1$, $w = 1$ and $u = 2$. According to Eq.(3.20), the amount of salt $s(t)$ after t minutes is given by

$$
\begin{aligned}
s(t) &= \left(\alpha(u-w)(v+(u-w)t)^{\frac{u}{u-w}} + (c - \alpha(u-w))\, v^{\frac{u}{u-w}}\right)\left(\frac{1}{(v+(u-w)t)}\right)^{\frac{w}{u-w}} \\
&= \frac{200t + t^2}{100 + t}
\end{aligned}
$$

Again, the time required for making the liquid volume in the container equal to 150 gallons is $t = 150 - 100 = 50$ minutes. Hence the concentration of the container at $t = 50$ minutes is given by

$$
\begin{aligned}
s(50) &= \frac{200 \times 50 + 50^2}{100 + 50} \\
&= 83.3 \text{ pounds}
\end{aligned}
$$

Exercises 3.4

1. If water containing 2 pounds of salt per gallon enters a tank containing s_0 pounds of salt dissolved in 200 gallons of water at the rate of 4 gallon per minute, and a well stirred solution leaves the tank at the same rate, then find the concentration of salt in the tank at any time t.
2. If pure water enters a tank containing 100 gallons of water in which 50 pounds of salt is dissolved at the rate of 1 gallon per minute, and a well stirred solution leaves the tank at the same rate, then find the concentration of salt in the tank at any time t.
3. A tank contains 50 gallons of water in which 25 pounds of salt is dissolved. Again, pure water runs into the tank at the rate of 2 gallons per minute, and the mixture flows out at the same rate through a second tank of infinite capacity initially containing 50 gallons of pure water. Find the time when the second tank contains the maximum amount of salt.

4. The initial concentration of salt in a 10-litre cubical tank is 0.02 gram per litre. If brine flows into the tank at a rate of 2 litre per second with a concentration of 0.01 gram per litre and the outflow equals the inflow, then find the time necessary to reach a concentration of 0.011 gram per litre in the tank.
5. A tank contains 1000 gallons of brine in which 500 pounds of salt are dissolved. If fresh water runs into the tank at the rate of 100 gallons per minute and at the same rate the mixture runs out after uniform stirring, then how long will it be before 50 pounds of salt are left in the tank?
6. A tank initially contains 50 gallons of fresh water. If brine containing 2 pounds per gallon of salt flows into the tank at the rate of 2 gallons per minute, and the mixture runs out at the same rate after uniform stirring, then how long will it take for the quantity of salt in the tank to increase from 40 pounds to 80 pounds?

3.5 Electrical Circuit

In this section, we are transferring a L-R-electrical circuit containing a resistance R, inductance L and voltage source $E(t)$ in series to a first order nonhomogeneous differential equation using Kirchhoff's voltage law. Let $i(t)$ be the current in the circuit at time t. Let a known electromotive force $E(t)$ be impressed across the circuit. Our problem is to find the current in the system as a function of time when the switch is put on. The experiments show that the voltage drop across the inductor is $L\left(\frac{di}{dt}\right)$ and across the resistor is $Ri(t)$. According to Kirchhoff's voltage law, the voltage impressed on a circuit is equal to the sum of the voltage drops in the rest of the circuit. Hence we have

$$L\,i' + R\,i = E(t) \tag{3.23}$$

Let $p(t) = \frac{R}{l}$ and $f(t) = \frac{E(t)}{L}$. According to Eq.(2.8), the solution of Eq.(3.23) is given by

$$\begin{aligned}
i(t) &= e^{-\int p(t)\,dt}\left(\int e^{\int p(t)\,dt} f(t)\,dt + k\right) \\
&= e^{-\int\left(\frac{R}{L}\right)dt}\left(\frac{1}{L}\int e^{\int\left(\frac{R}{L}\right)dt}E(t)\,dt + k\right) \\
&= e^{-\left(\frac{R}{L}\right)\int dt}\left(\frac{1}{L}\int e^{\left(\frac{R}{L}\right)\int dt}E(t)\,dt + k\right) \\
&= e^{-\left(\frac{R}{L}\right)t}\left(\frac{1}{L}\int e^{\left(\frac{R}{L}\right)t}E(t)\,dt + k\right)
\end{aligned} \tag{3.24}$$

Using the initial current i_0 in the circuit at $t = 0$, one can determine the constant k in the solution given in Eq.(3.24).

EXAMPLE 3.11 If an electromotive force of $160\cos 5t$ V is impressed on a series circuit composed of a $20\,\Omega$ resistor and a 10^{-1} H inductor, then find the current in the circuit.

Solution: Clearly, $L = 10^{-1}$ H, $R = 20\,\Omega$ and $E(t) = 160\cos 5t$ V. According to Eq.(3.24), we have

$$\begin{aligned}
i(t) &= e^{-\left(\frac{R}{L}\right)t}\left(\frac{1}{L}\int e^{\left(\frac{R}{L}\right)t}E(t)\,dt + k\right)\\
&= e^{-\left(\frac{20}{10^{-1}}\right)t}\left(\frac{1}{10^{-1}}\int e^{\left(\frac{20}{10^{-1}}\right)t}160\cos 5t\,dt + k\right)\\
&= e^{-200t}\left(10\int e^{200t}160\cos 5t\,dt + k\right)\\
&= e^{-200t}\left(1600\int e^{200t}\cos 5t\,dt + k\right)\\
&= e^{-200t}\left(\frac{12800}{1601}e^{200t}\cos 5t + \frac{320}{1601}e^{200t}\sin 5t + k\right)\\
&= \frac{12800}{1601}\cos 5t + \frac{320}{1601}\sin 5t + k\,e^{-200t}
\end{aligned}$$

EXAMPLE 3.12 If a electromotive force of $10\sin 2t$ V is impressed on a series circuit composed of a $8\,\Omega$ resistor and a 2 H inductor, then find the current in the circuit.

Solution: Clearly, $L = 2$ H, $R = 8\,\Omega$ and $E(t) = 10\sin 2t$ V. According to Eq.(3.24), we have

$$\begin{aligned}
i(t) &= e^{-\left(\frac{R}{L}\right)t}\left(\frac{1}{L}\int e^{\left(\frac{R}{L}\right)t}E(t)\,dt + k\right)\\
&= e^{-\frac{8}{2}t}\left(\frac{1}{2}\int e^{\frac{8}{2}t}10\sin 2t\,dt + k\right)\\
&= e^{-4t}\left(5\int e^{4t}\sin 2t\,dt + k\right)\\
&= e^{-4t}\left(\frac{1}{5}e^{4t}\sin 2t - \frac{1}{10}e^{4t}\cos 2t + k\right)\\
&= \frac{1}{5}\sin 2t - \frac{1}{10}\cos 2t + k\,e^{-4t}
\end{aligned}$$

Exercises 3.5

1. If a constant voltage of 12 V is impressed on a series circuit composed of a $10\,\Omega$ resistor and a 10^{-4} H inductor and current in the circuit is zero initially, then find the current after 2 seconds.
2. If a constant voltage of 10 V is impressed on a series circuit composed of a $25\,\Omega$ resistor and a 10^{-1} H inductor, and the current in the circuit is zero initially, then find the current after 1 second.
3. If a constant voltage of 100 V is impressed on a series circuit composed of a $100\,\Omega$ resistor and a 10^{-4} H inductor, and the current in the circuit is zero initially, then find the current after 2 seconds.

4. If a electromotive force of $15\sin(\omega t)$ V is impressed on a series circuit composed of a $25\,\Omega$ resistor and a 3×10^{-5} H inductor, and the current in the circuit is zero initially, then find the current after 4 seconds.

3.6 One-dimensional Heat Flow

If one or more hot bodies are placed in a colder medium, the heat from the hot bodies is disseminated by conduction (neglecting radiation as it is small at low temperature) to the surrounding medium. If the bodies are kept at a constant temperature, a steady state will be reached after a long period of time. According to Fourier's law, heat flux Q, the quantity of heat per unit area and time, is proportional to the greatest rate of decrease of temperature T. In other words, we have

$$Q = -\kappa\left(\frac{\partial T}{\partial n}\right) \tag{3.25}$$

where n is the normal to the surface and κ is the thermal conductivity of the fluid. The negative sign indicates that heat flows from points of higher temperature to points of lower temperature. In general, a description of temperature variation involves partial derivatives. However, in certain simple cases, the temperature depends only on the dependent variable x. If the heat flow is across a surface of area A, perpendicular to the x-direction, then Eq.(3.25) reduces to

$$Q = -\kappa\, A\left(\frac{\mathrm{d}T}{\mathrm{d}x}\right) \tag{3.26}$$

Clearly, the solution of Eq.(3.26) is given by

$$T(x) = -\kappa\, Ax + c \tag{3.27}$$

One can determine the unknown parameter c in Eq.(3.27) using the boundary condition $T(0)$ at $x = 0$.

EXAMPLE 3.13 A steam pipe of infinite length and diameter 1 metre has a cylindrical jacket, $\frac{1}{2}$ metre thick, made of an insulating material of thermal conductivity $\kappa = \frac{1}{3000}$. If the pipe is kept at 500 degrees centigrade and the outside of the jacket is at 100 degrees centigrade, then find the temperature distribution in the jacket.

Solution: Clearly, the temperature is a function of radius r of the jacket. In particular, $\frac{1}{2} \le r \le 1$. Again, $A = 2\pi r$. According to Eq.(3.26), we have

$$Q = -2\pi\kappa\, r\left(\frac{\mathrm{d}T}{\mathrm{d}r}\right) \tag{3.28}$$

Again, integration of Eq.(3.28) gives

$$\begin{aligned} T(r) &= C - \left(\frac{Q}{2\pi\kappa}\right)\ln r \\ &= C - \left(\frac{3000Q}{2\pi}\right)\ln r \\ &= C - \left(\frac{1500Q}{\pi}\right)\ln r \end{aligned} \tag{3.29}$$

One can determine the unknown parameters C and Q in Eq.(3.29), using the boundary conditions $T\left(\frac{1}{2}\right) = 500$ and $T(1) = 100$. Hence we have

$$T(r) = 100 - \left(\frac{400}{\ln 2}\right) \ln r$$

EXAMPLE 3.14 If the temperature of one end of a metallic rod having an area of 4 square metres is 100 degrees centigrade, the lateral surface of the rod is insulated, and the temperature 1 metre away from the heated end is 60 degrees centigrade, then find the temperature distribution inside the metallic rod.

Solution: Clearly, $A = 4$. According to Eq.(3.27), we have

$$\begin{aligned} T(x) &= -\kappa\, Ax + c \\ &= -4kx + c \end{aligned}$$

Again, $c = 100$ using the initial condition. Hence we have

$$T(x) = -4kx + 100$$

Again, $k = 10$ using the second condition. Hence the corresponding temperature distribution inside the metal rod is

$$T(x) = -40\, x + 100$$

Exercises 3.6

1. The temperature of one end of a metallic rod having an area of 4 square metres is maintained at T_0 degrees centigrade and the lateral surface of the rod is insulated. If the temperatures 2 metres and 3 metres away from the heated end are 60 degrees centigrade and 40 degrees centigrade respectively, then find the temperature distribution inside the metallic rod.
2. The temperature of one end of a metallic rod having an area of 4 square metres is maintained at T_0 degrees centigrade, and the lateral surface of the rod is insulated. If the temperatures 2 metres and 3 metres away from the heated end are 60 degrees centigrade and 40 degrees centigrade respectively, then find the initial temperature T_0.
3. If the temperature of one end having an area of 4 square metre of a metallic rod is at 100 degree centigrade and the lateral surface of the rod is insulated and temperature at 1 metre away from the heated end is 60 degree centigrade, then find the position on the metallic rod where the temperature is zero.
4. If at the right end at $x = 2$ of a metallic solid rod of cross-sectional area of 1200 square metres, 10 kilowatts heat is transferred and the left end is held at a constant temperature of 8 degree centigrade, then find the temperature distribution inside the metallic rod when the rod is laterally insulated.
5. A pipe, 20 centimetres in diameter, contains steam at 200 degrees centigrade and is protected with a covering 6 centimetres thick for which $k = 0.0003$. If the temperature of the outer surface of the covering is 30 degrees centigrade, then find the temperature distribution in the covering region.

6. A pipe 20 centimetres in diameter contains steam at 150 degrees centigrade and is protected with a covering 5 centimetres thick, for which $k = 0.0025$. If the temperature of the outer surface of the covering is 40 degrees centigrade, then find the temperature distribution in the covering region.

3.7 Law of Cooling

A body of temperature y_0 at time $t = 0$, is kept in a place having constant temperature α, which is maintained externally. Let $T(t)$ be the temperature of the body at any time t. If the rate at which the body cools is proportional to the difference between it and its surroundings, then we have

$$\begin{aligned} T' &= -k(T - \alpha) \\ &= -kT + k\alpha \end{aligned} \tag{3.30}$$

Clearly, Eq.(3.30) is a first order linear nonhomogeneous differential equation. Let $p(t) = k$ and $f(t) = k\alpha$. According to Eq.(2.8), we have

$$\begin{aligned} T(t) &= e^{-\int p(t)\,\mathrm{d}t}\left(\int e^{\int p(t)\,\mathrm{d}t} f(t)\,\mathrm{d}t + c\right) \\ &= e^{-\int k\,\mathrm{d}t}\left(\int e^{\int k\,\mathrm{d}t} k\alpha\,\mathrm{d}t + c\right) \\ &= e^{-kt}\left(\alpha \int k\,e^{kt}\,\mathrm{d}t + c\right) \\ &= e^{-kt}\left(\alpha e^{kt} + c\right) \\ &= ce^{-kt} + \alpha \end{aligned} \tag{3.31}$$

The parameter c in Eq.(3.31) can be determined using the temperature y_0 of the body at time $t = 0$.

EXAMPLE 3.15 A copper ball is heated initially to a temperature of 100 degrees centigrade and placed in water which is maintained at a temperature of 30 degrees centigrade. If at the end of 3 minutes, the temperature of the ball is reduced to 70 degrees centigrade, then find the temperature of the ball after 10 minutes.

Solution: Clearly, $\alpha = 30$. According to Eq.(3.31), we have

$$\begin{aligned} T(t) &= ce^{-kt} + \alpha \\ &= ce^{-kt} + 30 \end{aligned}$$

Again, $c = 70$ using the initial condition. Hence we have

$$T(t) = 70\,e^{-kt} + 30$$

Using the second condition, one can produce

$$70 = 70\, e^{-3k} + 30$$

In other words, we have $k = -\frac{\ln 3 - \ln 7}{3}$. Again, the temperature of the ball after 10 minutes is

$$T(10) = 70\, e^{\left(\frac{\ln 3 - \ln 7}{3}\right)10} + 30$$

EXAMPLE 3.16 If a pot of carrot-and-garlic soup cooling in air at 0 degree centigrade initially boils at 100 degrees centigrade and cools to 20 degrees centigrade during the first 30 minutes, then how much will it cool during the next 30 minutes?

Solution: Clearly, $\alpha = 0$. According to Eq.(3.31), we have

$$\begin{aligned} T(t) &= c\, e^{-kt} + \alpha \\ &= c\, e^{-kt} \end{aligned}$$

Again, $c = 100$ using the initial condition. Hence we have

$$T(t) = 100\, e^{-kt}$$

Using the second condition, one can produce

$$80 = 100\, e^{-30k}$$

In other words, we have $k = -\frac{\ln 4 - \ln 5}{30}$. Again, the temperature of the ball after 60 minutes is

$$\begin{aligned} T(60) &= 100\, e^{\left(\frac{\ln 4 - \ln 5}{30}\right)60} \\ &= 100\, e^{2\,(\ln 4 - \ln 5)} \\ &= 64 \end{aligned}$$

The temperature by which the pot cooled in the next 30 minutes is given by

$$\begin{aligned} T(60) - T(30) &= 80 - 64 \\ &= 16 \end{aligned}$$

Exercises 3.7

1. A body at an initial temperature of 100 degrees centigrade is in a room whose temperature is 20 degrees centigrade. If after 10 minutes its temperature is 40 degrees centigrade, then how long will it take for the body to reach a temperature of 25 degrees centigrade?
2. A body at an initial temperature of 100 degrees centigrade is in a room whose temperature is 20 degrees centigrade. If after 10 minutes its temperature is 40 degrees centigrade, then find its temperature at time t.
3. A body of unknown temperature is placed in a freezer which is kept at a constant temperature of 0 degrees centigrade. If the temperature of the body is 30 degrees centigrade after 15 minutes and 15 degrees centigrade after 30 minutes, then what is the initial temperature of the body?

4. A body of unknown temperature is placed in a hot chamber which is kept at a constant temperature of 100 degrees centigrade. If the temperature of the body is 30 degrees centigrade after 15 minutes and 75 degrees centigrade after 30 minutes, then what is the initial temperature of the body?
5. A metal ball is heated to a temperature of 100 degrees centigrade and placed at time $t = 0$ in water which is maintained at 40 degrees centigrade. If the temperature of the ball is reduced to 60 degrees centigrade in 4 minutes, then find the time at which the temperature of the ball is 50 degrees centigrade.
6. The President and the Prime Minister order coffee and receive cups of equal temperature at the same time. The President adds a small amount of cool cream immediately, but does not drink his coffee till 10 minutes later. The Prime Minister waits for 10 minutes, and then adds the same amount of cool cream and begins to drink. Who drinks the hotter coffee?

Chapter 4

Solution Existence and Uniqueness

In Chapter 2, we presented a few methods for determining the solutions of some first order differential equations. It is possible that we may not be able to solve the differential equation

$$y' = f(x, y), \qquad y(x_0) = y_0$$

explicitly. In other words, we may not exhibit a function $\phi(x)$, such that $y = \phi(x)$ satisfies the initial value problem

$$y' = f(x, y), \qquad y(x_0) = y_0$$

Therefore, the following questions arises:

1. How could we know that there is a solution of

$$y' = f(x, y), \qquad y(x_0) = y_0$$

   even though we cannot exhibit it?
2. How can we assure that there is exactly one solution to

$$y' = f(x, y), \qquad y(x_0) = y_0$$

These questions are important, because in approximations we are interested in finding approximate values of the solution $y(x)$ for a range of values of x, which is usually done by using digital computers.

4.1 Successive Approximation Method

Theorem 4.1.1 The differential equation $y' = f(x, y)$ with $y(x_0) = y_0$ has a solution if and only if the integral equation $y = y_0 + \int_{x_0}^{x} f(u, y)\, du$ has a solution.

Proof: Let $\phi(x)$ be a solution of the differential equation $y' = f(x, y)$ with $y(x_0) = y_0$. Hence we have $\phi'(x) = f(x, \phi(x))$ with $\phi(x_0) = y_0$. Again, we have

$$\begin{aligned}\phi(x) &= \phi(x_0) + \phi(x) - \phi(x_0) \\ &= \phi(x_0) + \int_{x_0}^{x} \phi'(u)\,\mathrm{d}u \\ &= \phi(x_0) + \int_{x_0}^{x} f(u, \phi(u))\,\mathrm{d}u \\ &= y_0 + \int_{x_0}^{x} f(u, \phi(u))\,\mathrm{d}u\end{aligned}$$

In other words, $\phi(x)$ is a solution of the integral equation $y = y_0 + \int_{x_0}^{x} f(u, y)\,\mathrm{d}u$

Let $\phi(x)$ be a solution of the integral equation $y = y_0 + \int_{x_0}^{x} f(u, y)\,\mathrm{d}u$. Hence we have

$$\phi(x) = y_0 + \int_{x_0}^{x} f(u, \phi(u))\,\mathrm{d}u \tag{4.1}$$

Clearly, $\phi(x_0) = y_0$. Again, differentiation of Eq. (4.1) gives

$$\phi'(x) = f(x, \phi(x))$$

In other words, $\phi(x)$ is a solution of the differential equation $y' = f(x, y)$ with $y(x_0) = y_0$.

Theorem 4.1.2 If the function $f(x, y)$ appearing in the differential equation $y' = f(x, y)$ with $y(x_0) = y_0$ is continuous over the region $\mathbb{W} = \{(x, y) : |x - x_0| \le a, |y - y_0| \le b\}$, then the function $\phi_n(x)$ defined by

$$\phi_n(x) = y_0 + \int_{x_0}^{x} f(v, \phi_{n-1}(v))\,\mathrm{d}v$$

with $\phi_0(x) = y_0$ exists on the interval I and satisfies $|\phi_n(x) - y_0| \le \alpha|x - x_0|$ for all $x \in I$, where $I = \{x : |x - x_0| \le h\}$, $h = \min\left\{a, \frac{b}{\alpha}\right\}$ and $\alpha = \max\{|f(x, y)| : (x, y) \in \mathbb{W}\}$.

Proof: Clearly, $f(x, y_0)$ is continuous on $\mathbb{W}$. Hence $\phi_1(x)$ is defined by

$$\phi_1(x) = y_0 + \int_{x_0}^{x} f(v, y_0)\,\mathrm{d}v$$

is continuous. Again, we have

$$|\phi_1(x) - y_0| = \left| y_0 + \int_{x_0}^{x} f(v,\, y_0)\,\mathrm{d}v - y_0 \right|$$

$$= \left| \int_{x_0}^{x} f(v,\, y_0)\,\mathrm{d}v \right|$$

$$= \alpha|x - x_0|$$

In other words, the theorem is true for $n = 1$. We will prove the theorem by the method of induction on n. Assume that $\phi_n(x)$ exists, is continuous on I and satisfies the condition $|\phi_n(x) - y_0| \leq \alpha|x - x_0|$. Clearly, we have $|x - x_0| \leq h \leq a$ and

$$|\phi_n(x) - y_0| \leq \alpha|x - x_0|$$

$$\leq \alpha h$$

$$\leq b$$

Hence $(x,\, \phi_n(x)) \in \mathbb{W}$ for all $x \in I$. In other words, $f(x,\, \phi_n(x))$ exists. In other words, $f(x,\, \phi_n(x))$ is continuous. Hence $\phi_{n+1}(x)$, defined by

$$\phi_{n+1}(x) = y_0 + \int_{x_0}^{x} f(v,\, \phi_n(v))\,\mathrm{d}v$$

exists and is continuous on I. Again, we have

$$|\phi_{n+1}(x) - y_0| = \left| y_0 + \int_{x_0}^{x} f(v,\, \phi_n(v))\,\mathrm{d}v - y_0 \right|$$

$$= \left| \int_{x_0}^{x} f(v,\, \phi_n(v))\,\mathrm{d}v \right|$$

$$= \alpha|x - x_0|$$

Theorem 4.1.3 If the function $f(x,\, y)$ appearing in the differential equation $y' = f(x,\, y)$ with $y(x_0) = y_0$ is continuous over the region $\mathbb{W} = \{(x,\, y) \,:\, |x - x_0| \leq a,\; |y - y_0| \leq b\}$ and $|f(x,\, y) - f(x,\, u)| \leq k|y - u|$ for all $(x,\, y),\; (x,\, u) \in \mathbb{W}$, then the function $\phi_n(x)$ defined by

$$\phi_n(x) = y_0 + \int_{x_0}^{x} f(v,\, \phi_{n-1}(v))\,\mathrm{d}v$$

with $\phi_0(x) = y_0$ satisfies $|\phi_n(x) - \phi_{n-1}(x)| \leq \left(\frac{\alpha k^{n-1}}{n!}\right)|x - x_0|^n$ for all $x \in I$, where $k > 0$, $I = \{x \,:\, |x - x_0| \leq h\}$, $h = \min\left\{a,\, \frac{b}{\alpha}\right\}$ and $\alpha = \max\{|f(x,\, y)| \,:\, (x,\, y) \in \mathbb{W}\}$.

Proof: Clearly, we have

$$\begin{aligned}
|\phi_1(x) - \phi_0(x)| &= |\phi_1(x) - y_0| \\
&= \left| y_0 + \int_{x_0}^{x} f(v,\, y_0)\, \mathrm{d}v - y_0 \right| \\
&= \left| \int_{x_0}^{x} f(v,\, y_0)\, \mathrm{d}v \right| \\
&= \alpha |x - x_0|
\end{aligned}$$

In other words, the theorem is true for $n = 1$. We will prove the theorem by the method of induction on n. Assume that

$$|\phi_n(x) - \phi_{n-1}(x)| \leq \left(\frac{\alpha k^{n-1}}{n!} \right) |x - x_0|^n$$

for all $x \in I$. Again, we have

$$\begin{aligned}
|\phi_{n+1}(x) - \phi_n(x)| &= \left| y_0 + \int_{x_0}^{x} f(v,\, \phi_n(v))\, \mathrm{d}v - y_0 - \int_{x_0}^{x} f(v,\, \phi_{n-1}(v))\, \mathrm{d}v \right| \\
&= \left| \int_{x_0}^{x} f(v,\, \phi_n(v))\, \mathrm{d}v - \int_{x_0}^{x} f(v,\, \phi_{n-1}(v))\, \mathrm{d}v \right| \\
&= \left| \int_{x_0}^{x} (f(v,\, \phi_n(v)) - f(v,\, \phi_{n-1}(v))\, \mathrm{d}v \right| \\
&\leq \int_{x_0}^{x} |f(v,\, \phi_n(v)) - f(v,\, \phi_{n-1}(v))|\, \mathrm{d}v \\
&\leq k \int_{x_0}^{x} |\phi_n(v) - \phi_{n-1}(v)|\, \mathrm{d}v \\
&\leq k \int_{x_0}^{x} \left(\frac{\alpha k^{n-1}}{n!} \right) |v - x_0|^n\, \mathrm{d}v
\end{aligned}$$

$$\leq \frac{\alpha k^n}{n!} \int_{x_0}^{x} |v - x_0|^n \, dv$$

$$= \frac{\alpha k^n}{(n+1)!} |x - x_0|^{n+1}$$

Theorem 4.1.4 If the function $f(x, y)$ appearing in the differential equation $y' = f(x, y)$ with $y(x_0) = y_0$ is continuous over the region $\mathbb{W} = \{(x, y) : |x - x_0| \leq a, |y - y_0| \leq b\}$ and $|f(x, y) - f(x, u)| \leq k|y - u|$ for all $(x, y), (x, u) \in \mathbb{W}$, then the sequence of functions $\{\phi_n(x)\}$ defined by

$$\phi_n(x) = y_0 + \int_{x_0}^{x} f(v, \phi_{n-1}(v)) \, dv$$

with $\phi_0(x) = y_0$ converges uniformly to a continuous function $\phi(x)$ for all $x \in I$, where $k > 0$, $I = \{x : |x - x_0| \leq h\}$, $h = \min\left\{a, \frac{b}{\alpha}\right\}$ and $\alpha = \max\{|f(x, y)| : (x, y) \in \mathbb{W}\}$.

Proof: One can easily verify that

$$\begin{aligned}
\phi_n(x) &= \phi_0(x) + (\phi_1(x) - \phi_0(x)) + (\phi_2(x) - \phi_1(x)) + \cdots + (\phi_n(x) - \phi_{n-1}(x)) \\
&= y_0 + (\phi_1(x) - \phi_0(x)) + (\phi_2(x) - \phi_1(x)) + \cdots + (\phi_n(x) - \phi_{n-1}(x)) \\
&= y_0 + \sum_{i=1}^{n} (\phi_i(x) - \phi_{i-1}(x))
\end{aligned}$$

In other words, the sequence $\{\phi_n(x)\}$ is the sequence of partial sums of the series

$$y_0 + \sum_{i=1}^{\infty} (\phi_i(x) - \phi_{i-1}(x)) \tag{4.2}$$

Hence the sequence $\{\phi_n(x)\}$ converges if and only if the series given in Eq.(4.2) converges. According to Theorem 4.1.3, we have

$$\begin{aligned}
y_0 + \sum_{i=1}^{\infty} (\phi_i(x) - \phi_{i-1}(x)) &\leq y_0 + \sum_{i=1}^{\infty} |(\phi_i(x) - \phi_{i-1}(x))| \\
&\leq y_0 + \sum_{i=1}^{\infty} \left(\frac{\alpha k^{i-1}}{i!}\right) |x - x_0|^i \\
&\leq y_0 + \frac{\alpha}{k} \sum_{i=1}^{\infty} \left(\frac{\alpha k^i}{i!}\right) |x - x_0|^i \\
&= y_0 + \frac{\alpha}{k} \sum_{i=1}^{\infty} \left(\frac{\alpha k^i}{i!}\right) h^i
\end{aligned}$$

$$= y_0 + \frac{\alpha}{k}\sum_{i=1}^{\infty}\left(\frac{\alpha(hk)^i}{i!}\right)$$

$$= y_0 + \frac{\alpha}{k}\left(e^{hk} - 1\right)$$

Clearly, the series given in Eq.(4.2) converges absolutely. In other words, the series given in Eq.(4.2) converges uniformly for all $x \in I$. Hence the associated sequence $\{\phi_n(x)\}$ converges uniformly. Let $\lim\limits_{n\to\infty} \phi_n(x) = \phi(x)$. Since the sequence $\{\phi_n(x)\}$ converges absolutely to $\phi(x)$, and $\phi_n(x)$ are continuous, therefore, $\phi(x)$ is continuous.

Theorem 4.1.5 If the function $f(x, y)$ appearing in the differential equation $y' = f(x, y)$ with $y(x_0) = y_0$ is continuous over the region $\mathbb{W} = \{(x, y) \ : \ |x - x_0| \le a, \ |y - y_0| \le b\}$ and $|f(x, y) - f(x, u)| \le k|y - u|$ for all $(x, y), \ (x, u) \in \mathbb{W}$, where $k > 0$, then the limiting function $\phi(x)$ of the sequence $\{\phi_n(x)\}$ defined by

$$\phi_n(x) = y_0 + \int_{x_0}^{x} f(v, \phi_{n-1}(v))\, dv$$

with $\phi_0(x) = y_0$ is a solution of the integral equation

$$\phi(x) = y_0 + \int_{x_0}^{x} f(v, \phi(v))\, dv$$

Proof: Define $\alpha = \max\{|f(x, y)| \ : \ (x, y) \in \mathbb{W}\}$, $h = \min\left\{a, \frac{b}{\alpha}\right\}$ and $I = \{x \ : \ |x - x_0| \le h\}$. According to Theorem 4.1.4, the sequence $\{\phi_n(x)\}$ converges uniformly to $\phi(x)$. Hence for given $\epsilon > 0$, there exists m, such that $|\phi(x) - \phi_n(x)| \le \frac{\epsilon}{kh}$ for all $x \in I$ and $n \ge m$. It is easy to verify that

$$\left|\int_{x_0}^{x} f(v, \phi(v))\, dv - \int_{x_0}^{x} f(v, \phi_n(v))\, dv\right| = \left|\int_{x_0}^{x} f(v, \phi(v))\, dv - f(v, \phi_n(v))\, dv\right|$$

$$\le \int_{x_0}^{x} |f(v, \phi(v)) - f(v, \phi_n(v))|\, dv$$

$$\le \int_{x_0}^{x} k|\phi(v) - \phi_n(v)|\, dv$$

$$\le k|x - x_0|\, (\max\{|\phi(v) - \phi_n(v)| \ : \ v \in I\})$$

$$\le kh\left(\frac{\epsilon}{kh}\right)$$

$$= \epsilon$$

In other words, we have

$$\lim_{n\to\infty} \int\limits_{x_0}^{x} f(v,\, \phi_n(v))\, dv = \int\limits_{x_0}^{x} f(v,\, \phi(v))\, dv$$

It is easy to verify that

$$\begin{aligned}
\phi(x) &= \lim_{n\to\infty} \phi_n(x) \\
&= \lim_{n\to\infty} \left(y_0 + \int\limits_{x_0}^{x} f(v,\, \phi_n(v))\, dv \right) \\
&= y_0 + \lim_{n\to\infty} \int\limits_{x_0}^{x} f(v,\, \phi_n(v))\, dv \\
&= y_0 + \int\limits_{x_0}^{x} \lim_{n\to\infty} f(v,\, \phi_n(v))\, dv \\
&= y_0 + \int\limits_{x_0}^{x} f(v,\, \phi(v))\, dv
\end{aligned}$$

In other words, $\phi(x)$ is a solution of the integral equation

$$\phi(x) = y_0 + \int\limits_{x_0}^{x} f(v,\, \phi(v))\, dv$$

Theorem 4.1.6 If the function $f(x,\, y)$ appearing in the differential equation $y' = f(x,\, y)$ with $y(x_0) = y_0$ is continuous over the region $\mathbb{W} = \{(x,\, y) \,:\, |x - x_0| \le a,\ |y - y_0| \le b\}$ and $|f(x,\, y) - f(x,\, u)| \le k|y - u|$ for all $(x,\, y),\ (x,\, u) \in \mathbb{W}$ where $k > 0$, then there exists a unique solution to the differential equation $y' = f(x,\, y)$ with $y(x_0) = y_0$ over the interval $I = \{x \,:\, |x - x_0| \le h\}$ where $h = \min\left\{a,\, \frac{b}{\alpha}\right\}$ and $\alpha = \max\{|f(x,\, y)| \,:\, (x,\, y) \in \mathbb{W}\}$.

Proof: Let $\psi(x)$ and $\phi(x)$ be two different solutions of the differential equation $y' = f(x,\, y)$ with $y(x_0) = y_0$. In other words, we have

$$\psi'(x) = f(x,\, \psi(x)) \qquad \psi(x_0) = y_0$$

and

$$\phi'(x) = f(x,\, \phi(x)) \qquad \phi(x_0) = y_0$$

According to Theorem 4.1.1, we have

$$\psi(x) = y_0 + \int\limits_{x_0}^{x} f(v,\, \psi(v))\, dv$$

and

$$\phi(x) = y_0 + \int_{x_0}^{x} f(v,\, \phi(v))\, dv$$

Again, $\psi(x) - \phi(x)$ is continuous in the interval I by Theorem 4.1.4. Hence $|\psi(x) - \phi(x)| \leq c$ for all $x \in I$. It is easy to verify that for $x \in I$, we have

$$\begin{aligned}
|\psi(x) - \phi(x)| &= \left| y_0 + \int_{x_0}^{x} f(v,\, \psi(v))\, dv - y_0 - \int_{x_0}^{x} f(v,\, \phi(v))\, dv \right| \\
&= \left| \int_{x_0}^{x} f(v,\, \psi(v))\, dv - \int_{x_0}^{x} f(v,\, \phi(v))\, dv \right| \\
&= \left| \int_{x_0}^{x} (f(v,\, \psi(v)) - f(v,\, \phi(v)))\, dv \right| \\
&\leq \int_{x_0}^{x} |f(v,\, \psi(v)) - f(v,\, \phi(v))|\, dv| \\
&\leq k \int_{x_0}^{x} |\psi(x) - \phi(x)|\, dv
\end{aligned} \tag{4.3}$$

and

$$\begin{aligned}
|\psi(x) - \phi(x)| &\leq kc|x - x_0| \\
&\leq kch
\end{aligned} \tag{4.4}$$

Clearly, Eqs.(4.3) and (4.4) are coupled. One can show by repeated use of Eqs.(4.3) and (4.4) that

$$|\psi(x) - \phi(x)| \leq \frac{ck^n h^n}{n!}$$

for any positive integer n. In other words, we have

$$\begin{aligned}
\lim_{n\to\infty} |\psi(x) - \phi(x)| &\leq \lim_{n\to\infty} \left(\frac{ck^n h^n}{n!} \right) \\
&= 0
\end{aligned}$$

Hence the given differential equation has a unique solution.

Theorem 4.1.7 If the function $f(x,\, y)$ of the initial value problem $y' = f(x,\, y)$ with $y(x_0) = y_0$ is continuous and satisfy the condition $|f(x,\, y) - f(x,\, u)| \leq k|y - u|$ where $k > 0$ over the region $\mathbb{W} = \{(x,\, y) \,:\, |x - x_0| \leq a,\ |y - y_0| \leq b\}$, then a small change $\overline{y}_0$ to the initial value y_0 induces a small change $\overline{\phi}(x)$ to the existing solution $\phi(x)$.

Proof: Let $|y_0 - \overline{y}_0| \leq \delta$ and $\epsilon = \delta e^{kh}$, where $|x - x_0| \leq h$. Clearly, $\phi(x)$ and $\overline{\phi}(x)$ are solutions of the initial value problems

$$y' = f(x, y) \qquad y(x_0) = y_0$$

and

$$y' = f(x, y) \qquad y(x_0) = \overline{y}_0$$

respectively. In other words, we have

$$\phi(x) = y_0 + \int_{x_0}^{x} f(v, \phi(v))\, dv \tag{4.5}$$

and

$$\overline{\phi}(x) = \overline{y}_0 + \int_{x_0}^{x} f(v, \overline{\phi}(v))\, dv \tag{4.6}$$

One can have, using Eqs.(4.5) and (4.6)

$$\begin{aligned} |\phi(x) - \overline{\phi}(x)| &= \left| y_0 + \int_{x_0}^{x} f(v, \phi(v))\, dv - \overline{y}_0 - \int_{x_0}^{x} f(v, \overline{\phi}(v))\, dv \right| \\ &= \left| y_0 - \overline{y}_0 + \int_{x_0}^{x} \left(f(v, \phi(v)) - f(v, \overline{\phi}(v)) \right) dv \right| \\ &\leq |y_0 - \overline{y}_0| + \int_{x_0}^{x} \left| f(v, \phi(v)) - f(v, \overline{\phi}(v)) \right| dv \\ &\leq |y_0 - \overline{y}_0| + k \int_{x_0}^{x} \left| \phi(v) - \overline{\phi}(v) \right| dv \end{aligned} \tag{4.7}$$

Let $\psi(x) = \int_{x_0}^{x} |\phi(v) - \overline{\phi}(v)| dv$. Clearly, Eq.(4.7) reduces to

$$\begin{aligned} \psi'(x) - k\psi(x) &\leq |y_0 - \overline{y}_0| \\ &\leq \delta \end{aligned} \tag{4.8}$$

One can multiply Eq.(4.8) by $e^{-k(x-x_0)}$ to have

$$\begin{aligned} \delta e^{-k(x-x_0)} &\geq e^{-k(x-x_0)} \left(\psi'(x) - k\psi(x) \right) \\ &= \left(\psi(x) e^{-k(x-x_0)} \right)' \end{aligned} \tag{4.9}$$

Again, the integration of Eq.(4.9) from x_0 to x gives

$$\psi e^{-k(x-x_0)} \leq \frac{\delta}{k}\left(1 - e^{-k(x-x_0)}\right)$$

In other words, we have

$$\psi(x) \leq \frac{\delta}{k}\left(e^{k(x-x_0)} - 1\right) \tag{4.10}$$

One can have, using Eqs.(4.7) and (4.10),

$$\begin{aligned}
|\phi(x) - \overline{\phi}(x)| &\leq |y_0 - \overline{y}_0| + k\left(\frac{\delta}{k}\right)\left(e^{k(x-x_0)} - 1\right) \\
&= \delta e^{k(x-x_0)} \\
&= \delta e^{kh} \\
&= \epsilon
\end{aligned}$$

Similarly, integration of Eq.(4.9) by replacing $x - x_0$ by $x_0 - x$ from x to x_0 gives

$$\psi(x) \leq \left(\frac{\delta}{k}\right)\left(e^{k(x_0-x)} - 1\right) \tag{4.11}$$

One can have, using Eq.(4.7) by changing the order of integration limit and Eq. (4.11),

$$\begin{aligned}
|\phi(x) - \overline{\phi}(x)| &\leq |y_0 - \overline{y}_0| + k\left(\frac{\delta}{k}\right)\left(e^{k(x_0-x)} - 1\right) \\
&= \delta e^{k(x_0-x)} \\
&= \delta e^{kh} \\
&= \epsilon
\end{aligned}$$

EXAMPLE 4.1 Show that the solution $y(x)$ of the initial value problem $y' = y^2 + \cos x^2$ with $y(0) = 0$ exists for $|x| \leq \frac{1}{2}$.

Solution: Clearly, $f(x, y) = y^2 + \cos x^2$, $a = \frac{1}{2}$, $x_0 = 0$ and $y_0 = 0$. Hence the region is $\mathbb{W} = \{(x, y) \,:\, |x| \leq \frac{1}{2},\ |y| \leq b\}$. Again, $f(x, y)$ is continuous at every point of $\mathbb{W}$. It is easy to verify that $\alpha = \max\{|f(x, y)| \,:\, (x, y) \in \mathbb{W}\} = 1 + b^2$. Again, we have

$$\begin{aligned}
h &= \min\left\{a, \frac{b}{1+b^2}\right\} \\
&= \min\left\{\frac{1}{2}, \frac{b}{1+b^2}\right\} \\
&= \min\left\{\frac{1}{2}, \frac{1}{2}\right\} \\
&= \frac{1}{2}
\end{aligned}$$

EXAMPLE 4.2 Show that the solution $y(x)$ of the initial value problem $y' = y + \cos x^2$ with $y(0) = 0$ exists for $|x| \le \frac{1}{2}$.

Solution: Clearly, $f(x, y) = y + \cos x^2$, $a = \frac{1}{2}$, $x_0 = 0$ and $y_0 = 0$. Hence the region is $\mathbb{W} = \{(x, y) : |x| \le \frac{1}{2}, |y| \le b\}$. Again, $f(x, y)$ is continuous at every point of $\mathbb{W}$. It is easy to verify that $\alpha = \max\{|f(x, y)| : (x, y) \in \mathbb{W}\} = 1 + b$. Again, we have

$$\begin{aligned} h &= \min\left\{a, \frac{b}{1+b}\right\} \\ &= \min\left\{\frac{1}{2}, \frac{b}{1+b}\right\} \\ &= \min\left\{\frac{1}{2}, \frac{1}{2}\right\} \\ &= \frac{1}{2} \end{aligned}$$

EXAMPLE 4.3 Find the largest interval on which the initial value problem $y' = e^{2y}$ with $y(0) = 0$ has a unique solution.

Solution: Clearly, $f(x, y) = e^{2y}$, $x_0 = 0$ and $y_0 = 0$. Hence the region is $\mathbb{W} = \{(x, y) : |x| \le a, |y| \le b\}$. Again, $f(x, y)$ is continuous at every point of $\mathbb{W}$. It is easy to verify that $\alpha = \max\{|f(x, y)| : (x, y) \in \mathbb{W}\} = e^{2b}$. Again, we have

$$\begin{aligned} h &= \min\left\{a, \frac{b}{e^{2b}}\right\} \\ &= \min\left\{a, \frac{1}{2e}\right\} \end{aligned}$$

In other words, the maximum interval is possible when $a = \frac{1}{2e}$ at $b = \frac{1}{2}$. The maximum interval is given by $-\frac{1}{2e} \le x \le \frac{1}{2e}$.

EXAMPLE 4.4 Find the interval on which the initial value problem $y' = e^{(y-x)^2}$ with $y(0) = 1$ has a unique solution.

Solution: Clearly, $f(x, y) = e^{(y-x)^2}$, $x_0 = 0$ and $y_0 = 1$. Hence the region is $\mathbb{W} = \{(x, y) : |x| \le a, |y - 1| \le b\}$. Again, $f(x, y)$ is continuous at every point of $\mathbb{W}$. It is easy to verify that $\alpha = \max\{|f(x, y)| : (x, y) \in \mathbb{W}\} = e^{(1+b)^2}$. Again, we have

$$\begin{aligned} h &= \min\left\{a, \frac{b}{e^{(1+b)^2}}\right\} \\ &= \min\left\{a, \left(\frac{\sqrt{3}-1}{2}\right) e^{\ 1+\frac{\sqrt{3}}{2}}\right\} \end{aligned}$$

In other words, the maximum interval is possible when $a = \left(\frac{\sqrt{3}-1}{2}\right) e^{\ 1+\frac{\sqrt{3}}{2}}$ at $b = \frac{1+\sqrt{3}}{2}$. The interval on which the given initial value problem has a unique solution is given by

$$\left(\frac{1-\sqrt{3}}{2}\right) e^{\ 1+\frac{\sqrt{3}}{2}} \le x \le \left(\frac{\sqrt{3}-1}{2}\right) e^{\ 1+\frac{\sqrt{3}}{2}}$$

EXAMPLE 4.5 Find the largest interval in order to have a unique solution to the initial value problem $y' = x^2 + y^2$ with $y(0) = 0$.

Solution: Clearly, $f(x, y) = x^2 + y^2$, $x_0 = 0$ and $y_0 = 0$. Hence the region is $\mathbb{W} = \{(x, y) : |x| \leq a,\ |y| \leq b\}$. Again, $f(x, y)$ is continuous at every point of $\mathbb{W}$. It is easy to verify that $\alpha = \max\{|f(x, y)| : (x, y) \in \mathbb{W}\} = a^2 + b^2$. Again, we have

$$\begin{aligned} h &= \min\left\{a,\ \frac{b}{a^2 + b^2}\right\} \\ &= \min\left\{a,\ \frac{1}{2a}\right\} \end{aligned}$$

In other words, the maximum interval is possible when $a = \frac{1}{2a}$ at $b = a$. The maximum interval is given by $-\frac{1}{\sqrt{2}} \leq x \leq \frac{1}{\sqrt{2}}$.

EXAMPLE 4.6 Find the largest interval in order to have a unique solution to the initial value problem $y' = y^2$ with $y(1) = -1$.

Solution: Clearly, $f(x, y) = y^2$, $x_0 = 1$ and $y_0 = -1$. Hence the region is $\mathbb{W} = \{(x, y) : |x - 1| \leq a,\ |y + 1| \leq b\}$. Again, $f(x, y)$ is continuous at every point of $\mathbb{W}$. It is easy to verify that $\alpha = \max\{|f(x, y)| : (x, y) \in \mathbb{W}\} = (b - 1)^2$. Again, we have

$$\begin{aligned} h &= \min\left\{a,\ \frac{b}{(b + 1)^2}\right\} \\ &= \min\left\{a,\ \frac{1}{4}\right\} \end{aligned}$$

In other words, the maximum interval is possible when $a = \frac{1}{4}$ at $b = 1$. The maximum interval is given by $-\frac{3}{3} \leq x \leq \frac{5}{4}$.

EXAMPLE 4.7 Solve the initial value differential equation $y' = x$ with $y(0) = 1$ by approximated method.

Solution: Clearly, $f(x, y) = x$. The corresponding integral equation is

$$y(x) = 1 + \int_0^x f(u,\ y(u))\, du$$

Define the approximated solution $\phi_n(x)$ to the integral equation by

$$\phi_n(x) = 1 + \int_0^x f(u,\ \phi_{n-1}(u))\, du$$

with $\phi_0(x) = 1$. Clearly, we have

$$\phi_n(x) = 1 + \int_0^x f(u,\, \phi_{n-1}(u))\, du$$

$$= 1 + \int_0^x u\, du$$

$$= 1 + \frac{x^2}{2}$$

Since $\phi_n(x)$ is independent of $\phi_{n-1}(x)$, therefore, the approximated solution to the given initial value differential equation is $y(x) = 1 + \frac{x^2}{2}$.

EXAMPLE 4.8 Solve the initial value differential equation $y' = xy$ with $y(0) = 1$ by approximated method.

Solution: Clearly, $f(x,\, y) = xy$. The corresponding integral equation is

$$y(x) = 1 + \int_0^x f(u,\, y(u))\, du$$

Define the approximated solution $\phi_n(x)$ to the integral equation by

$$\phi_n(x) = 1 + \int_0^x f(u,\, \phi_{n-1}(u))\, du$$

with $\phi_0(x) = 1$. Clearly, we have

$$\phi_1(x) = 1 + \int_0^x f(u,\, \phi_0(u))\, du$$

$$= 1 + \int_0^x u\, du$$

$$= 1 + \frac{x^2}{2}$$

$$\phi_2(x) = 1 + \int_0^x f(u,\, \phi_1(u))\, du$$

$$= 1 + \int_0^x u\left(1 + \frac{u^2}{2}\right) du$$

$$= 1 + \frac{x^2}{2} + \frac{x^4}{8}$$

$$\phi_3(x) = 1 + \int_0^x f(u,\, \phi_2(u))\, du$$

$$= 1 + \int_0^x u\left(1 + \frac{u^2}{2} + \frac{u^4}{8}\right) du$$

$$= 1 + \frac{x^2}{2} + \frac{x^4}{8} + \frac{x^6}{48}$$

and so on. Since $\phi_n(x)$ is dependent on $\phi_{n-1}(x)$, therefore, the approximated solution to the given initial value differential equation is the limiting value of the sequence $\{\phi_n(x)\}$ obtained, which is $y(x) = e^{\frac{x^2}{2}}$.

EXAMPLE 4.9 Solve the initial value differential equation $y' = y$ with $y(0) = 1$ by approximated method.

Solution: Clearly, $f(x,\, y) = y$. The corresponding integral equation is

$$y(x) = 1 + \int_0^x f(u,\, y(u))\, du$$

Define the approximated solution $\phi_n(x)$ to the integral equation by

$$\phi_n(x) = 1 + \int_0^x f(u,\, \phi_{n-1}(u))\, du$$

with $\phi_0(x) = 1$. Clearly, we have

$$\phi_1(x) = 1 + \int_0^x f(u,\, \phi_0(u))\, du$$

$$= 1 + \int_0^x du$$

$$= 1 + x$$

$$\phi_2(x) = 1 + \int_0^x f(u,\, \phi_1(u))\, du$$

$$= 1 + \int_0^x (1 + u)\, du$$

$$= 1 + x + \frac{x^2}{2}$$

$$\phi_3(x) = 1 + \int_0^x f(u, \phi_2(u))\, du$$

$$= 1 + \int_0^x \left(1 + u + \frac{u^2}{2}\right) du$$

$$= 1 + x + \frac{x^2}{2} + \frac{x^3}{6}$$

and, so on. Since $\phi_n(x)$ is dependent on $\phi_{n-1}(x)$, therefore, the approximated solution to the given initial value differential equation is the limiting value of the sequence $\{\phi_n(x)\}$ obtained, which is $y(x) = e^x$.

Exercises 4.1

1. Find the approximated solution of the initial value problem $y' = x + y$ with $y(0) = 1$.
2. Find the approximated solution of the initial value problem $y' = y$ with $y(0) = 0$ and $x \geq 0$.
3. Find the approximated solution of the initial value problem $y' = e^x + y^2$ with $y(0) = 0$.
4. Find the approximated solution of the initial value problem $y' = 1 + xy^2$ with $y(0) = 0$.
5. Find the approximated solution of the initial value problem $y' = 1 + xy$ with $y(1) = 2$.
6. Find the integral equation corresponding to the initial value problem $y' = x + y$ with $y(0) = 1$.
7. Find the integral equation corresponding to the initial value problem $y' = -y$ with $y(0) = 1$ and $x \geq 0$.
8. Find the integral equation corresponding to the initial value problem $y' = e^x + y^2$ with $y(0) = 0$.
9. Find the integral equation corresponding to the initial value problem $y' = 1 + xy^2$ with $y(0) = 0$.
10. Find the integral equation corresponding to the initial value problem $y' = 1 + xy$ with $y(1) = 2$.
11. Find the largest interval in order to have a unique solution to the initial value problem $y' = e^x + y$ with $y(0) = 0$.
12. Find the largest interval in order to have a unique solution to the initial value problem $y' = x^2 + y$ with $y(1) = 3$.
13. Find the largest interval in order to have a unique solution to the initial value problem $y' = 1 + y^2$ with $y(0) = 0$.
14. Find the largest interval in order to have a unique solution to the initial value problem $y' = -y$ with $y(0) = -1$ and $x \geq 0$.

Chapter 5

Second Order Linear Equations

Many of the differential equations that describe physical phenomena are linear differential equations. Among these, the second order equation is the most common and the most important special case. In this section, we present certain aspects of general theory of second order equation with constant coefficients, and variable coefficients.

5.1 Independent Functions

The general solution of any linear differential equation $y''+p(x)y'+q(x)y=0$ can be expressed in the form $y(x)=c_1y_1(x)+c_2y_2(x)$ where $y_1(x)$ and $y_2(x)$ are two independent solutions of the said differential equation. Hence it is necessary to introduce the concepts of the independent solutions. In general, two functions $y_1(x)$ and $y_2(x)$ are *independent* over an interval $(a,\ b)$ if and only if one of them cannot be expressed as some constant multiple of the other over the said interval. One can make this concept very clear by the following examples.

EXAMPLE 5.1 Show that $y_1(x)=|x|x$ and $y_2(x)=x^2$ are independent over the interval $(-1,\ 1)$.

Solution: Let $y_1(x)=\alpha y_2(x)$ over the interval $(-1,\ 1)$. Again, it is easy to verify that

$$y_1(x)=\begin{cases} -x^2, \text{ when } & x<0 \\ x^2, \text{ when } & x>0 \end{cases}$$

Clearly, the constant $\alpha=-1$ for $x<0$ and $\alpha=1$ for $x>0$. Hence $y_1(x)$ is not a constant multiple of $y_2(x)$. In other words, $y_1(x)$ and $y_2(x)$ are independent.

EXAMPLE 5.2 Show that $y_1(x)=|x|x$ and $y_2(x)=x^2$ are not independent over the interval $(0,\ 1)$.

Solution: Let $y_1(x)=\alpha y_2(x)$ over the interval $(0,\ 1)$. Again, it is easy to verify that $y_1(x)=x^2$. Clearly, the constat $\alpha=1$ for $x>0$. Hence $y_1(x)$ is a constant multiple of $y_2(x)$. In other words, $y_1(x)$ and $y_2(x)$ are not independent.

EXAMPLE 5.3 Show that $y_1(x)=|x|x$ and $y_2(x)=x^2$ are not independent over the interval $(-1,\ 0)$.

Solution: Let $y_1(x) = \alpha y_2(x)$ over the interval $(-1, 0)$. Again, it is easy to verify that $y_1(x) = -x^2$. Clearly, the constat $\alpha = -1$ for $x < 0$. Hence $y_1(x)$ is a constant multiple of $y_2(x)$. In other words, $y_1(x)$ and $y_2(x)$ are not independent.

Exercises 5.1

1. Show that $y_1(x) = \sin(x)$ and $y_2(x) = \cos(x)$ are independent over the interval $(-\infty, \infty)$.
2. Show that $y_1(x) = \sin^2(x)$ and $y_2(x) = \cos^2(x)$ are independent over the interval $(-\infty, \infty)$.
3. Show that $y_1(x) = x + 1$ and $y_2(x) = x + 2$ are independent over the interval $(-\infty, \infty)$.
4. Show that $y_1(x) = x^2$ and $y_2(x) = x$ are independent over the interval $(-\infty, \infty)$.
5. Show that $y_1(x) = x$ and $y_2(x) = e^x$ are independent over the interval $(-\infty, \infty)$.
6. Show that $y_1(x) = x$ and $y_2(x) = e^{-x}$ are independent over the interval $(-\infty, \infty)$.
7. Show that $y_1(x) = xe^x$ and $y_2(x) = e^x$ are independent over the interval $(-\infty, \infty)$.
8. Show that $y_1(x) = xe^{-x}$ and $y_2(x) = e^{-x}$ are independent over the interval $(-\infty, \infty)$.

5.2 Second Order Homogeneous Equation

We know that every general solution of any second order homogeneous differential equation will contain two arbitrary parameters. For this purpose, we look for two independent solutions $y_1(x)$ and $y_2(x)$. Again, two solutions $y_1(x)$ and $y_2(x)$ are said to be *independent* if and only if $y_1'y_2 - y_1y_2' \neq 0$ for all x in the required domain. If one has two independent solutions of any second order differential equation, the following theorem tells how to construct a general solution of that equation:

Theorem 5.2.1 If $y_1(x)$ and $y_2(x)$ are two independent solutions of any second order homogeneous equation $y'' + p(x)y' + q(x)y = 0$, then $y(x) = c_1y_1(x) + c_2y_2(x)$ is the general solution of the given equation.

Proof: If $y_1(x)$ and $y_2(x)$ are two solutions of the given differential equation $y'' + p(x)y' + q(x)y = 0$, then we have

$$y_1'' + p(x)y_1' + q(x)y_1 = 0 \tag{5.1}$$

and

$$y_2'' + p(x)y_2' + q(x)y_2 = 0 \tag{5.2}$$

Again, we have from Eqs.(5.1) and (5.2)

$$\begin{aligned} y'' + p(x)y' + q(x)y &= c_1y_1'' + c_2y_2'' + p(x)\left(c_1y_1' + c_2y_2'\right) + q(x)(c_1y_1 + c_2y_2) \\ &= c_1\left(y_1'' + p(x)y_1' + q(x)y_1\right) + c_2\left(y_2'' + p(x)y_2' + q(x)y_2\right) \\ &= 0 \end{aligned}$$

Hence $y(x) = c_1y_1(x)+c_2y_2(x)$ is a solution of the given differential equation with two arbitrary constants. In other words, $y(x) = c_1y_1(x) + c_2y_2(x)$ is the general solution of the differential equation $y'' + p(x)y' + q(x)y = 0$, provided $y_1(x)$ and $y_2(x)$ are two independent solutions of the given differential equation. The unknowns c_1 and c_2 present in the general solution $y(x) = c_1y_1(x)+c_2y_2(x)$ can be determined if and only if two initial conditions are given. This is discussed in Section 5.5.

EXAMPLE 5.4 Show that $y(x) = c_1y_1(x)+c_2y_2(x)$ is a solution of the differential equation $(1-x)y'' + xy' - y = 0$, where $y_1(x) = x$ and $y_2(x) = e^x$.

Solution: It is easy to verify that $y_1(x) = x$ and $y_2(x) = e^x$ are two independent solutions of the given differential equation. According to Theorem 5.2.1, $y(x) = c_1y_1(x) + c_2y_2(x)$ is a solution of the given differential equation.

EXAMPLE 5.5 Show that $y(x) = c_1y_1(x)+c_2y_2(x)$ is a solution of the differential equation $xy'' - (1+x)y' + y = 0$, where $y_1(x) = 1 + x$ and $y_2(x) = e^x$.

Solution: It is easy to verify that $y_1(x) = 1+x$ and $y_2(x) = e^x$ are two independent solutions of the given differential equation. According to Theorem 5.2.1, $y(x) = c_1y_1(x) + c_2y_2(x)$ is a solution of the given differential equation.

EXAMPLE 5.6 Show that $y(x) = c_1y_1(x)+c_2y_2(x)$ is a solution of the differential equation $x^2y'' - 2y = 0$, where $y_1(x) = x^2$ and $y_2(x) = x^{-1}$.

Solution: It is easy to verify that $y_1(x) = x^2$ and $y_2(x) = x^{-1}$ are two solutions of the given differential equation. According to Theorem 5.2.1, $y(x) = c_1y_1(x) + c_2y_2(x)$ is a solution of the given differential equation.

Theorem 5.2.2 The Wronskian $w(y_1,\, y_2)$ of any two independent solutions $y_1(x)$ and $y_2(x)$ of the differential equation $y'' + p(x)y' + q(x)y = 0$ satisfies the identity

$$w(y_1,\, y_2;\, x) = w(y_1,\, y_2;\, a)\, e^{-\int_a^x p(u)\,du}$$

Proof: Clearly, we have

$$y_1'' + p(x)y_1' + q(x)y_1 = 0 \tag{5.3}$$

and

$$y_2'' + p(x)y_2' + q(x)y_2 = 0 \tag{5.4}$$

If Eq.(5.3) is multiplied by y_2 and Eq.(5.4) is multiplied by y_1, and the first equation is subtracted from the second, then we have

$$\begin{aligned} 0 &= y_1y_2'' + p(x)y_1y_2' - y_1''y_2 - p(x)y_1'y_2 \\ &= (y_1y_2' - y_1'y_2)' + p(x)(y_1y_2' - y_1'y_2) \\ &= w' + p(x)w \end{aligned} \tag{5.5}$$

Clearly, Eq.(5.5) is a first order homogeneous differential equation. According to Eq.(2.2), the solution of Eq.(5.5) is

$$w(y_1,\, y_2;\, x) = w(y_1,\, y_2;\, a)e^{-\int_a^x p(u)\,du} \tag{5.6}$$

by integrating from a to x.

Corollary 5.2.1 If $y_1(x)$ and $y_2(x)$ are any two independent solutions of the differential equation $y''+p(x)y'+q(x)y=0$, then $w(y_1, y_2) = y_1y_2' - y_1'y_2$, the Wronskian of the differential equation, is unique up to an arbitrary constant multiple.

Proof: According to Eq.(5.6), the conclusion follows because a is arbitrary.

Corollary 5.2.2 The Wronskian $w(y_1,\ y_2)$ of any two independent solutions $y_1(x)$ and $y_2(x)$ of the differential equation $y'' + p(x)y' + q(x)y = 0$ is identically positive, identically negative or identically zero.

Proof: According to Eq.(5.6), the conclusion follows because $e^{-\int_a^x p(u)\,du}$ is positive.

EXAMPLE 5.7 Find the Wronskian of the differential equation $y'' + xy' + x^2y = 0$.

Solution: Clearly, $p(x) = x$. According to Eq.(5.6), the Wronskian $w(y_1,\ y_2;\ x)$ of the given differential equation is given by

$$\begin{aligned} w(y_1,\ y_2;\ x) &= w(y_1,\ y_2;\ a)e^{-\int_a^x p(u)\,du} \\ &= ce^{-\int_a^x u\,du} \\ &= ce^{\frac{a^2}{2}}e^{-\frac{x^2}{2}} \\ &= ke^{-\frac{x^2}{2}} \end{aligned}$$

EXAMPLE 5.8 Find the Wronskian of the differential equation $y'' + xy' + xy = 0$.

Solution: Clearly, $p(x) = x$. According to Eq.(5.6), the Wronskian $w(y_1,\ y_2;\ x)$ of the given differential equation is given by

$$\begin{aligned} w(y_1,\ y_2;\ x) &= w(y_1,\ y_2;\ a)e^{-\int_a^x p(u)\,du} \\ &= ce^{-\int_a^x u\,du} \\ &= ce^{\frac{a^2}{2}}e^{-\frac{x^2}{2}} \\ &= ke^{-\frac{x^2}{2}} \end{aligned}$$

Theorem 5.2.3 Any two solutions $y_1(x)$ and $y_2(x)$ of the differential equation $y'' + p(x)y' + q(x) = 0$ are dependent if and only if $w(y_1,\ y_2) = 0$.

Proof: Let $w(y_1,\ y_2) = 0$. In other words, we have

$$y_1y_2' - y_1'y_2 = 0$$

Hence we have

$$\frac{y_2'}{y_2} = \frac{y_1'}{y_1}$$

In other words, we have

$$\frac{dy_2}{y_2} = \frac{dy_1}{y_1}$$

It is easy to verify that

$$\begin{aligned} y_1 &= e^{\ln y_1} \\ &= e^{\int \frac{dy_1}{y_1}} \\ &= e^{\int \frac{dy_2}{y_2}} \\ &= ce^{\ln y_2} \\ &= cy_2 \end{aligned}$$

Let $y_1(x)$ and $y_2(x)$ be dependent. Clearly, $y_1 = cy_2$. Hence we have

$$\begin{aligned} w(y_1,\, y_2) &= y_1 y_2' - y_1' y_2 \\ &= cy_2 y_2' - cy_2' y_2 \\ &= 0 \end{aligned}$$

EXAMPLE 5.9 Show that $y_1(x) = x$ and $y_2(x) = e^x$ are two independent solutions of the differential equation $(1 - x)y'' + xy' - y = 0$.

Solution: It is easy to verify that $w(y_1,\, y_2) = e^x(x-1)$ which is never zero for all x. According to Theorem 5.2.3, the conclusion follows.

EXAMPLE 5.10 Show that $y_1(x) = 1+x$ and $y_2(x) = e^x$ are two independent solutions of the differential equation $xy'' - (1 + x)y' + y = 0$.

Solution: It is easy to verify that $w(y_1,\, y_2) = xe^x$ which is never zero for all x. According to Theorem 5.2.3, the conclusion follows.

Exercises 5.2

1. Find the Wronskian of the differential equation $y'' + y' + y = x$.

   HINT: Clearly, $p(x) = 1$. According to Eq.(5.6), $w(y_1,\, y_2) = ce^{-\int p(x)\,dx} = ce^{-x}$.

2. Find the Wronskian of the differential equation $y'' + xy' + y = x$.

   HINT: Clearly, $p(x) = x$. According to Eq.(5.6), $w(y_1,\, y_2) = ce^{-\int p(x)\,dx} = ce^{-\frac{x^2}{2}}$.

3. Find the Wronskian of the differential equation $y'' + x^2y' + y = x$.

   HINT: Clearly, $p(x) = x^2$. According to Eq.(5.6), $w(y_1,\, y_2) = ce^{-\int p(x)\,dx} = ce^{-\frac{x^3}{3}}$.

4. Find the Wronskian of the differential equation $xy'' + y' + y = x$.

   HINT: Clearly, $p(x) = \frac{1}{x}$. According to Eq.(5.6), $w(y_1,\, y_2) = ce^{-\int p(x)\,dx} = \frac{c}{x}$.

5. Find the Wronskian of the differential equation $xy'' + xy' + y = x$.
   HINT: Clearly, $p(x) = 1$. According to Eq.(5.6), $w(y_1, y_2) = c\,e^{-\int p(x)\,dx} = c\,e^{-x}$.
6. Find the Wronskian of the differential equation $xy'' + x^2y' + y = x$.
   HINT: Clearly, $p(x) = x$. According to Eq.(5.6), $w(y_1, y_2) = c\,e^{-\int p(x)\,dx} = c\,e^{-\frac{x^2}{2}}$.

5.2.1 Second Order Constant Coefficient Equations

If $p(x)$ and $q(x)$ are variables, it becomes difficult to find two independent solutions $y_1(x)$ and $y_2(x)$. In this section, we have considered and developed the necessary theory for finding the two independent solutions of any second order homogeneous differential equation when $p(x) = a$ and $q(x) = b$ are constants.

Theorem 5.2.4 The differential equation $y' + ay = 0$ admits a solution of the form $y(x) = e^{rx}$.

Proof: Clearly, the given equation is a first order homogeneous equation. One can easily verify by using the formula given in Eq.(2.2) that

$$\begin{aligned} y(x) &= e^{\int -a\,dx} \\ &= e^{-ax} \\ &= e^{rx} \end{aligned}$$

where $r = -a$.

Theorem 5.2.5 $y(x) = e^{rx}$ is a solution of the differential equation $y'' + ay' + by = 0$ if and only if $r^2 + ar + b = 0$.

Proof: Let $y(x) = e^{rx}$ be a solution of the given equation $y'' + ay' + by = 0$. Hence we have

$$\begin{aligned} 0 &= y'' + ay' + by \\ &= r^2e^{rx} + are^{rx} + be^{rx} \\ &= (r^2 + ar + b)e^{rx} \end{aligned}$$

Again, $e^{rx} \neq 0$. Hence $r^2 + ar + b = 0$.

Let $r^2 + ar + b = 0$. Let $y(x) = e^{rx}$. Hence we have

$$\begin{aligned} y'' + ay' + by &= r^2e^{rx} + are^{rx} + be^{rx} \\ &= (r^2 + ar + b)e^{rx} \\ &= 0 \end{aligned}$$

In other words, $y(x) = e^{rx}$ is a solution of the differential equation $y'' + ay' + by = 0$.

Theorem 5.2.6 If r is a multiple root of the equation $r^2 + ar + b = 0$, then $y(x) = x\,e^{rx}$ is a solution of the differential equation $y'' + ay' + by = 0$.

Proof: According to Theorem 5.2.5, $y_1(x) = e^{rx}$ is a solution of the differential equation $y'' + ay' + by = 0$. Let $y(x) = u(x)y_1(x)$ be a solution of the differential equation. Hence we have

$$\begin{aligned}
0 &= y'' + ay' + by \\
&= u''y_1 + 2u'y_1' + uy_1'' + a(u'y_1 + uy_1') + buy_1 \\
&= u(y_1'' + ay_1' + by_1) + u''y_1 + 2u'y_1' + au'y_1 \\
&= u''y_1 + 2u'y_1' + au'y_1 \\
&= u''e^{rx} + 2ru'e^{rx} + au'e^{rx} \\
&= (u'' + 2ru' + au')\, e^{rx} \\
&= (u'' + (2r + a)u')\, e^{rx} \\
&= (u'' + (2r + a)u')\, y_1(x)
\end{aligned}$$

Again, $y_1(x) = e^{rx} \neq 0$. Hence we have

$$u'' + (2r + a)u' = 0$$

Again, r is a multiple root of the equation $r^2 + ar + b = 0$, gives $r = -\frac{a}{2}$. Hence we have

$$u'' = 0$$

In other words, $u(x) = x$. Hence $y(x) = u(x)y_1(x) = xe^{rx}$ is a solution of the differential equation $y'' + ay' + by = 0$.

EXAMPLE 5.11 If $r = \alpha \pm i\beta$ are the roots of the equation $r^2 + ar + b = 0$, then show that $y_1(x) = e^{\alpha x} \cos \beta x$ and $y_2(x) = e^{\alpha x} \sin \beta x$ are two independent solutions of the differential equation $y'' + ay' + by = 0$.

Solution: Clearly, we have

$$\begin{aligned}
y_1(x) &= e^{\alpha x + i\beta x} \\
&= e^{\alpha x} e^{i\beta x} \\
&= e^{\alpha x} (\cos \beta x + i \sin \beta x)
\end{aligned}$$

and

$$\begin{aligned}
y_2(x) &= e^{\alpha x - i\beta x} \\
&= e^{\alpha x} e^{-i\beta x} \\
&= e^{\alpha x} (\cos \beta x - i \sin \beta x)
\end{aligned}$$

are two independent solutions of the given differential $y'' + ay' + by = 0$. Hence the general solution is given by

$$\begin{aligned}
y(x) &= c_1 y_1(x) + c_2 y_2(x) \\
&= c_1 e^{\alpha x} (\cos \beta x + i \sin \beta x) + c_2 e^{\alpha x} (\cos \beta x - i \sin \beta x) \\
&= (c_1 + c_2) e^{\alpha x} \cos \beta x + i(c_1 - c_2) e^{\alpha x} \sin \beta x \\
&= k_1 e^{\alpha x} \cos \beta x + i k_2 e^{\alpha x} \sin \beta x
\end{aligned}$$

Again, the real part of $y(x)$ and the imaginary part of $y(x)$ are solutions of the given differential equation $y''+ay'+by = 0$. Hence $y_1(x) = e^{\alpha x}\cos \beta x$ and $y_2(x) = e^{\alpha x}\sin \beta x$ are two independent solutions of the differential equation $y'' + ay' + by = 0$.

EXAMPLE 5.12 Find the general solution of the differential equation $y'' + ay' + by = 0$.

Solution: Let $y(x) = e^{rx}$. Clearly, we have

$$\begin{aligned}
0 &= y'' + ay' + by \\
&= r^2 e^{rx} + are^{rx} + be^{rx} \\
&= (r^2 + ar + b)e^{rx}
\end{aligned}$$

The general solution of the given differential equation depends upon the nature of the roots of the equation $r^2 + ar + b = 0$. The type of the general solution is given by the following:

1. If r_1 and r_2 are two distinct real roots of the equation $r^2 + ar + b = 0$, then the type of the general solution is $y(x) = c_1 e^{r_1 x} + c_2 e^{r_2 x}$.
2. If r is the only real root of the equation $r^2 + ar + b = 0$, then the type of the general solution is $y(x) = (c_1 + c_2 x)\, e^{rx}$.
3. If $\alpha \pm i\beta$ are two complex roots of the equation $r^2 + ar + b = 0$, then the type of the general solution is $y(x) = e^{\alpha x} (c_1 \sin \beta x + c_2 \cos \beta x)$.

EXAMPLE 5.13 Find the general solution of the differential equation $y'' + 5y' + 6y = 0$.

Solution: Let $y(x) = e^{rx}$ be the solution of the given differential equation. Hence we have

$$r^2 + 5r + 6 = 0 \tag{5.7}$$

Again, $r = -2, \ -3$ are the real roots of Eq.(5.7). Hence we have $y_1(x) = e^{-3x}$ and $y_2(x) = e^{-2x}$. The general solution of the given differential equation is $y(x) = c_1 e^{-3x} + c_2 e^{-2x}$.

EXAMPLE 5.14 Find the general solution of the differential equation $y'' + 4y' + 4y = 0$.

Solution: Let $y(x) = e^{rx}$ be the solution of the given differential equation. Hence we have

$$r^2 + 4r + 4 = 0 \tag{5.8}$$

Again, $r = -2$ is the only real root of Eq.(5.8). Hence we have $y_1(x) = e^{-2x}$ and $y_2(x) = xe^{-2x}$. The general solution of the given differential equation is $y(x) = c_1 e^{-2x} + c_2 x e^{-2x}$.

EXAMPLE 5.15 Find the general solution of the differential equation $y'' + 4y = 0$.
Solution: Let $y(x) = e^{rx}$ be the solution of the given differential equation. Hence we have

$$r^2 + 4 = 0 \tag{5.9}$$

Again, $r = \pm 2i$ are the complex roots of Eq.(5.9). Hence we have $y_1(x) = \cos 2x$ and $y_2(x) = \sin 2x$. The general solution of the given differential equation is $y(x) = c_1 \cos 2x + c_2 \sin 2x$.

EXAMPLE 5.16 Find the general solution of the differential equation $y'' - 2y' + 2y = 0$.
Solution: Let $y(x) = e^{rx}$ be the solution of the given differential equation. Hence we have

$$r^2 - 2r + 2 = 0 \tag{5.10}$$

Again, $r = 1 \pm i$ are the complex roots of Eq.(5.10). Hence we have $y_1(x) = e^x \cos x$ and $y_2(x) = e^x \sin x$. The general solution of the given differential equation is $y(x) = c_1 e^x \cos x + c_2 e^x \sin x$.

Exercises 5.2.1

1. Find the general solution of the differential equation $y'' + 5y' - 6y = 0$.

   HINT: Clearly, the characteristic equation is $r^2 + 5r - 6 = 0$. In other words, $r = 1, -6$. Hence the general solution can be given by $y(x) = c_1 e^x + c_2 e^{-6x}$.
2. Find the general solution of the differential equation $y'' - 5y' - 6y = 0$.

   HINT: Clearly, the characteristic equation is $r^2 - 5r - 6 = 0$. In other words, $r = -1, 6$. Hence the general solution can be given by $y(x) = c_1 e^{-x} + c_2 e^{6x}$.
3. Find the general solution of the differential equation $y'' - 9y = 0$.

   HINT: Clearly, the characteristic equation is $r^2 - 9 = 0$. In other words, $r = \pm 3$. Hence the general solution can be given by $y(x) = c_1 e^{3x} + c_2 e^{-3x}$.
4. Find the general solution of the differential equation $y'' + 9y = 0$.

   HINT: Clearly, the characteristic equation is $r^2 + 9 = 0$. In other words, $r = \pm 3i$. Hence the general solution can be given by $y(x) = c_1 \cos 3x + c_2 \sin 3x$.
5. Find the general solution of the differential equation $y'' + 4y = 0$.

   HINT: Clearly, the characteristic equation is $r^2 + 4 = 0$. In other words, $r = \pm 2i$. Hence the general solution can be given by $y(x) = c_1 \cos 2x + c_2 \sin 2x$.
6. Find the general solution of the differential equation $y'' + 4y' + 4y = 0$.

   HINT: Clearly, the characteristic equation is $r^2 + 4r + 4 = 0$. In other words, $r = -2, -2$. Hence the general solution can be given by $y(x) = (c_1 + c_2 x)e^{-5x}$.
7. Find the general solution of the differential equation $y'' - 4y' - 4y = 0$.

   HINT: Clearly, the characteristic equation is $r^2 - 4r - 4 = 0$. In other words, $r = 2 \pm 2\sqrt{2}$. Hence the general solution can be given by $y(x) = e^{2x}(c_1 e^{-2\sqrt{2}x} + c_2 e^{2\sqrt{2}x})$.
8. Find the general solution of the differential equation $y'' - 4y' + 8y = 0$.

   HINT: Clearly, the characteristic equation is $r^2 - 4r + 8 = 0$. In other words, $r = 2 \pm 2i$. Hence the general solution can be given by $y(x) = e^{2x}(c_1 \cos 2x + c_2 \sin 2x)$.

9. Find the general solution of the differential equation $y'' + 6y' + 5y = 0$.

   HINT: Clearly, the characteristic equation is $r^2+6r+5 = 0$. In other words, $r = -1, -5$. Hence the general solution can be given by $y(x) = c_1\, e^{-x} + c_2\, e^{-5x}$.

10. Find the general solution of the differential equation $y'' + 4y' + 3y = 0$.

    HINT: Clearly, the characteristic equation is $r^2+4r+3 = 0$. In other words, $r = -1, -3$. Hence the general solution can be given by $y(x) = c_1\, e^{-x} + c_2\, e^{-3x}$.

11. Find the general solution of the differential equation $y'' + 3y' - 4y = 0$.

    HINT: Clearly, the characteristic equation is $r^2 + 3r - 4 = 0$. In other words, $r = 1, -4$. Hence the general solution can be given by $y(x) = c_1\, e^{x} + c_2\, e^{-4x}$.

12. Find the general solution of the differential equation $y'' + 2y' + 5y = 0$.

    HINT: Clearly, the characteristic equation is $r^2+2r+5 = 0$. In other words, $r = -1\pm 2i$. Hence the general solution can be given by $y(x) = e^{-x}(c_1 \cos 2x + c_2 \sin 2x)$.

13. Find the general solution of the differential equation $y'' + 6y' + 9y = 0$.

    HINT: Clearly, the characteristic equation is $r^2+6r+9 = 0$. In other words, $r = -3, -3$. Hence the general solution can be given by $y(x) = (c_1 + c_2\, x)e^{-3x}$.

14. Find the general solution of the differential equation $y'' - 6y' + 9y = 0$.

    HINT: Clearly, the characteristic equation is $r^2 - 6r + 9 = 0$. In other words, $r = 3, 3$. Hence the general solution can be given by $y(x) = (c_1 + c_2\, x)e^{3x}$.

15. Find the general solution of the differential equation $y'' + 6y' = 0$.

    HINT: Clearly, the characteristic equation is $r^2 + 6r = 0$. In other words, $r = 0, -6$. Hence the general solution can be given by $y(x) = c_1 + c_2\, e^{-6x}$.

16. Find the general solution of the differential equation $y'' + 5y' = 0$.

    HINT: Clearly, the characteristic equation is $r^2 + 5r = 0$. In other words, $r = 0, -5$. Hence the general solution can be given by $y(x) = c_1 + c_2\, e^{-5x}$.

17. Find the general solution of the differential equation $y'' + 4y' = 0$.

    HINT: Clearly, the characteristic equation is $r^2 + 4r = 0$. In other words, $r = 0, -4$. Hence the general solution can be given by $y(x) = c_1 + c_2\, e^{-4x}$.

18. Find the general solution of the differential equation $y'' = 0$.

    HINT: Clearly, the characteristic equation is $r^2 = 0$. In other words, $r = 0, 0$. Hence the general solution can be given by $y(x) = c_1 + c_2\, x$.

5.3 Second Order Nonhomogeneous Equation

In the previous section, we have discussed how to solve the equation of the type $y'' + p(x)y' + q(x)y = 0$ where $p(x)$ and $q(x)$ are constants. The present section deals with equations of the type $y'' + p(x)y' + q(x)y = f(x)$, where $p(x)$ and $q(x)$ are constants. One can break the given equation into two parts as follows:

1. The homogeneous solution of the equation $y'' + p(x)y' + q(x)y = 0$ which is discussed in the previous section.

2. The solution $h(x)$ of the equation $y'' + p(x)y' + q(x)y = f(x)$, called the *particular integral*, which is a new concept.

There are three different methods for finding the particular integral, and they are as follows

1. Method of variational parameters
2. Method of undetermined coefficients
3. Operator method.

5.3.1 Method of Variational Parameters

This method determines the particular integral of any nonhomogeneous differential equation $r(x)y'' + p(x)y' + q(x)y = f(x)$, when $y_1(x)$ and $y_2(x)$ are two independent solutions of the corresponding homogeneous differential equation. In other words, we have

$$r(x)y_1'' + p(x)y_1' + q(x)y_1 = 0$$

and

$$r(x)y_2'' + p(x)y_2' + q(x)y_2 = 0$$

Let $h(x) = v_1(x)y_1(x) + v_2(x)y_2(x)$ be the particular integral of the given differential equation. Again, we have

$$h'(x) = v_1y_1' + v_2y_2' + v_1'y_1 + v_2'y_2$$

In order to avoid the second derivative of the unknown quantities $v_1(x)$ and $v_2(x)$, assume that

$$v_1'y_1 + v_2'y_2 = 0 \tag{5.11}$$

Again, we have

$$h''(x) = v_1y_1'' + v_2y_2'' + v_1'y_1' + v_2'y_2'$$

Substituting the expressions for $h(x)$, $h'(x)$ and $h''(x)$ in the given differential equation, we have

$$\begin{aligned}
f(x) &= r(x)h'' + p(x)h' + q(x)h \\
&= r(x)\left(v_1y_1'' + v_2y_2'' + v_1'y_1' + v_2'y_2'\right) + p(x)\left(v_1y_1' + v_2y_2'\right) + q(x)\left(v_1y_1 + v_2y_2\right) \\
&= v_1\left(r(x)y_1'' + p(x)y_1' + q(x)y_1\right) + v_2\left(r(x)y_2'' + p(x)y_2' + q(x)y_2\right) \\
&\quad + r(x)\left(v_1'y_1' + v_2'y_2'\right) \\
&= r(x)\left(v_1'y_1' + v_2'y_2'\right)
\end{aligned} \tag{5.12}$$

Using Eqs.(5.11) and (5.12), one can conclude that

$$v_1(x) = -\int \frac{f(x)y_2(x)}{r(x)w(y_1,\, y_2)}\mathrm{d}x$$

and

$$v_2(x) = \int \frac{f(x)y_1(x)}{r(x)w(y_1,\, y_2)}\mathrm{d}x$$

where $w(y_1, y_2) = y_1y_2' - y_1'y_2$, called the *Wronskian* of two independent solutions $y_1(x)$ and $y_2(x)$. Hence the particular integral of the given differential equation is

$$\begin{aligned}
h(x) &= y_1(x)v_1(x) + y_2(x)v_2(x) \\
&= -y_1(x)\int \frac{f(x)y_2(x)}{r(x)w(y_1, y_2)}\mathrm{d}x + y_2(x)\int \frac{f(x)y_1(x)}{r(x)w(y_1, y_2)}\mathrm{d}x \\
&= \int_0^x f(\gamma)\left(\frac{y_1(\gamma)y_2(x) - y_1(x)y_2(\gamma)}{r(\gamma)w(y_1, y_2; \gamma)}\right)\mathrm{d}\gamma \qquad (5.13)
\end{aligned}$$

EXAMPLE 5.17 Find the general solution of the differential equation $y'' - 2y' + y = x$.

Solution: Clearly, $y_1(x) = e^x$ and $y_2(x) = xe^x$ are two independent solutions of the corresponding homogeneous equations. Again, $w(y_1, y_2) = e^x$. Clearly, $r(x) = 1$ and $f(x) = x$. Hence we have

$$\begin{aligned}
v_1(x) &= -\int \frac{f(x)y_2(x)}{r(x)w(y_1, y_2)}\mathrm{d}x \\
&= -\int \frac{x^2}{e^x}\,\mathrm{d}x \\
&= \frac{x^2}{e^x} + \frac{2x}{e^x} + \frac{2}{e^x}
\end{aligned}$$

and

$$\begin{aligned}
v_2(x) &= \int \frac{f(x)y_1(x)}{r(x)w(y_1, y_2)}\mathrm{d}x \\
&= \int \frac{x}{e^x}\,\mathrm{d}x \\
&= -\frac{x}{e^x} - \frac{1}{e^x}
\end{aligned}$$

According to Eq.(5.13), the particular integral is

$$\begin{aligned}
h(x) &= y_1(x)v_1(x) + y_2(x)v_2(x) \\
&= e^x\left(\frac{x^2}{e^x} + \frac{2x}{e^x} + \frac{2}{e^x}\right) + xe^x\left(-\frac{x}{e^x} - \frac{1}{e^x}\right) \\
&= x + 2
\end{aligned}$$

Hence the general solution is given by

$$\begin{aligned}
y(x) &= c_1y_1(x) + c_2y_2(x) + h(x) \\
&= c_1e^x + c_2xe^x + x + 2
\end{aligned}$$

EXAMPLE 5.18 Find the general solution of the differential equation $y'' + 4y' + y = 8x^2$.

Solution: Clearly, $y_1(x) = \cos 2x$ and $y_2(x) = \sin 2x$ are two independent solutions of the corresponding homogeneous equation. Again, $w(y_1, y_2) = 2$. Clearly, $r(x) = 1$ and $f(x) = 8x^2$. Hence we have

$$\begin{aligned} v_1(x) &= -\int \left(\frac{f(x)y_2(x)}{r(x)w(y_1, y_2)} \right) \mathrm{d}x \\ &= -4 \int x^2 \sin 2x \, \mathrm{d}x \\ &= 2x^2 \cos 2x - 2x \sin 2x - \cos 2x \end{aligned}$$

and

$$\begin{aligned} v_2(x) &= \int \left(\frac{f(x)y_1(x)}{r(x)w(y_1, y_2)} \right) \mathrm{d}x \\ &= 4 \int x^2 \cos 2x \, \mathrm{d}x \\ &= 2x^2 \sin 2x + 2x \cos 2x - \sin 2x \end{aligned}$$

According to Eq.(5.13), the particular integral is

$$\begin{aligned} h(x) &= y_1(x)v_1(x) + y_2(x)v_2(x) \\ &= \left(2x^2 \cos 2x - 2x \sin 2x - \cos 2x\right) \cos 2x \\ &\quad + \left(2x^2 \sin 2x + 2x \cos 2x - \sin 2x\right) \sin 2x \\ &= 2x^2 - 1 \end{aligned}$$

Hence the general solution is given by

$$\begin{aligned} y(x) &= c_1 y_1(x) + c_2 y_2(x) + h(x) \\ &= c_1 \cos 2x + c_2 \sin 2x + 2x^2 - 1 \end{aligned}$$

EXAMPLE 5.19 Find the general solution of the differential equation $y'' - 2y' + y = e^x$.

Solution: Clearly, $y_1(x) = e^x$ and $y_2(x) = xe^x$ are two independent solutions of the corresponding homogeneous equations. Again, $w(y_1, y_2) = e^x$. Clearly, $r(x) = 1$ and $f(x) = e^x$. Hence we have

$$\begin{aligned} v_1(x) &= -\int \frac{f(x)y_2(x)}{r(x)w(y_1, y_2)} \mathrm{d}x \\ &= -\int x \, \mathrm{d}x \\ &= -\frac{x^2}{2} \end{aligned}$$

and

$$
\begin{aligned}
v_2(x) &= \int \frac{f(x)y_1(x)}{r(x)w(y_1,\, y_2)} \mathrm{d}x \\
&= \int \mathrm{d}x \\
&= x
\end{aligned}
$$

According to Eq.(5.13), the particular integral is

$$
\begin{aligned}
h(x) &= y_1(x)v_1(x) + y_2(x)v_2(x) \\
&= e^x \left(-\frac{x^2}{2}\right) + xe^x x \\
&= \left(\frac{x^2}{2}\right) e^x
\end{aligned}
$$

Hence the general solution is given by

$$
\begin{aligned}
y(x) &= c_1y_1(x) + c_2y_2(x) + h(x) \\
&= c_1e^x + c_2xe^x + \left(\frac{x^2}{2}\right) e^x
\end{aligned}
$$

EXAMPLE 5.20 Find the general solution of the differential equation $y'' - 5y' + 6y = e^{4x}$.

Solution: Clearly, $y_1(x) = e^{2x}$ and $y_2(x) = e^{3x}$ are two independent solutions of the corresponding homogeneous equations. Again, $w(y_1,\, y_2) = e^{5x}$. Clearly, $r(x) = 1$ and $f(x) = e^{4x}$. Hence we have

$$
\begin{aligned}
v_1(x) &= -\int \frac{f(x)y_2(x)}{r(x)w(y_1,\, y_2)} \mathrm{d}x \\
&= -\int e^{2x}\, \mathrm{d}x \\
&= -\frac{e^{2x}}{2}
\end{aligned}
$$

and

$$
\begin{aligned}
v_2(x) &= \int \frac{f(x)y_1(x)}{r(x)w(y_1,\, y_2)} \mathrm{d}x \\
&= \int e^x\, \mathrm{d}x \\
&= e^x
\end{aligned}
$$

According to Eq.(5.13), the particular integral is

$$\begin{aligned} h(x) &= y_1(x)v_1(x) + y_2(x)v_2(x) \\ &= e^{2x}\left(-\frac{e^{2x}}{2}\right) + e^{3x}e^{x} \\ &= \frac{e^{4x}}{2} \end{aligned}$$

Hence the general solution is given by

$$\begin{aligned} y(x) &= c_1y_1(x) + c_2y_2(x) + h(x) \\ &= c_1e^{2x} + c_2e^{3x} + \frac{e^{4x}}{2} \end{aligned}$$

EXAMPLE 5.21 Find the general solution of the differential equation $y'' + y = \sec x$.

Solution: Clearly, $y_1(x) = \cos x$ and $y_2(x) = \sin x$ are two independent solutions of the corresponding homogeneous equations. Again, $w(y_1, y_2) = 1$. Clearly, $r(x) = 1$ and $f(x) = \sec x$. Hence we have

$$\begin{aligned} v_1(x) &= -\int \frac{f(x)y_2(x)}{r(x)w(y_1,\, y_2)}\mathrm{d}x \\ &= -\int \sin x\, \sec x\, \mathrm{d}x \\ &= -\int \tan x\, \mathrm{d}x \\ &= \ln(\cos x) \end{aligned}$$

and

$$\begin{aligned} v_2(x) &= \int \frac{f(x)y_1(x)}{r(x)w(y_1,\, y_2)}\mathrm{d}x \\ &= \int \cos x\, \sec x\, \mathrm{d}x \\ &= \int \mathrm{d}x \\ &= x \end{aligned}$$

According to Eq.(5.13), the particular integral is

$$\begin{aligned} h(x) &= y_1(x)v_1(x) + y_2(x)v_2(x) \\ &= \cos x\, \ln(\cos x) + x\, \sin x \end{aligned}$$

Hence the general solution is given by

$$\begin{aligned} y(x) &= c_1 y_1(x) + c_2 y_2(x) + h(x) \\ &= c_1 \cos x + c_2 \sin x + \cos x \ln(\cos x) + x \sin x \end{aligned}$$

EXAMPLE 5.22 Find the general solution of the differential equation $y'' + y = \operatorname{cosec} x$.

Solution: Clearly, $y_1(x) = \cos x$ and $y_2(x) = \sin x$ are two independent solutions of the corresponding homogeneous equations. Again, $w(y_1, y_2) = 1$. Clearly, $r(x) = 1$ and $f(x) = \operatorname{cosec} x$. Hence we have

$$\begin{aligned} v_1(x) &= -\int \frac{f(x) y_2(x)}{r(x) w(y_1, y_2)} \mathrm{d}x \\ &= -\int \sin x \operatorname{cosec} x \, \mathrm{d}x \\ &= -\int \mathrm{d}x \\ &= -x \end{aligned}$$

and

$$\begin{aligned} v_2(x) &= \int \frac{f(x) y_1(x)}{r(x) w(y_1, y_2)} \mathrm{d}x \\ &= \int \cos x \operatorname{cosec} x \, \mathrm{d}x \\ &= \int \cot x \, \mathrm{d}x \\ &= \ln(\sin x) \end{aligned}$$

According to Eq.(5.13), the particular integral is

$$\begin{aligned} h(x) &= y_1(x) v_1(x) + y_2(x) v_2(x) \\ &= -x \cos x + \sin x \ln(\sin x) \end{aligned}$$

Hence the general solution is given by

$$\begin{aligned} y(x) &= c_1 y_1(x) + c_2 y_2(x) + h(x) \\ &= c_1 \cos x + c_2 \sin x - x \cos x + \sin x \ln(\sin x) \end{aligned}$$

EXAMPLE 5.23 Find the general solution of the differential equation $xy'' - y' = (3+x)x^2 e^x$.

Solution: Clearly, $y_1(x) = 1$ and $y_2(x) = x^2$ are two independent solutions of the corresponding homogeneous equation. Again, $w(y_1, y_2) = 2x$. Clearly, $r(x) = x$ and $f(x) = (3+x)x^2 e^x$.

Hence we have

$$\begin{aligned} v_1(x) &= -\int\left(\frac{f(x)y_2(x)}{r(x)w(y_1,\, y_2)}\right)\mathrm{d}x \\ &= -\frac{1}{2}\int(3+x)x^2e^x\,\mathrm{d}x \\ &= -\frac{x^3e^x}{2} \end{aligned}$$

and

$$\begin{aligned} v_2(x) &= \int\left(\frac{f(x)y_1(x)}{r(x)w(y_1,\, y_2)}\right)\mathrm{d}x \\ &= \frac{1}{2}\int(3+x)e^x\,\mathrm{d}x \\ &= \frac{3xe^x}{2} \end{aligned}$$

According to Eq.(5.13), the particular integral is

$$\begin{aligned} h(x) &= y_1(x)v_1(x) + y_2(x)v_2(x) \\ &= x^3e^x \end{aligned}$$

Hence the general solution is given by

$$\begin{aligned} y(x) &= c_1y_1(x) + c_2y_2(x) + h(x) \\ &= c_1 + c_2x^2 + x^3\,e^x \end{aligned}$$

Exercises 5.3.1

1. Find the general solution of the differential equation $y'' - 2y' + 2y = e^x$.

   HINT: Clearly, $y_1(x) = e^x\cos x$ and $y_2(x) = e^x\sin x$. Again, $w(y_1,\, y_2) = e^{2x}$. Hence $y(x) = c_1\,e^x\cos x + c_2\,e^x\sin x + \cos^2(x)\,e^x + \sin^2(x)\,e^x = c_1\,e^x\cos x + c_2\,e^x\sin x + e^x = (c_1\cos x + c_2\sin x + 1)e^x$ is the general solution.

2. Find the general solution of the differential equation $y'' - 5y' + 6y = e^{2x}$.

   HINT: Clearly, $y_1(x) = e^{2x}$ and $y_2(x) = e^{3x}$. Again, $w(y_1,\, y_2) = e^{5x}$. Hence $y(x) = c_1\,e^{2x} + c_2\,e^{3x} - (x+1)e^{2x}$ is the general solution.

3. Find the general solution of the differential equation $y'' - 5y' + 6y = e^{3x}$.

   HINT: Clearly, $y_1(x) = e^{2x}$ and $y_2(x) = e^{3x}$. Again, $w(y_1,\, y_2) = e^{5x}$. Hence $y(x) = c_1\,e^{2x} + c_2\,e^{3x} + (x-1)e^{3x}$ is the general solution.

4. Find the general solution of the differential equation $y'' - 2y' + 2y = \sin x$.

   HINT: Clearly, $y_1(x) = e^x\cos x$ and $y_2(x) = e^x\sin x$. Again, $w(y_1,\, y_2) = e^{2x}$. Hence $y(x) = c_1\,e^x\cos x + c_2\,e^x\sin x + \cos x\,\left(\frac{1}{5}\sin 2x - \frac{1}{10}\cos 2x\right) - \sin x\,\left(\frac{1}{10}\sin 2x + \frac{1}{5}\cos 2x\right) + \frac{1}{2}\cos x$ is the general solution.

5. Find the general solution of the differential equation $y'' - 5y' + 6y = \sin x$.

   HINT: Clearly, $y_1(x) = e^{2x}$ and $y_2(x) = e^{3x}$. Again, $w(y_1, y_2) = e^{5x}$. Hence $y(x) = c_1 e^{2x} + c_2 e^{3x} + \frac{1}{10}(\sin x + \cos x)$ is the general solution.

6. Find the general solution of the differential equation $y'' + 2y' + 2y = 4\,e^{-x}\sec^3 x$.

   HINT: Clearly, $y_1(x) = e^{-x}\cos x$ and $y_2(x) = e^{-x}\sin x$. Again, $w(y_1, y_2) = e^{-2x}$. Hence $y(x) = (c_1\cos x + c_2 \sin x)e^{-x} + 2\tan x \sin(x)\, e^{-x}$ is the general solution.

7. Find the general solution of the differential equation $y'' + 2y' + y = e^{-x}\cos x$.

   HINT: Clearly, $y_1(x) = e^{-x}$ and $y_2(x) = x\,e^{-x}$. Again, $w(y_1, y_2) = e^{-2x}$. Hence $y(x) = (c_1 + c_2 x)e^{-x} - e^{-x}\cos x$ is the general solution.

8. Find the general solution of the differential equation $y'' + 2y' + 2y = e^{-x}\cos x$.

   HINTS: Clearly, $y_1(x) = e^{-x}\cos x$ and $y_2(x) = e^{-x}\sin x$ are two independent solutions of the corresponding homogeneous differential equation. Again, $w(y_1, y_2) = e^{-2x}$. Hence $y(x) = c_1\,e^{-x}\cos x + c_2\,e^{-x}\sin x + e^{-x}\cos^2 x + \frac{x\,e^{-x}\sin x}{2} + \frac{e^{-x}\sin 2x \sin x}{4}$.

9. Find the general solution of the differential equation $y'' - 2y' + 4y = e^{x}\sin x$.

   HINT: Clearly, $y_1(x) = e^{x}\cos(\sqrt{3}x)$ and $y_2(x) = e^{x}\sin(\sqrt{3}x)$. Again, $w(y_1, y_2) = \sqrt{3}\,e^{2x}$. In other words, $y(x) = \frac{e^{x}\cos(\sqrt{3}x)}{2\sqrt{3}}(\sqrt{3}\sin(x)\cos(\sqrt{3}x) - \sin(\sqrt{3}x)\cos x) + c_1\,e^{x}\cos(\sqrt{3}x) + c_2\,e^{x}\sin(\sqrt{3}x) + \frac{e^{x}\sin(\sqrt{3}x)}{2\sqrt{3}}(\cos x\cos(\sqrt{3}x) + \sqrt{3}\sin(\sqrt{3}x)\sin x)$ is the general solution.

10. Find the general solution of the differential equation $y'' - 2y' + 4y = e^{x}\cos x$.

    HINT: Clearly, $y_1(x) = e^{x}\cos(\sqrt{3}x)$ and $y_2(x) = e^{x}\sin(\sqrt{3}x)$. Again, $w(y_1, y_2) = \sqrt{3}\,e^{2x}$. In other words, $y(x) = \frac{e^{x}\cos(\sqrt{3}x)}{2\sqrt{3}}(\sqrt{3}\cos(x)\cos(\sqrt{3}x) + \sin(\sqrt{3}x)\sin x) + c_1\,e^{x}\cos(\sqrt{3}x) + c_2\,e^{x}\sin(\sqrt{3}x) - \frac{e^{x}\sin(\sqrt{3}x)}{2\sqrt{3}}(\cos x\sin(\sqrt{3}x) - \sqrt{3}\sin(\sqrt{3}x)\cos x)$ is the general solution.

11. Find the general solution of the differential equation $y'' - y' - 6y = e^{3x}\cos x$.

    HINT: Clearly, $y_1(x) = e^{-2x}$ and $y_2(x) = e^{3x}$. Again, $w(y_1, y_2) = 5\,e^{x}$. Hence $y(x) = c_1\,e^{-2x} + c_2\,e^{3x} + \left(\frac{5\sin x - \cos x}{26}\right)e^{3x}$ is the general solution.

12. Find the general solution of the differential equation $y'' - y' - 6y = e^{2x}\cos x$.

    HINT: Clearly, $y_1(x) = e^{-2x}$ and $y_2(x) = e^{3x}$. Again, $w(y_1, y_2) = 5\,e^{x}$. Hence $y(x) = c_1\,e^{-2x} + c_2\,e^{3x} + \frac{1}{34}(3\sin x - 5\cos x)\,e^{2x}$ is the general solution.

13. Find the general solution of the differential equation $y'' - y' - 6y = e^{3x}\sin x$.

    HINT: Clearly, $y_1(x) = e^{-2x}$ and $y_2(x) = e^{3x}$. Again, $w(y_1, y_2) = 5\,e^{x}$. Hence $y(x) = c_1\,e^{-2x} + c_2\,e^{3x} - \left(\frac{\sin x + 5\cos x}{26}\right)e^{3x}$ is the general solution.

14. Find the general solution of the differential equation $y'' - y' - 6y = e^{2x}\sin x$.

    HINT: Clearly, $y_1(x) = e^{-2x}$ and $y_2(x) = e^{3x}$. Again, $w(y_1, y_2) = 5\,e^{x}$. Hence $y(x) = c_1\,e^{-2x} + c_2\,e^{3x} - \frac{1}{34}(5\sin x + 3\cos x)\,e^{2x}$ is the general solution.

15. Find the general solution of the differential equation $y'' - 4y' + 4y = e^{x}\cos x$.

    HINT: Clearly, $y_1(x) = e^{2x}$ and $y_2(x) = x\,e^{2x}$. Again, $w(y_1, y_2) = e^{4x}$. Hence $y(x) = c_1\,e^{2x} + c_2\,x\,e^{2x} - \frac{\sin x\, e^{x}}{2}$ is the general solution.

16. Find the general solution of the differential equation $y'' - 5y' + 6y = e^{3x}\cos x$.

    HINT: Clearly, $y_1(x) = e^{2x}$ and $y_2(x) = e^{3x}$. Again, $w(y_1, y_2) = e^{5x}$. Hence $y(x) = c_1 e^{2x} + c_2 e^{3x} + \frac{1}{2}(\sin x - \cos x)e^{3x}$ is the general solution.

17. Find the general solution of the differential equation $y'' - 5y' + 6y = e^{2x}\cos x$.

    HINT: Clearly, $y_1(x) = e^{2x}$ and $y_2(x) = e^{3x}$. Again, $w(y_1, y_2) = e^{5x}$. Hence $y(x) = c_1 e^{2x} + c_2 e^{3x} - \frac{1}{2}(\sin x + \cos x)e^{2x}$ is the general solution.

18. Find the general solution of the differential equation $y'' - 5y' + 6y = e^{3x}\sin x$.

    HINT: Clearly, $y_1(x) = e^{2x}$ and $y_2(x) = e^{3x}$. Again, $w(y_1, y_2) = e^{5x}$. Hence $y(x) = c_1 e^{2x} + c_2 e^{3x} - \frac{1}{2}(\sin x + \cos x)e^{3x}$ is the general solution.

19. Find the general solution of the differential equation $y'' - 5y' + 6y = e^{2x}\sin x$.

    HINT: Clearly, $y_1(x) = e^{2x}$ and $y_2(x) = e^{3x}$. Again, $w(y_1, y_2) = e^{5x}$. Hence $y(x) = c_1 e^{2x} + c_2 e^{3x} + \frac{1}{2}(\cos x - \sin x)e^{2x}$ is the general solution.

20. Find the general solution of the differential equation $y'' - 3y' + 2y = e^{2x} + e^{x}$.

    HINT: Clearly, $y_1(x) = e^{x}$ and $y_2(x) = e^{2x}$. Again, $w(y_1, y_2) = e^{3x}$. Hence $y(x) = c_1 e^{x} + c_2 e^{2x} + (x-1)e^{2x} - (x+1)e^{x} = (x+k_2)e^{2x} - (x-k_1)e^{x} = x(e^{2x} - e^{x}) + k_1 e^{x} + k_2 e^{2x}$ is the general solution.

21. Find the general solution of the differential equation $y'' + 3y' + 2y = e^{2x} + e^{x}$.

    HINT: Clearly, $y_1(x) = e^{-2x}$ and $y_2(x) = e^{-x}$. Again, $w(y_1, y_2) = e^{-3x}$. Hence $y(x) = c_1 e^{-2x} + c_2 e^{-x} + \frac{e^{2x}}{12} + \frac{e^{x}}{6}$ is the general solution.

22. Find the general solution of the differential equation $y'' + 4y = \tan 2x$.

    HINT: Clearly, $y_1(x) = \cos 2x$ and $y_2(x) = \sin 2x$. Again, $w(y_1, y_2) = 2$. Hence $y(x) = c_1\cos 2x + c_2\sin 2x - \frac{1}{4}\cos 2x\ln(\sec 2x + \tan 2x)$ is the general solution.

23. Find the general solution of the differential equation $y'' + 4y = \cot 2x$.

    HINT: Clearly, $y_1(x) = \cos 2x$ and $y_2(x) = \sin 2x$. Again, $w(y_1, y_2) = 2$. Hence $y(x) = c_1\cos 2x + c_2\sin 2x - \frac{1}{4}\sin 2x\ln(\operatorname{cosec} 2x + \cot 2x)$ is the general solution.

24. Find the general solution of the differential equation $y'' + 9y = \sec 3x$.

    HINT: Clearly, $y_1(x) = \cos 3x$ and $y_2(x) = \sin 3x$. Again, $w(y_1, y_2) = 3$. Hence $y(x) = c_1\cos 3x + c_2\sin 3x + x\sin 3x + \frac{1}{3}\cos 3x\ln(\cos 3x)$ is the general solution.

25. Find the general solution of the differential equation $y'' + 2y' + y = x\,e^{x}$.

    HINT: Clearly, $y_1(x) = e^{-x}$ and $y_2(x) = x\,e^{-x}$. Again, $w(y_1, y_2) = e^{-2x}$. Hence $y(x) = (c_1 + c_2 x)e^{-x} + \frac{1}{4}(x-1)e^{x}$ is the general solution.

26. Find the general solution of the differential equation $y'' + 2y' + y = e^{-x}\ln x$.

    HINT: Clearly, $y_1(x) = e^{-x}$ and $y_2(x) = x\,e^{-x}$. Again, $w(y_1, y_2) = e^{-2x}$. Hence $y(x) = (c_1 + c_2 x)e^{-x} + \left(\frac{x^2\ln x}{2} - \frac{3x^2}{4}\right)e^{-x}$ is the general solution.

27. Find the general solution of the differential equation $y'' + 2y' + 5y = e^{-x}\sec 2x$.

    HINT: Clearly, $y_1(x) = e^{-x}\cos 2x$ and $y_2(x) = e^{-x}\sin 2x$. Again, $w(y_1, y_2) = 2\,e^{-2x}$. Hence $y(x) = (c_1\cos 2x + c_2\sin 2x)e^{-x} + \left(\frac{\cos 2x\ln(\cos 2x)}{4} + \frac{x\sin(2x)}{2}\right)e^{-x}$ is the general solution.

5.3.2 Method of Undetermined Coefficients

This is one of the methods by which one can determine the particular integral when the function $f(x)$ of the differential equation $y'' + ay' + by = f(x)$ takes some particular form. In particular, we have

1. $f(x) = \sum_{i=0}^{n} a_i x^i$
2. $f(x) = \cos \alpha x$
3. $f(x) = \sin \alpha x$
4. $f(x) = e^{\alpha x}\left(\sum_{i=0}^{n} a_i x^i\right)$

One has to assume the type of the particular solution $h(x)$ looking at the types of differentials in the given equation and the nature of $f(x)$ with some unknowns which will be determined in course of time. We have presented the type of $h(x)$ corresponding to different types of $f(x)$ and differentials in Table 5.1.

Table 5.1: Type of $h(x)$ corresponding to different types of $f(x)$ and differentials

Type of differentials	$f(x)$	*Characteristic root*	*Type of* $h(x)$
$y'' + ay' + by$	$\sum_{i=0}^{n} a_i x^i$	$b \neq 0$	$\sum_{i=0}^{n} \alpha_i x^i$
y''	$\sum_{i=0}^{n} a_i x^i$	-	$\sum_{i=2}^{n+2} \alpha_i x^i$
$y'' + ay'$	$\sum_{i=0}^{n} a_i x^i$	$a \neq 0$	$\sum_{i=1}^{n+1} \alpha_i x^i$
$y'' + ay' + by$	$\sin \alpha x$	$i\alpha$ is not a root	$c_1 \sin \alpha x + c_2 \cos \alpha x$
$y'' + ay' + by$	$\cos \alpha x$	$i\alpha$ is not a root	$c_1 \sin \alpha x + c_2 \cos \alpha x$
$y'' + b^2 y$	$\sin \alpha x$	$i\alpha$ is not a root	$c \sin \alpha x$
$y'' + b^2 y$	$\cos \alpha x$	$i\alpha$ is not a root	$c \cos \alpha x$
$y'' + ay' + by$	$e^{\alpha x}$	α is not a root	$c\, e^{\alpha x}$
$y'' + ay' + by$	$e^{\alpha x}$	α is a simple root	$c\, x\, e^{\alpha x}$
$y'' + ay' + by$	$e^{\alpha x}$	α is a multiple root	$c\, x^2\, e^{\alpha x}$
$y'' + b^2 y$	$\cos bx$	-	$c\, x \sin bx$
$y'' + b^2 y$	$\sin bx$	-	$c\, x \cos bx$
$y'' + ay' + by$	$e^{\alpha x} \sum_{i=0}^{n} a_i x^i$	α is not a root	$e^{\alpha x} \sum_{i=0}^{n} \alpha_i x^i$
$y'' + ay' + by$	$e^{\alpha x} \sum_{i=0}^{n} a_i x^i$	α is a simple root	$e^{\alpha x} \sum_{i=1}^{n+1} \alpha_i x^i$
$y'' + ay' + by$	$e^{\alpha x} \sum_{i=0}^{n} a_i x^i$	α is a multiple root	$e^{\alpha x} \sum_{i=2}^{n+2} \alpha_i x^i$

EXAMPLE 5.24 Find the general solution of the differential equation $y''+4y'-5y=x+3$.

Solution: Clearly, $y_1(x) = e^x$ and $y_2(x) = e^{-5x}$ are two independent solutions of the corresponding homogeneous equation. Again, $f(x) = x+3$. Let $h(x) = \alpha x + \beta$. Clearly, we have

$$\begin{aligned} x+3 &= 4\alpha - 5(\alpha x+\beta) \\ &= 4\alpha - 5\beta - 5\alpha x \end{aligned} \tag{5.14}$$

Comparing both sides of Eq.(5.14), we have

$$4\alpha - 5\beta = 3$$

$$-5\alpha = 1$$

Clearly, $\alpha = -\frac{1}{5}$ and $\beta = -\frac{19}{25}$. Hence the particular integral is $h(x) = -\frac{1}{25}(5x+19)$. The corresponding general solution is given by

$$\begin{aligned} y(x) &= c_1 y_1(x) + c_2 y_2(x) + h(x) \\ &= c_1 e^x + c_2 e^{-5x} - \frac{1}{25}(19x+5) \end{aligned}$$

EXAMPLE 5.25 Find the general solution of the differential equation $y''+4y = 4\cos 2x$.

Solution: Clearly, $y_1(x) = \cos 2x$ and $y_2(x) = \sin 2x$ are two independent solutions of the corresponding homogeneous equation. Again, $f(x) = 4\cos 2x$. Let $h(x) = \alpha\, x \sin 2x$. Clearly, we have

$$\begin{aligned} 4\cos 2x &= 4\alpha\cos 2x - 4\alpha\, x\sin 2x + 4\alpha\, x \sin 2x \\ &= 4\alpha \cos 2x \end{aligned} \tag{5.15}$$

Comparing both sides of Eq.(5.15), we have

$$\alpha = 1$$

Hence the particular integral is $h(x) = x\sin 2x$. The corresponding general solution is given by

$$\begin{aligned} y(x) &= c_1 y_1(x) + c_2 y_2(x) + h(x) \\ &= c_1 \cos 2x + c_2 \sin 2x + x \sin 2x \end{aligned}$$

EXAMPLE 5.26 Find the general solution of the differential equation $y''+4y = 4\sin 2x$.

Solution: Clearly, $y_1(x) = \cos 2x$ and $y_2(x) = \sin 2x$ are two independent solutions of the corresponding homogeneous equation. Again, $f(x) = 4\sin 2x$. Let $h(x) = \alpha\, x\cos 2x$. Clearly, we have

$$\begin{aligned} 4\sin 2x &= -4\alpha\sin 2x - 4\alpha\, x\cos 2x + 4\alpha\, x\cos 2x \\ &= -4\alpha \sin 2x \end{aligned} \tag{5.16}$$

Comparing both sides of Eq.(5.16), we have

$$\alpha = -1$$

Hence the particular integral is $h(x) = -x\cos 2x$. The corresponding general solution is given by

$$\begin{aligned} y(x) &= c_1 y_1(x) + c_2 y_2(x) + h(x) \\ &= c_1 \cos 2x + c_2 \sin 2x - x\cos 2x \end{aligned}$$

EXAMPLE 5.27 Find the general solution of the differential equation $y'' - 5y' + 4y = 2e^{3x}$.
Solution: Clearly, $y_1(x) = e^x$ and $y_2(x) = e^{4x}$ are two independent solutions of the corresponding homogeneous equation. Again, $f(x) = 2e^{3x}$. Let $h(x) = \alpha e^{3x}$. Clearly, we have

$$\begin{aligned} 2e^{3x} &= 9\alpha\, e^{3x} - 5(3\alpha\, e^{3x}) + 4\alpha\, e^{3x} \\ &= -2\alpha\, e^{3x} \end{aligned} \tag{5.17}$$

Comparing both sides of Eq.(5.17), we have

$$\alpha = -1$$

Hence the particular integral is $h(x) = -e^{3x}$. The corresponding general solution is given by

$$\begin{aligned} y(x) &= c_1 y_1(x) + c_2 y_2(x) + h(x) \\ &= c_1 e^x + c_2 e^{4x} - e^{3x} \end{aligned}$$

EXAMPLE 5.28 Find the general solution of the differential equation $y'' - 5y' + 4y = 3e^x$.
Solution: Clearly, $y_1(x) = e^x$ and $y_2(x) = e^{4x}$ are two independent solutions of the corresponding homogeneous equation. Again, $f(x) = 3e^x$. Let $h(x) = \alpha\, x\, e^x$. Clearly, we have

$$\begin{aligned} 3e^x &= 2\alpha\, e^x + \alpha\, x\, e^x - 5(\alpha\, e^x + \alpha\, x\, e^x) + 4\alpha\, x\, e^x \\ &= -3\alpha\, e^x \end{aligned} \tag{5.18}$$

Comparing both sides of Eq.(5.18), we have

$$\alpha = -1$$

Hence the particular integral is $h(x) = -x\, e^x$. The corresponding general solution is given by

$$\begin{aligned} y(x) &= c_1 y_1(x) + c_2 y_2(x) + h(x) \\ &= c_1 e^x + c_2 e^{4x} - x\, e^x \end{aligned}$$

EXAMPLE 5.29 Find the general solution of the differential equation $y'' - y = 4xe^x$.

Solution: Clearly, $y_1(x) = e^x$ and $y_2(x) = e^{-x}$ are two independent solutions of the corresponding homogeneous equation. Again, $f(x) = 4xe^x$. Let $h(x) = (\alpha x + \beta x^2)\, e^x$. Clearly, we have

$$\begin{aligned} 4xe^x &= (2\alpha + 2\beta + 4\beta)e^x + (\alpha x + \beta x^2)e^x - (\alpha x + \beta x^2)e^x \\ &= (2\alpha + 2\beta + 4\beta x)\, e^x \qquad (5.19) \end{aligned}$$

Comparing both sides of Eq.(5.19), we have

$$2\alpha + 2\beta = 0$$
$$4\beta = 4$$

In other words, we have

$$\alpha = -1$$
$$\beta = 1$$

Hence the particular integral is $h(x) = (x-1)x\, e^x$. The corresponding general solution is given by

$$\begin{aligned} y(x) &= c_1 y_1(x) + c_2 y_2(x) + h(x) \\ &= c_1 e^x + c_2 e^{-x} + (x-1)x\, e^x \end{aligned}$$

EXAMPLE 5.30 Find the general solution of the differential equation $y'' - 4y = 9xe^x$.

Solution: Clearly, $y_1(x) = e^{2x}$ and $y_2(x) = e^{-2x}$ are two independent solutions of the corresponding homogeneous equation. Again, $f(x) = 9xe^x$. Let $h(x) = (\alpha + \beta x)\, e^x$. Clearly, we have

$$\begin{aligned} 9xe^x &= (\alpha + 2\beta + \beta x)e^x - 4(\alpha + \beta x)e^x \\ &= (2\beta - 3\alpha - 3\beta x)\, e^x \qquad (5.20) \end{aligned}$$

Comparing both sides of Eq.(5.20), we have

$$2\beta - 3\alpha = 0$$
$$-3\beta = 9$$

In other words, we have

$$\alpha = -2$$
$$\beta = -3$$

Hence the particular solution is $h(x) = -(2 + 3x)\, e^x$. The corresponding general solution is given by

$$\begin{aligned} y(x) &= c_1 y_1(x) + c_2 y_2(x) + h(x) \\ &= c_1 e^{2x} + c_2 e^{-2x} - (2 + 3x)\, e^x \end{aligned}$$

Exercises 5.3.2

1. Find the general solution of the differential equation $y'' + 2y = 2x$.

   HINT: Let $y(x) = ax + b$. In other words, $a = 1$ and $b = 0$. Hence $y(x) = x$.

2. Find the general solution of the differential equation $y'' + y' + 2y = 2x$.

   HINT: Let $y(x) = ax + b$. In other words, $a = 1$ and $b = -\frac{1}{2}$. Hence $y(x) = x - \frac{1}{2}$.

3. Find the general solution of the differential equation $y'' + 4y' + 4y = x^2 + x + 4$.

   HINT: Let $y(x) = ax^2 + bx + c$. In other words, $a = \frac{1}{4}$ and $b = -\frac{1}{4}$ and $c = \frac{9}{8}$. Hence $y(x) = \frac{1}{8}(2x^2 - 2x + 9)$.

4. Find the general solution of the differential equation $y'' + 4y' + 4y = x^2$.

   HINT: Let $y(x) = ax^2 + bx + c$. In other words, $a = \frac{1}{4}$ and $b = -\frac{1}{2}$ and $c = \frac{5}{8}$. Hence $y(x) = \frac{1}{8}(2x^2 - 4x + 5)$.

5. Find the general solution of the differential equation $y'' + 3y' + 2y = x^2$.

   HINT: Let $y(x) = ax^2 + bx + c$. In other words, $a = \frac{1}{2}$ and $b = -\frac{3}{2}$ and $c = \frac{7}{4}$. Hence $y(x) = \frac{1}{4}(2x^2 - 6x + 7)$.

6. Find the general solution of the differential equation $y'' + 4y' - y = x^2 + 5$.

   HINT: Let $y(x) = ax^2 + bx + c$. In other words, $a = \frac{1}{2}$ and $b = -\frac{3}{2}$ and $c = \frac{17}{4}$. Hence $y(x) = \frac{1}{4}(2x^2 - 6x + 17)$.

7. Find the general solution of the differential equation $y'' + 9y = 5\sin x$.

   HINT: Let $y(x) = a\sin x$. In other words, $a = \frac{5}{8}$. Hence $y(x) = \frac{5\sin x}{8}$.

8. Find the general solution of the differential equation $y'' + y = \operatorname{cosec}^3 x$.

   HINT: Let $y(x) = a\operatorname{cosec} x$. In other words, $a = \frac{1}{2}$. Hence $y(x) = \frac{\operatorname{cosec} x}{2}$.

9. Find the general solution of the differential equation $y'' + 4y = \sec^3 2x$.

   HINT: Let $y(x) = a\sec 2x$. In other words, $a = \frac{1}{8}$. Hence $y(x) = \frac{\operatorname{cosec} 2x}{8}$.

10. Find the general solution of the differential equation $y'' + 9y = 5\sin 3x$.

    HINT: Let $y(x) = ax\cos 3x$. In other words, $a = -\frac{5}{6}$. Hence $y(x) = -\frac{5x\cos 3x}{6}$.

11. Find the general solution of the differential equation $y'' + 9y = 5\cos 3x$.

    HINT: Let $y(x) = ax\sin 3x$. In other words, $a = \frac{5}{6}$. Hence $y(x) = \frac{5x\sin 3x}{6}$.

12. Find the general solution of the differential equation $y'' + 4y = \cos 2x$.

    HINT: Let $y(x) = ax\sin 2x$. In other words, $a = \frac{1}{4}$. Hence $y(x) = \frac{x\sin 2x}{4}$.

13. Find the general solution of the differential equation $y'' + 4y = \sin 2x$.

    HINT: Let $y(x) = ax\cos 2x$. In other words, $a = -\frac{1}{4}$. Hence $y(x) = -\frac{x\cos 2x}{4}$.

14. Find the general solution of the differential equation $y'' + 4y = 2\cos x$.

    HINT: Let $y(x) = a\cos x$. In other words, $a = \frac{2}{3}$. Hence $y(x) = \frac{2\cos x}{3}$.

15. Find the general solution of the differential equation $y'' + 4y = 2\sin x$.

    HINT: Let $y(x) = a\sin x$. In other words, $a = \frac{2}{3}$. Hence $y(x) = \frac{2\sin x}{3}$.

16. Find the general solution of the differential equation $y'' + 9y = x^2 + \sin 3x$.

    HINT: Let $y(x) = a + bx + cx^2 + d\,x\sin 3x$. In other words, $a = -\frac{2}{81}$, $b = 0$, $c = \frac{1}{9}$ and $d = -\frac{1}{6}$. Hence $y(x) = -\frac{2}{81} + \frac{x^2}{9} - \frac{x\sin 3x}{6}$.

17. Find the general solution of the differential equation $y'' + 9y = x^2 + \sin 2x$.

    HINT: Let $y(x) = a + bx + cx^2 + d\sin 2x$. In other words, $a = -\frac{2}{81}$, $b = 0$, $c = \frac{1}{9}$ and $d = \frac{1}{5}$. Hence $y(x) = -\frac{2}{81} + \frac{x^2}{9} + \frac{\sin 2x}{5}$.

18. Find the general solution of the differential equation $y'' + 3y' + 4y = 2\cos x$.

    HINT: Let $y(x) = a\cos x + b\sin x$. In other words, $a = -1$ and $b = 1$. Hence $y(x) = \sin x - \cos x$.

19. Find the general solution of the differential equation $y'' + y = e^{-x}$.

    HINT: Let $y(x) = a\,e^{-x}$. In other words, $a = \frac{1}{2}$. Hence $y(x) = \frac{e^{-x}}{2}$.

20. Find the general solution of the differential equation $y'' + 2y' - 5y = 2e^x$.

    HINT: Let $y(x) = a\,e^x$. In other words, $a = -1$. Hence $y(x) = -e^x$.

21. Find the general solution of the differential equation $y'' - y = e^x$.

    HINT: Let $y(x) = a\,x\,e^x$. In other words, $a = \frac{1}{2}$. Hence $y(x) = \frac{x\,e^x}{2}$.

22. Find the general solution of the differential equation $y'' + y = e^{2x}$.

    HINT: Let $y(x) = a\,e^{2x}$. In other words, $a = \frac{1}{5}$. Hence $y(x) = \frac{e^x}{5}$.

23. Find the general solution of the differential equation $y'' - 16y = e^{4x}$.

    HINT: Let $y(x) = a\,x\,e^{4x}$. In other words, $a = \frac{1}{8}$. Hence $y(x) = \frac{x\,e^x}{8}$.

24. Find the general solution of the differential equation $y'' - 16y = 2e^{4x}$.

    HINT: Let $y(x) = a\,x\,e^{4x}$. In other words, $a = \frac{1}{4}$. Hence $y(x) = \frac{x\,e^x}{4}$.

25. Find the general solution of the differential equation $y'' + 4y' + 4y = e^{-2x}$.

    HINT: Let $y(x) = a\,x^2\,e^{-2x}$. In other words, $a = \frac{1}{2}$. Hence $y(x) = \frac{x^2\,e^x}{2}$.

5.3.3 Operator Method

Any second nonhomogeneous linear differential equation with constant coefficient can be written in the form $F(D)y = f(x)$, where $F(D)y = (D^2 + aD + b)y = y'' + ay' + by$. This method is somewhat easier than the two other methods discussed in Section 5.3. The motivation behind this method is when the operator $\frac{1}{F(D)}$, called the *inverse operator*, is operated to any continuous function $f(x)$, and gives the particular integral $y(x) = \left(\frac{1}{F(D)}\right) f(x)$ of the equation $F(D)y = f(x)$. In other words, we have

$$
\begin{aligned}
F(D)\left(\frac{1}{F(D)}f(x)\right) &= F(D)y \\
&= f(x)
\end{aligned}
$$

Accordingly, we have

$$Dy = D\left(\frac{1}{D}f(x)\right)$$
$$= f(x)$$

In other words, we have

$$y(x) = \frac{1}{D}f(x)$$

and

$$y(x) = \int f(x)\,\mathrm{d}x$$

Hence we have

$$\frac{1}{D}f(x) = \int f(x)\,\mathrm{d}x$$

Theorem 5.3.1 If α is any constant, then $F(D)e^{\alpha x} = F(\alpha)e^{\alpha x}$.

Proof: It is easy to verify that $D^n(e^{\alpha x}) = \alpha^n e^{\alpha x}$ for $n \geq 0$ by induction on n. Hence $F(D)e^{\alpha x} = F(\alpha)e^{\alpha x}$.

Theorem 5.3.2 If α is any constant and $g(x)$ is any function, then $F(D)e^{\alpha x}g(x) = e^{\alpha x}\left(F(D+\alpha)g(x)\right)$.

Proof: It is easy to verify that $D(e^{\alpha x}g(x)) = \alpha e^{\alpha x}g(x) + e^{\alpha x}D(g(x)) = e^{\alpha x}(\alpha + D)g(x)$. In general, one can show by induction on $n \geq 0$ that $D^n(e^{\alpha x}g(x)) = e^{\alpha x}(D+\alpha)^n g(x)$. In other words, we have $F(D)e^{\alpha x}g(x) = e^{\alpha x}\left(F(D+\alpha)g(x)\right)$.

Theorem 5.3.3 If α and β are arbitrary constants, then

$$F\left(D^2\right)\sin(\alpha x+\beta) = F\left(-\alpha^2\right)\sin(\alpha x+\beta)$$

Proof: It is easy to verify that $D^2\sin(\alpha x+\beta) = -\alpha^2\sin(\alpha x+\beta)$. Hence $F\left(D^2\right)\sin(\alpha x+\beta) = F\left(-\alpha^2\right)\sin(\alpha x+\beta)$.

Theorem 5.3.4 If α and β are arbitrary constants, then

$$F\left(D^2\right)\cos(\alpha x+\beta) = F\left(-\alpha^2\right)\cos(\alpha x+\beta)$$

Proof: It is easy to verify that $D^2\cos(\alpha x+\beta) = -\alpha^2\cos(\alpha x+\beta)$. Hence $F\left(D^2\right)\cos(\alpha x+\beta) = F\left(-\alpha^2\right)\cos(\alpha x+\beta)$.

Theorem 5.3.5 If α is any constant and $F(\alpha) \neq 0$, then $\left(\frac{1}{F(D)}\right)e^{\alpha x} = \frac{e^{\alpha x}}{F(\alpha)}$.

Proof: According to Theorem 5.3.1, we have

$$F(D)e^{\alpha x} = F(\alpha)e^{\alpha x} \tag{5.21}$$

Using Eq.(5.21), we have

$$\begin{aligned}\frac{e^{\alpha x}}{F(\alpha)} &= \left(\frac{1}{F(\alpha)}\right)\left(\frac{1}{F(D)}\right)F(D)e^{\alpha x}\\ &= \left(\frac{1}{F(\alpha)}\right)\left(\frac{1}{F(D)}\right)F(\alpha)e^{\alpha x}\\ &= \frac{1}{F(D)}e^{\alpha x}\end{aligned}$$

EXAMPLE 5.31 Find the particular integral of the differential equation $(D^2 - 3D + 2)y = e^{3x}$.

Solution: Clearly, $F(D) = D^2 - 3D + 2$ and $\alpha = 3$. According to Theorem 5.3.5, the particular integral is

$$\begin{aligned}h(x) &= \left(\frac{1}{F(D)}\right)e^{\alpha x}\\ &= \frac{e^{\alpha x}}{F(\alpha)}\\ &= \frac{e^{3x}}{F(3)}\\ &= \frac{e^{3x}}{2}\end{aligned}$$

EXAMPLE 5.32 Find the particular integral of the differential equation $(D^2 - 3D + 2)y = 2e^{3x}$.

Solution: Clearly, $F(D) = D^2 - 3D + 2$ and $\alpha = 3$. According to Theorem 5.3.5, the particular integral is

$$\begin{aligned}h(x) &= \left(\frac{1}{F(D)}\right)(2e^{\alpha x})\\ &= 2\left(\frac{e^{\alpha x}}{F(\alpha)}\right)\\ &= 2\left(\frac{e^{3x}}{F(3)}\right)\\ &= 2\left(\frac{e^{3x}}{2}\right)\\ &= e^{3x}\end{aligned}$$

Theorem 5.3.6 If α is any constant and $g(x)$ is any function, then $\left(\frac{1}{F(D)}\right)e^{\alpha x}g(x) = e^{\alpha x}\left(\frac{1}{F(D+\alpha)}\right)g(x)$.

Proof: According to Theorem 5.3.2, we have

$$\begin{aligned} F(D)\left(e^{\alpha x}\left(\frac{1}{F(D+\alpha)}\right)g(x)\right) &= F(D)\left(e^{\alpha x}\left(\left(\frac{1}{F(D+\alpha)}\right)g(x)\right)\right) \\ &= e^{\alpha x}F(D+\alpha)\left(\left(\frac{1}{F(D+\alpha)}\right)g(x)\right) \\ &= e^{\alpha x}g(x) \end{aligned}$$

In other words, $e^{\alpha x}\left(\frac{1}{F(D+\alpha)}\right)g(x)$ is the particular integral of the equation $F(D)y = e^{\alpha x}\,g(x)$. Again, we know that $\left(\frac{1}{F(D)}\right)e^{\alpha x}g(x)$ is the particular integral of the equation $F(D)y = e^{\alpha x}\,g(x)$. Hence we have

$$\left(\frac{1}{F(D)}\right)e^{\alpha x}g(x) = e^{\alpha x}\left(\frac{1}{F(D+\alpha)}\right)g(x)$$

EXAMPLE 5.33 Find the particular integral of the differential equation $(D^2 + D)y = 2$.

Solution: Clearly, $F(D) = D^2 + D$, $\alpha = 0$ and $g(x) = 2$. According to Theorem 5.3.6, the particular integral is

$$\begin{aligned} h(x) &= \left(\frac{1}{F(D)}\right)e^{\alpha x}g(x) \\ &= e^{\alpha x}\left(\frac{1}{F(D+\alpha)}\right)g(x) \\ &= \left(\frac{1}{D^2+D}\right)g(x) \\ &= \left(\frac{1}{D}\right)(1+D)^{-1}\,g(x) \\ &= \left(\frac{1}{D}\right)(1-\cdots)\,g(x) \\ &= \left(\frac{1}{D}\right)g(x) \\ &= 2x \end{aligned}$$

EXAMPLE 5.34 Find the particular integral of the differential equation $(D^2+2)y = x\,e^{3x}$.

Solution: Clearly, $F(D) = D^2 + 2$, $\alpha = 3$ and $g(x) = x$. According to Theorem 5.3.6, the particular integral is

$$\begin{aligned} h(x) &= \left(\frac{1}{F(D)}\right)e^{3x}g(x) \\ &= e^{3x}\left(\frac{1}{F(D+3)}\right)g(x) \end{aligned}$$

$$= e^{3x}\left(\frac{1}{F(D+3)}\right)x$$

$$= e^{3x}\left(\frac{1}{(D+3)^2+2}\right)x$$

$$= e^{3x}\left(\frac{1}{D^2+6D+11}\right)x$$

$$= \frac{e^{3x}}{11}\left(1+\frac{D^2+6D}{11}\right)^{-1}x$$

$$= \frac{e^{3x}}{11}\left(1-\frac{D^2+6D}{11}+\cdots\right)x$$

$$= \frac{e^{3x}}{11}\left(x-\frac{6}{11}\right)$$

EXAMPLE 5.35 Find the particular integral of the differential equation $(D^2+2)y = x^2\, e^{3x}$.

Solution: Clearly, $F(D) = D^2 + 2$, $\alpha = 3$ and $g(x) = x^2$. According to Theorem 5.3.6, the particular integral is

$$h(x) = \left(\frac{1}{F(D)}\right)e^{3x}g(x)$$

$$= e^{3x}\left(\frac{1}{F(D+3)}\right)x^2$$

$$= e^{3x}\left(\frac{1}{(D+3)^2+2}\right)x^2$$

$$= e^{3x}\left(\frac{1}{D^2+6D+11}\right)x^2$$

$$= \frac{e^{3x}}{11}\left(1+\frac{D^2+6D}{11}\right)^{-1}x^2$$

$$= \frac{e^{3x}}{11}\left(1-\frac{D^2+6D}{11}+\left(\frac{D^2+6D}{11}\right)^2-\cdots\right)x^2$$

$$= \frac{e^{3x}}{11}\left(x^2-\frac{2+12x}{11}+\frac{72}{121}\right)$$

$$= \frac{e^{3x}}{11}\left(x^2-\frac{12x}{11}+\frac{50}{121}\right)$$

Theorem 5.3.7 If α is any constant and $F(D) = (D-\alpha)^r\phi(D)$, such that $\phi(\alpha) \neq 0$, then $\left(\frac{1}{F(D)}\right)e^{\alpha x} = \left(\frac{e^{\alpha x}}{F(\alpha)}\right)\left(\frac{x^r}{r!}\right)$.

Proof: Let $g(x) = 1$. According to Theorems 5.3.5 and 5.3.6, it is easy to verify that

$$\begin{aligned}
\left(\frac{1}{F(D)}\right)e^{\alpha x} &= \left(\frac{1}{(D-\alpha)^r}\right)\left(\frac{1}{\phi(D)}\right)e^{\alpha x} \\
&= \left(\frac{1}{\phi(\alpha)}\right)\left(\frac{1}{(D-\alpha)^r}\right)e^{\alpha x} \\
&= \left(\frac{1}{\phi(\alpha)}\right)\left(\frac{1}{(D-\alpha)^r}\right)e^{\alpha x}g(x) \\
&= \left(\frac{1}{\phi(\alpha)}\right)e^{\alpha x}\left(\frac{1}{(D-\alpha+\alpha)^r}\right)g(x) \\
&= \left(\frac{e^{\alpha x}}{\phi(\alpha)}\right)\left(\frac{1}{D^r}\right)g(x) \\
&= \left(\frac{e^{\alpha x}}{\phi(\alpha)}\right)\left(\frac{x^r}{r!}\right)
\end{aligned}$$

EXAMPLE 5.36 Find the particular integral of the differential equation $(D-2)^2y = e^{2x}$.
Solution: Clearly, $F(D) = (D-2)^2$, $\alpha = 2$, $\phi(D) = 1$ and $r = 2$. According to Theorem 5.3.7, the particular integral is

$$\begin{aligned}
h(x) &= \left(\frac{1}{F(D)}\right)e^{2x} \\
&= \left(\frac{e^{2x}}{\phi(2)}\right)\left(\frac{x^2}{2!}\right) \\
&= e^{2x}\left(\frac{x^2}{2}\right)
\end{aligned}$$

EXAMPLE 5.37 Find the particular integral of the differential equation $(D+2)(D-2)^2y = e^{2x}$.
Solution: Clearly, $F(D) = (D+2)(D-2)^2$, $\alpha = 2$, $\phi(D) = D+2$ and $r = 2$. According to Theorem 5.3.7, the particular integral is

$$\begin{aligned}
h(x) &= \left(\frac{1}{F(D)}\right)e^{2x} \\
&= \left(\frac{e^{2x}}{\phi(2)}\right)\left(\frac{x^2}{2!}\right) \\
&= \left(\frac{e^{2x}}{4}\right)\left(\frac{x^2}{2}\right)
\end{aligned}$$

Theorem 5.3.8 If α and β are arbitrary constants, then $\left(\frac{1}{F(D^2)}\right)\sin(\alpha x+\beta)=\frac{\sin(\alpha x+\beta)}{F(-\alpha^2)}$.

Proof: According to Theorem 5.3.3, we have

$$F(D^2)\sin(\alpha x+\beta) = F(-\alpha^2)\sin(\alpha x+\beta) \tag{5.22}$$

One can operate $\frac{1}{F(-\alpha^2)F(D^2)}$ to Eq. (5.22) to have

$$\begin{aligned}\frac{\sin(\alpha x+\beta)}{F(-\alpha^2)} &= \left(\frac{1}{F(-\alpha^2)}\right)F(D^2)\left(\frac{1}{F(D^2)}\right)\sin(\alpha x+\beta)\\ &= \left(\frac{1}{F(-\alpha^2)}\right)\left(\frac{1}{F(D^2)}\right)(F(D^2)\sin(\alpha x+\beta))\\ &= \left(\frac{1}{F(-\alpha^2)}\right)\left(\frac{1}{F(D^2)}\right)(F(-\alpha^2)\sin(\alpha x+\beta))\\ &= \frac{1}{F(D^2)}\sin(\alpha x+\beta)\end{aligned}$$

EXAMPLE 5.38 Find the particular integral of the differential equation $(D^2+2)y=\sin 2x$.

Solution: Clearly, $F(D) = D^2+2$, $\alpha = 2$ and $\beta = 0$. According to Theorem 5.3.8, the particular integral is given by

$$\begin{aligned}h(x) &= \left(\frac{1}{F(D^2)}\right)\sin(\alpha x+\beta)\\ &= \frac{\sin(\alpha x+\beta)}{F(-\alpha^2)}\\ &= \frac{\sin 2x}{F(-2^2)}\\ &= \frac{\sin 2x}{-2}\\ &= -\frac{\sin 2x}{2}\end{aligned}$$

EXAMPLE 5.39 Find the particular integral of the differential equation $(D^2-2)y = 6\sin 2x$.

Solution: Clearly, $F(D) = D^2-2$, $\alpha = 2$ and $\beta = 0$. According to Theorem 5.3.8, the particular integral is given by

$$\begin{aligned}h(x) &= \frac{1}{F(D^2)}(6\sin(\alpha x+\beta))\\ &= 6\left(\frac{\sin(\alpha x+\beta)}{F(-\alpha^2)}\right)\end{aligned}$$

$$= 6\left(\frac{\sin 2x}{F(-2^2)}\right)$$

$$= 6\left(\frac{\sin 2x}{-6}\right)$$

$$= -\sin 2x$$

Theorem 5.3.9 If α and β are arbitrary constants, then $\left(\frac{1}{F(D^2)}\right)\cos(\alpha x+\beta) = \frac{\cos(\alpha x+\beta)}{F(-\alpha^2)}$.

Proof: According to Theorem 5.3.4, we have

$$F(D^2)\cos(\alpha x+\beta) = F(-\alpha^2)\cos(\alpha x+\beta) \tag{5.23}$$

One can operate $\frac{1}{F(-\alpha^2)F(D^2)}$ to Eq.(5.23) to have

$$\frac{\cos(\alpha x+\beta)}{F(-\alpha^2)} = \left(\frac{1}{F(-\alpha^2)}\right)F(D^2)\left(\frac{1}{F(D^2)}\right)\cos(\alpha x+\beta)$$

$$= \left(\frac{1}{F(-\alpha^2)}\right)\left(\frac{1}{F(D^2)}\right)(F(D^2)\cos(\alpha x+\beta))$$

$$= \left(\frac{1}{F(-\alpha^2)}\right)\left(\frac{1}{F(D^2)}\right)(F(-\alpha^2)\cos(\alpha x+\beta))$$

$$= \left(\frac{1}{F(D^2)}\right)\cos(\alpha x+\beta)$$

EXAMPLE 5.40 Find the particular integral of the differential equation $(D^2+2)y = \sin 2x$.

Solution: Clearly, $F(D) = D^2 + 2$, $\alpha = 2$ and $\beta = 0$. According to Theorem 5.3.9, the particular integral is given by

$$h(x) = \left(\frac{1}{F(D^2)}\right)\cos(\alpha x+\beta)$$

$$= \frac{\cos(\alpha x+\beta)}{F(-\alpha^2)}$$

$$= \frac{\cos 2x}{F(-2^2)}$$

$$= \frac{\cos 2x}{-2}$$

$$= -\frac{\cos 2x}{2}$$

EXAMPLE 5.41 Find the particular integral of the differential equation $(D^2 - 2)y = 6\sin 2x$.

Solution: Clearly, $F(D) = D^2 - 2$, $\alpha = 2$ and $\beta = 0$. According to Theorem 5.3.9, the particular integral is given by

$$\begin{aligned} h(x) &= \left(\frac{1}{F(D^2)}\right)(6\cos(\alpha x + \beta)) \\ &= 6\left(\frac{\cos(\alpha x + \beta)}{F(-\alpha^2)}\right) \\ &= 6\left(\frac{\cos 2x}{F(-2^2)}\right) \\ &= 6\left(\frac{\cos 2x}{-6}\right) \\ &= -\cos 2x \end{aligned}$$

Theorem 5.3.10 If α and β are constants such that $F(-\alpha^2) = 0$, then $\left(\frac{1}{F(D^2)}\right)\cos(\alpha x + \beta) + i\left(\frac{1}{F(D^2)}\right)\sin(\alpha x + \beta) = e^{i(\alpha x+\beta)}\left(\frac{1}{F(D+i\alpha)^2}\right)g(x)$, where $g(x) = 1$.

Proof: Apply Theorem 5.3.6.

EXAMPLE 5.42 Find the particular integral of the differential equation $(D^2+1)y = \cos x$.
Solution: Clearly, $F(D) = D^2 + 1$, $\alpha = 1$ and $\beta = 0$. Let $g(x) = 1$. According to Theorem 5.3.10, we have

$$\begin{aligned} \left(\frac{1}{F(D^2)}\right)\cos(\alpha x + \beta) + i\left(\frac{1}{F(D^2)}\right)\sin(\alpha x + \beta) &= e^{i(\alpha x+\beta)}\left(\frac{1}{F(D+i\alpha)}\right)g(x) \\ &= e^{ix}\left(\frac{1}{(D+i)^2+1}\right)g(x) \\ &= e^{ix}\left(\frac{1}{D^2+2iD}\right)g(x) \\ &= e^{ix}\left(\frac{1}{2iD}\right)\left(1+\frac{D}{2i}\right)^{-1}g(x) \\ &= e^{ix}\left(\frac{1}{2iD}\right)(1-\cdots)\,g(x) \\ &= e^{ix}\left(\frac{x}{2i}\right) \\ &= -ix\left(\frac{\cos x}{2}\right) + x\left(\frac{\sin x}{2}\right) \end{aligned}$$

In other words, the particular integral is

$$\begin{aligned} h(x) &= \left(\frac{1}{F(D^2)}\right)\cos(\alpha x + \beta) \\ &= x\left(\frac{\sin x}{2}\right) \end{aligned}$$

EXAMPLE 5.43 Find the particular integral of the differential equation $(D^2+1)y = \sin x$.

Solution: Clearly, $F(D) = D^2+1$, $\alpha = 1$ and $\beta = 0$. Let $g(x) = 1$. According to Theorem 5.3.10, we have

$$\begin{aligned}
\left(\frac{1}{F(D^2)}\right)\cos(\alpha x+\beta) + i\left(\frac{1}{F(D^2)}\right)\sin(\alpha x+\beta) &= e^{i(\alpha x+\beta)}\left(\frac{1}{F(D+i\alpha)^2}\right)g(x)\\
&= e^{ix}\left(\frac{1}{(D+i)^2+1}\right)g(x)\\
&= e^{ix}\left(\frac{1}{D^2+2iD}\right)g(x)\\
&= e^{ix}\left(\frac{1}{2iD}\right)\left(1+\frac{D}{2i}\right)^{-1}g(x)\\
&= e^{ix}\left(\frac{1}{2iD}\right)(1-\cdots)\,g(x)\\
&= e^{ix}\left(\frac{x}{2i}\right)\\
&= -ix\left(\frac{\cos x}{2}\right)+x\left(\frac{\sin x}{2}\right)
\end{aligned}$$

In other words, the particular integral is

$$\begin{aligned}
h(x) &= \left(\frac{1}{F(D^2)}\right)\sin(\alpha x+\beta)\\
&= -x\left(\frac{\cos x}{2}\right)
\end{aligned}$$

Theorem 5.3.11 If $f(x)$ is any function of x, then $\left(\frac{1}{F(D)}\right)x\,f(x) = x\left(\frac{1}{F(D)}\right)f(x)$ $+\left(\frac{1}{F(D)}\right)' f(x)$.

Proof: It is easy to verify that $D^n(x\,g(x)) = xD^n(g(x)) + (D^n)'g(x)$. Clearly, $D(x\,g(x))$ $= g(x) + xD(g(x))$. We will verify $D^n(x\,g(x)) = xD^n(g(x)) + (D^n)'g(x)$ by the method of induction over $n \geq 1$. Assume that $D^n(x\,g(x)) = xD^n(g(x)) + (D^n)'g(x)$ is true for $n = r$. Hence we have $D^r(x\,g(x)) = xD^r(g(x)) + (D^r)'g(x)$. Again, one can verify that

$$\begin{aligned}
D^{r+1}(x\,g(x)) &= D(D^r(x\,g(x)))\\
&= D(xD^r(g(x)) + (D^r)'(g(x)))\\
&= xD^{r+1}(g(x)) + D^r(g(x)) + D((D^r)'(g(x)))\\
&= xD^{r+1}(g(x)) + D^r(g(x)) + D(rD^{r-1}(g(x)\\
&= xD^{r+1}(g(x)) + D^r(g(x)) + rD^r(g(x))
\end{aligned}$$

$$= xD^{r+1}(g(x)) + (r+1)D^r(g(x))$$

$$= xD^{r+1}(g(x)) + (D^{r+1})'(g(x))$$

In other words, we have

$$F(D)(x\,g(x)) = xF(D)(g(x)) + F'(D)(g(x)) \tag{5.24}$$

Let $F(D)g(x) = f(x)$. Clearly, we have

$$g(x) = \left(\frac{1}{F(D)}\right) f(x) \tag{5.25}$$

One can replace the operator expression given in Eq.(5.25) for $g(x)$ in Eq.(5.24) to have the requirement

$$F(D)\left(x\left(\frac{1}{F(D)}\right) f(x)\right) = xF(D)\left(\left(\frac{1}{F(D)}\right) f(x)\right) + F'(D)\left(\left(\frac{1}{F(D)}\right) f(x)\right)$$

$$= xf(x) + F'(D)\left(\left(\frac{1}{F(D)}\right) f(x)\right)$$

In other words, we have

$$xf(x) = F(D)\left(x\left(\frac{1}{F(D)}\right) f(x)\right) - F'(D)\left(\left(\frac{1}{F(D)}\right) f(x)\right) \tag{5.26}$$

One can operate $\frac{1}{F(D)}$ to Eq.(5.26) to have

$$\left(\frac{1}{F(D)}\right)(xf(x)) = x\left(\frac{1}{F(D)}\right) f(x) - \left(\frac{1}{F(D)}\right)\left(F'(D)\left(\left(\frac{1}{F(D)}\right) f(x)\right)\right)$$

$$= x\left(\frac{1}{F(D)}\right) f(x) - F'(D)\left(\left(\frac{1}{F(D)}\right)^2 f(x)\right)$$

$$= x\left(\frac{1}{F(D)}\right) f(x) + \left(\frac{1}{F(D)}\right)' f(x)$$

EXAMPLE 5.44 Find the particular integral of the differential equation $(D^2+4)y = x\sin x$.

Solution: Clearly, $F(D) = D^2 + 4$ and $f(x) = \sin x$. According to Theorem 5.3.11, the particular integral is

$$h(x) = \left(\frac{1}{F(D)}\right)(x\sin x)$$

$$= x\left(\frac{1}{F(D)}\right)\sin x + \left(\frac{1}{F(D)}\right)'\sin x$$

$$= x\left(\frac{1}{F(-1^2)}\right)\sin x - 2D\left(\frac{1}{(D^2+4)^2}\right)\sin x$$

$$= x\left(\frac{1}{3}\right)\sin x - 2D\left(\frac{1}{(-1^2+4)^2}\right)\sin x$$

$$= x\left(\frac{1}{3}\right)\sin x - 2D\left(\frac{1}{9}\right)\sin x$$

$$= \frac{x}{3}\sin x - \frac{2}{9}\cos x$$

EXAMPLE 5.45 Find the particular integral of the differential equation $(D^2+4)y = x\cos x$.

Solution: Clearly, $F(D) = D^2 + 4$ and $f(x) = \cos x$. According to Theorem 5.3.11, the particular integral is

$$h(x) = \left(\frac{1}{F(D)}\right)(x\cos x)$$

$$= x\left(\frac{1}{F(D)}\right)\cos x + \left(\frac{1}{F(D)}\right)'\cos x$$

$$= x\left(\frac{1}{F(-1^2)}\right)\cos x - 2D\left(\frac{1}{(D^2+4)^2}\right)\cos x$$

$$= x\left(\frac{1}{3}\right)\cos x - 2D\left(\frac{1}{(-1^2+4)^2}\right)\cos x$$

$$= x\left(\frac{1}{3}\right)\cos x - 2D\left(\frac{1}{9}\right)\cos x$$

$$= \frac{x}{3}\cos x + \frac{2}{9}\sin x$$

Exercises 5.3.3

1. Solve the differential equation $(D^2 + D - 2)y = x + \sin x$.
2. Solve the differential equation $(D^2 - 2D + 1)y = x\,e^x \sin x$.
3. Solve the differential equation $(D^2 + D)y = x\sin x$.
4. Solve the differential equation $(D^2 + D)y = x\cos x$.
5. Solve the differential equation $(D^2 - 2D + 4)y = e^x \sin x$.
6. Solve the differential equation $(D^2 - 2D + 4)y = e^x \cos x$.
7. Solve the differential equation $(D^2 - 3D + 2)y = e^x + e^{2x}$.
8. Solve the differential equation $(D^2 + D - 6)y = e^{3x}$.
9. Solve the differential equation $(D^2 + 5D - 6)y = \sin x$.
10. Solve the differential equation $(D^2 - 6D + 5)y = \sin x$.

11. Solve the differential equation $(D^2+4)y = \sin 3x + x^2 + e^x$.
12. Solve the differential equation $(D^2-4)y = \sin 3x + x^2 + e^x$.
13. Solve the differential equation $(D^2+1)y = \sinh 3x$.
14. Solve the differential equation $(D^2+1)y = \cosh 3x$.

5.4 Variable Coefficient Equation

All the methods discussed so far are applicable to second order constant coefficient linear differential equations. The method which is discussed in this section can be applicable for some of the second order differential equations with variable coefficients. The following methods are discussed in this section:

1. Method of Reduction Order
2. Removal of First Derivative
3. Change of Independent Variable.

5.4.1 Method of Reduction Order

This method is applicable to every differential equation of second order, provided one of the solutions to the corresponding homogeneous equation is known. Let $y'' + p(x)y' + q(x)y = 0$ be any homogeneous differential equation, and $y(x) = u(x)$ be one of the solutions of the given differential equation. Hence we have

$$u'' + p(x)u' + q(x)u = 0 \tag{5.27}$$

Assume that $y(x) = u(x)v(x)$ is the second independent solution of the given differential equation. Again, we have

$$\begin{aligned} y' &= uv' + u'v \\ y'' &= u''v + 2u'v' + uv'' \end{aligned}$$

Substituting the expressions for y, y' and y'' in the given differential equation, we have

$$\begin{aligned} 0 &= y'' + p(x)y' + q(x)y \\ &= u''v + 2u'v' + uv'' + p(x)\left(uv' + u'v\right) + q(x)uv \\ &= v\left(u'' + p(x)u' + q(x)u\right) + 2u'v' + uv'' + p(x)uv' \\ &= 2u'v' + uv'' + p(x)uv' \\ &= uv'' + 2u'v' + p(x)uv' \\ &= uv'' + \left(2u' + p(x)u\right)v' \end{aligned} \tag{5.28}$$

Again, the solution of Eq.(5.28) according to the formula given in Eq.(2.2) is given by

$$
\begin{aligned}
v(x) &= \int \mathrm{d}v \\
&= \int e^{-\int \ p(x)+2\frac{u'}{u} \ \mathrm{d}x} \mathrm{d}x \\
&= \int e^{-\int p(x)\,\mathrm{d}x} e^{-2\int \frac{u'}{u}\,\mathrm{d}x}\,\mathrm{d}x \\
&= \int \left(e^{-\int p(x)\,\mathrm{d}x} e^{-2\int \frac{\mathrm{d}u}{u}}\right)\mathrm{d}x \\
&= \int \left(e^{-\int p(x)\,\mathrm{d}x} u^{-2}\right)\mathrm{d}x \\
&= \int \left(\frac{e^{-\int p(x)\,\mathrm{d}x}}{u^2}\right)\,\mathrm{d}x
\end{aligned} \tag{5.29}
$$

EXAMPLE 5.46 Find the second solution of the differential equation $xy'' + 2y' + xy = 0$ using $u(x) = \frac{\sin x}{x}$ as one of the solution.

Solution: Clearly, $p(x) = \frac{2}{x}$ and $u(x) = \frac{\sin x}{x}$. According to the formula given in Eq.(5.29), we have

$$
\begin{aligned}
v(x) &= \int \left(\frac{e^{-\int p(x)\,\mathrm{d}x}}{u^2}\right)\,\mathrm{d}x \\
&= \int \operatorname{cosec}^2 x\,\mathrm{d}x \\
&= -\cot x
\end{aligned}
$$

Hence the required second solution is $y(x) = u(x)v(x) = -\left(\frac{\cos x}{x}\right)$.

EXAMPLE 5.47 Find the second solution of the differential equation $(x-1)y''-xy'+y=0$ using $u(x) = e^x$ as one of the solution.

Solution: Clearly, $p(x) = -\frac{x}{x-1}$ and $u(x) = e^x$. According to the formula given in Eq.(5.29), we have

$$
\begin{aligned}
v(x) &= \int \frac{e^{-\int p(x)\,\mathrm{d}x}}{u^2}\,\mathrm{d}x \\
&= \int \frac{e^{\int \left(\frac{x}{x-1}\right)\mathrm{d}x}}{e^{2x}}\,\mathrm{d}x
\end{aligned}
$$

$$= \int \frac{e^{\int \left(1+\frac{1}{x-1}\right) dx}}{e^{2x}} dx$$

$$= \int \frac{(x-1)e^x}{e^{2x}} dx$$

$$= \int (x-1)e^{-x} dx$$

$$= xe^{-x}$$

Hence the required second solution is $y(x) = u(x)v(x) = x$.

EXAMPLE 5.48 Find the second solution of the differential equation $(x-1)y'' - xy' + y = 0$ using $u(x) = x$ as one of the solutions.

Solution: Clearly, $p(x) = -\frac{x}{x-1}$ and $u(x) = x$. According to the formula given in Eq.(5.29), we have

$$v(x) = \int \frac{e^{-\int p(x)\, dx}}{u^2} dx$$

$$= \int \frac{e^{\int \left(\frac{x}{x-1}\right) dx}}{x^2} dx$$

$$= \int \frac{e^{\int \left(1+\frac{1}{x-1}\right) dx}}{x^2} dx$$

$$= \int \frac{(x-1)e^x}{x^2} dx$$

$$= \int (x^{-1} - x^{-2})e^x dx$$

$$= \frac{e^x}{x}$$

Hence the required second solution is $y(x) = u(x)v(x) = e^x$.

EXAMPLE 5.49 Find the second solution of the differential equation $xy'' + (x-1)y' - y = 0$ using $u(x) = x - 1$ as one of the solutions.

Solution: Clearly, $p(x) = \frac{x-1}{x}$ and $u(x) = x - 1$. According to the formula given in Eq.(5.29), we have

$$v(x) = \int \frac{e^{-\int p(x)\, dx}}{u^2} dx$$

$$= \int \frac{e^{\int \left(\frac{1-x}{x}\right) dx}}{(x-1)^2} dx$$

$$= \int \frac{e^{\int\left(\frac{1}{x}-1\right)\,\mathrm{d}x}}{(x-1)^2}\,\mathrm{d}x$$

$$= \int \frac{xe^{-x}}{(x-1)^2}\,\mathrm{d}x$$

$$= \int \left((x-1)^{-1} + (x-1)^{-2}\right) e^{-x}\,\mathrm{d}x$$

$$= \frac{e^{-x}}{1-x}$$

Hence the required second solution is $y(x) = u(x)v(x) = -e^{-x}$.

EXAMPLE 5.50 Find the second solution of the differential equation $xy''+(x-1)y'-y=0$ using $u(x) = -e^{-x}$ as one of the solution.

Solution: Clearly, $p(x) = \frac{x-1}{x}$ and $u(x) = -e^{-x}$. According to the formula given in Eq.(5.29), we have

$$v(x) = \int \frac{e^{-\int p(x)\,\mathrm{d}x}}{u^2}\mathrm{d}x$$

$$= \int \frac{e^{\int\left(\frac{1-x}{x}\right)\,\mathrm{d}x}}{e^{-2x}}\,\mathrm{d}x$$

$$= \int \frac{e^{\int\left(\frac{1}{x}-1\right)\,\mathrm{d}x}}{e^{-2x}}\,\mathrm{d}x$$

$$= \int \frac{xe^{-x}}{e^{-2x}}\,\mathrm{d}x$$

$$= \int xe^{x}\,\mathrm{d}x$$

$$= (x-1)e^{-x}$$

Hence the required second solution is $y(x) = u(x)v(x) = 1-x$.

EXAMPLE 5.51 Find the second solution of the differential equation $y'' + xy' - y = 0$ using $u(x) = x$ as one of the solution.

Solution: Clearly, $p(x) = x$ and $u(x) = x$. According to the formula given in Eq.(5.29), we have

$$v(x) = \int \frac{e^{-\int p(x)\,\mathrm{d}x}}{u^2}\,\mathrm{d}x$$

$$= \int \frac{e^{-\int x\,\mathrm{d}x}}{x^2}\,\mathrm{d}x$$

$$= \int \frac{e^{-\frac{x^2}{2}}}{x^2}\,\mathrm{d}x$$

Hence the required second solution is $y(x) = u(x)v(x) = x \int \frac{e^{-\frac{x^2}{2}}}{x^2}\,\mathrm{d}x$.

EXAMPLE 5.52 Find the solution of the differential equation $y'' + 3y' + 2y = 0$ taking $u(x) = e^{-x}$ as a solution of the homogeneous equation.

Solution: Clearly, $p(x) = 3$ and $u(x) = e^{-x}$. According to the formula given in Eq.(5.29), we have

$$\begin{aligned} v(x) &= \int \frac{e^{-\int p(x)\,\mathrm{d}x}}{u^2}\,\mathrm{d}x \\ &= \int \frac{e^{-\int 3\,\mathrm{d}x}}{e^{-2x}}\,\mathrm{d}x \\ &= \int \frac{e^{-3x}}{e^{-2x}}\,\mathrm{d}x \\ &= \int e^{-x}\,\mathrm{d}x \\ &= -e^{-x} \end{aligned}$$

Hence the required second solution is $y(x) = u(x)v(x) = e^{-x}(-e^{-x}) = -e^{-2x}$.

Let $y'' + p(x)y' + q(x)y = f(x)$ be any non-homogeneous differential equation, and $y(x) = u(x)$ is one of the solutions of the corresponding given homogeneous differential equation. Hence we have

$$u'' + p(x)u' + q(x)u = 0 \tag{5.30}$$

Assume that $y(x) = u(x)v(x)$ is the second independent solution of the given differential equation. Again, we have

$$\begin{aligned} y' &= uv' + u'v \\ y'' &= u''v + 2u'v' + uv'' \end{aligned}$$

Substituting the expressions for y, y' and y'' in the given differential equation, we have

$$\begin{aligned} f(x) &= y'' + p(x)y' + q(x)y \\ &= u''v + 2u'v' + uv'' + p(x)\,(uv' + u'v) + q(x)uv \\ &= v\,(u'' + p(x)u' + q(x)u) + 2u'v' + uv'' + p(x)uv' \\ &= 2u'v' + uv'' + p(x)uv' \\ &= uv'' + 2u'v' + p(x)uv' \\ &= uv'' + (2u' + p(x)u)\,v' \end{aligned} \tag{5.31}$$

Again, the particular solution of Eq.(5.31), according to the formula given in Eq.(2.4) is given by

$$\begin{aligned}
v(x) &= \int \mathrm{d}v \\
&= \int e^{-\int \ p(x)+2\ \frac{u'}{u}\ \ \mathrm{d}x} \left(e^{\int \ p(x)+2\ \frac{u'}{u}\ \ \mathrm{d}x} \left(\frac{f(x)}{u} \right) \mathrm{d}x \right) \mathrm{d}x \\
&= \int \left(e^{-\int p(x)\,\mathrm{d}x} e^{-2\int \ \frac{u'}{u}\ \mathrm{d}x} \right) \left(\int \left(e^{\int p(x)\,\mathrm{d}x} e^{2\int \ \frac{u'}{u}\ \mathrm{d}x} \right) \left(\frac{f(x)}{u} \right) \mathrm{d}x \right) \mathrm{d}x \\
&= \int \left(e^{-\int p(x)\,\mathrm{d}x} e^{-2\int \frac{\mathrm{d}u}{u}} \right) \left(\int \left(e^{\int p(x)\,\mathrm{d}x} e^{2\int \frac{\mathrm{d}u}{u}} \right) \left(\frac{f(x)}{u} \right) \mathrm{d}x \right) \mathrm{d}x \\
&= \int \left(e^{-\int p(x)\,\mathrm{d}x} u^{-2} \right) \left(\int \left(e^{\int p(x)\,\mathrm{d}x} u^{2} \left(\frac{f(x)}{u} \right) \mathrm{d}x \right) \right) \mathrm{d}x \\
&= \int \left(\frac{e^{-\int p(x)\,\mathrm{d}x}}{u^2} \right) \left(\int \left(u e^{\int p(x)\,\mathrm{d}x} f(x)\,\mathrm{d}x \right) \right) \mathrm{d}x \qquad (5.32)
\end{aligned}$$

EXAMPLE 5.53 Find the particular solution of the differential equation $y'' - x^2 y' + xy = x$ taking $u(x) = x$ as a solution of the homogeneous equation.

Solution: Clearly, $p(x) = -x^2$, $u(x) = x$ and $f(x) = x$. According to the formula given in Eq.(5.32), we have

$$\begin{aligned}
v(x) &= \int \frac{e^{-\int p(x)\,\mathrm{d}x}}{u^2} \left(\int \left(u e^{\int p(x)\,\mathrm{d}x} f(x)\,\mathrm{d}x \right) \right) \mathrm{d}x \\
&= \int \frac{e^{\int x^2\,\mathrm{d}x}}{x^2} \left(\int \left(x e^{\int -x^2\,\mathrm{d}x} x\,\mathrm{d}x \right) \right) \mathrm{d}x \\
&= \int \frac{e^{\int x^2\,\mathrm{d}x}}{x^2} \left(\int \left(x^2 e^{\int -x^2\,\mathrm{d}x}\,\mathrm{d}x \right) \right) \mathrm{d}x \\
&= \int \frac{e^{\frac{x^3}{3}}}{x^2} \left(\int \left(x^2 e^{-\ \frac{x^3}{3}}\ \mathrm{d}x \right) \right) \mathrm{d}x \\
&= \int \frac{e^{\frac{x^3}{3}}}{x^2} \left(-e^{-\ \frac{x^3}{3}} \right) \mathrm{d}x \\
&= \int -\frac{1}{x^2}\,\mathrm{d}x \\
&= \frac{1}{x}
\end{aligned}$$

Hence the required particular solution is $y(x) = u(x)v(x) = 1$.

EXAMPLE 5.54 Find the particular solution of the differential equation $y'' - xy' + y = 1$ taking $u(x) = x$ as a solution of the homogeneous equation.

Solution: Clearly, $p(x) = -x$, $u(x) = x$ and $f(x) = 1$. According to the formula given in Eq.(5.32), we have

$$\begin{aligned}
v(x) &= \int \frac{e^{-\int p(x)\,dx}}{u^2}\left(\int \left(ue^{\int p(x)\,dx} f(x)\,dx\right)\right) dx \\
&= \int \frac{e^{\int x\,dx}}{x^2}\left(\int \left(xe^{\int -x\,dx}\,dx\right)\right) dx \\
&= \int \frac{e^{\frac{x^2}{2}}}{x^2}\left(\int \left(xe^{-\frac{x^2}{2}}\,dx\right)\right) dx \\
&= \int \frac{e^{\frac{x^2}{2}}}{x^2}\left(-e^{-\frac{x^2}{2}}\right) dx \\
&= \int -\frac{1}{x^2}\,dx \\
&= \frac{1}{x}
\end{aligned}$$

Hence the required particular solution is $y(x) = u(x)v(x) = 1$.

EXAMPLE 5.55 Find the general solution of the differential equation $y'' + y = \sin x$ taking $u(x) = \sin x$ as a solution of the homogeneous equation.

Solution: Clearly, $p(x) = 0$ and $u(x) = \sin x$. According to the formula given in Eq.(5.29), we have

$$\begin{aligned}
v(x) &= \int \frac{e^{-\int p(x)\,dx}}{u^2}\,dx \\
&= \int \frac{1}{\sin^2 x}\,dx \\
&= \int \operatorname{cosec}^2 x\,dx \\
&= -\cot x
\end{aligned}$$

Hence the second independent solution of the homogeneous differential equation can be given by $w(x) = u(x)v(x) = -\cos x$. Again, $f(x) = \sin x$. According to the formula given in Eq.(5.32), we have

$$\begin{aligned}
z(x) &= \int \frac{e^{-\int p(x)\,dx}}{u^2}\left(\int \left(ue^{\int p(x)\,dx} f(x)\,dx\right)\right) dx \\
&= \int \frac{1}{u^2}\left(\int uf(x)\,dx\right) dx
\end{aligned}$$

$$
\begin{aligned}
&= \int \frac{1}{\sin^2 x}\left(\int \sin^2 x \, dx\right) dx \\
&= \int \operatorname{cosec}^2 x \left(\int \left(\frac{1-\cos 2x}{2}\right) dx\right) dx \\
&= \int \operatorname{cosec}^2 x \left(\frac{x}{2} - \frac{\sin 2x}{4}\right) dx \\
&= \frac{1}{2}\int x \operatorname{cosec}^2 x \, dx - \frac{1}{2}\int \left(\frac{\cos x}{\sin x}\right) dx \\
&= \frac{1}{2}\int x \operatorname{cosec}^2 x \, dx - \frac{1}{2}\int \cot x dx \\
&= -\frac{x\cot x}{2} + \frac{1}{2}\int \cot x \, dx - \frac{1}{2}\int \cot x dx \\
&= -\frac{x\cot x}{2}
\end{aligned}
$$

Hence the general solution can be given by

$$
\begin{aligned}
y(x) &= c_1 u(x) + c_2 w(x) + u(x)z(x) \\
&= c_1 \sin x - c_2 \cos x - \frac{x\cos x}{2} \\
&= \alpha \sin x + \beta \cos x - \frac{x\cos x}{2}
\end{aligned}
$$

EXAMPLE 5.56 Find the general solution of the differential equation $y'' + 9y = \sin 3x$ taking $u(x) = \cos 3x$ as a solution of the homogeneous equation.

Solution: Clearly, $p(x) = 0$ and $u(x) = \cos 3x$. Let $y = u(x)v(x) = v(x)\cos 3x$ be the particular solution of the given differential equation. Accordingly, one can have

$$
\begin{aligned}
v'' - 6\tan(3x)v' &= z' - 6\tan(3x)z \\
&= 1
\end{aligned}
\tag{5.33}
$$

Clearly, Eq.(5.33) is a first order non-homogeneous linear differential equation. According to the formula given in Eq.(2.8), we have

$$
\begin{aligned}
v(x) &= \int \left(e^{\int 6\tan 3x \, dx} \int e^{-\int 6\tan 3x \, dx} dx + c\right) dx \\
&= \int \left(\sec^2 3x \int \cos^2 3x dx + c\right) dx \\
&= \int \left(\sec^2 3x \left(\frac{x}{2} + \frac{\sin 6x}{12} + c\right)\right) dx \\
&= \int \left(\frac{x\sec^2 3x}{2} + \frac{\tan 3x}{6} + c\sec^2 3x\right) dx \\
&= \frac{x\tan 3x}{6} + c\left(\frac{\tan 3x}{3}\right) + k
\end{aligned}
$$

In other words, one can have

$$\begin{aligned} y(x) &= u(x)v(x) \\ &= \cos 3x\left(\frac{x\tan 3x}{6}+c\left(\frac{\tan 3x}{3}\right)+k\right) \\ &= \frac{x\sin 3x}{6}+c\left(\frac{\sin 3x}{3}\right)+k\cos 3x \\ &= \frac{x\sin 3x}{6}+\beta\sin 3x+k\cos 3x \end{aligned}$$

as the general solution to the given differential equation as there are two arbitrary constants.

EXAMPLE 5.57 Find the general solution of the differential equation $y''-3y'+2y=e^x$ taking $u(x)=e^x$ as a solution of the homogeneous equation.

Solution: Clearly, $p(x)=0$ and $u(x)=e^x$. Let $y=u(x)v(x)=v(x)\,e^x$ be the particular solution of the given differential equation. Accordingly, one can have

$$\begin{aligned} v''-v' &= z'-z \\ &= 1 \end{aligned} \tag{5.34}$$

Clearly, Eq.(5.34) is a first order non-homogeneous linear differential equation. According to the formula given in Eq.(2.8), we have

$$\begin{aligned} v(x) &= \int\left(e^{\int \mathrm{d}x}\int e^{-\mathrm{d}x}\mathrm{d}x+c\right)\mathrm{d}x \\ &= \int\left(e^x\int e^{-x}\mathrm{d}x+c\right)\mathrm{d}x \\ &= \int\left(e^x\left(-e^{-x}+c\right)\right)\mathrm{d}x \\ &= \int\left(-1+ce^x\right)\mathrm{d}x \\ &= -x+c\,e^x+k \end{aligned}$$

In other words, one can have

$$\begin{aligned} y(x) &= u(x)v(x) \\ &= \left(-x+c\,e^x+k\right)e^x \\ &= -x\,e^x+ce^{2x}+k\,e^x \end{aligned}$$

as the general solution to the given differential equation as there are two arbitrary constants.

EXAMPLE 5.58 Find the general solution of the differential equation $x^2y''-2xy'+2y=\ln(x)$ taking $u(x)=x$ as a solution of the homogeneous equation.

Solution: Clearly, $p(x)=-2x$ and $u(x)=x$. Let $y=u(x)v(x)=v(x)\,x$ be the particular solution of the given differential equation. Accordingly, one can have

$$x^3v''=\ln(x) \tag{5.35}$$

Clearly, Eq.(5.35) is a first order non-homogeneous linear differential equation. According to the formula given in Eq.(2.8), we have

$$\begin{aligned} v(x) &= \int \left(\int \left(\frac{\ln x}{x^2} \right) dx + c \right) dx \\ &= \int \left(-\frac{\ln x}{x^2} - \frac{1}{4x^2} + c \right) dx + k \\ &= \frac{\ln x}{2x} + \frac{3}{4x} + cx + k \end{aligned}$$

In other words, one can have

$$\begin{aligned} y(x) &= u(x)v(x) \\ &= \left(\frac{\ln x}{2x} + \frac{3}{4x} + cx + k \right) x \\ &= \frac{\ln x}{2} + \frac{3}{4} + cx^2 + kx \end{aligned}$$

as the general solution to the given differential equation as there are two arbitrary constants.

Exercises 5.4.1

1. Find the general solution of the differential equation $xy'' - (2x + 1)y' + (x + 1)y = 0$ taking $u(x) = e^x$ as a solution.
2. Find the general solution of the differential equation $(3-x)y'' - (9-4x)y' + (6-3x)y = 0$ taking $u(x) = e^x$ as a solution.
3. Find the general solution of the differential equation $(1 - x^2)y'' - 2xy' + 2y = 0$ taking $u(x) = x$ as a solution.
4. Find the general solution of the differential equation $x^2y'' + 2x(1 - x)y' + x(x - 2)y = 0$ taking $u(x) = e^x$ as a solution.
5. Find the general solution of the differential equation $xy'' - (2x - 1)y' + (x - 1)y = 0$ taking $u(x) = e^x$ as a solution.
6. Find the general solution of the differential equation $(x \sin x + \cos x)y'' - x\cos(x)y' + \cos(x)y = 0$ taking $u(x) = x$ as a solution.
7. Find the general solution of the differential equation $(x^2 \cos x - 2x \sin x)y'' + (x^2 + 2)\sin(x)y' - 2(x \sin x + \cos x)y = 0$ taking $u(x) = x^2$ as a solution.
8. Find the general solution of the differential equation $\tan^2(x)y'' - 2\tan(x)y' + (2 + \tan^2(x))y = 0$ taking $u(x) = \sin x$ as a solution.
9. Find the general solution of the differential equation $(x-1)y'' - xy' + y = (x-1)^2$ taking $u(x) = x$ as a solution of the homogeneous equation.
10. Find the general solution of the differential equation $xy'' + (1 - x)y' + y = e^x$ taking $u(x) = 1 - x$ as a solution of the homogeneous equation.
11. Find the general solution of the differential equation $xy'' + (1 + x)y' - y = e^x$ taking $u(x) = 1 + x$ as a solution of the homogeneous equation.

12. Find the general solution of the differential equation $xy'' + (1-x)y' - y = e^x$ taking $u(x) = e^x$ as a solution of the homogeneous equation.
13. Find the general solution of the differential equation $xy'' - y' + (1-x)y = xe^{-x}$ taking $u(x) = e^{-x}$ as a solution of the homogeneous equation.
14. Find the general solution of the differential equation $(x+2)y'' - (2x+5)y' + 2y = (x+1)e^x$ taking $u(x) = 2x+5$ as a solution of the homogeneous equation.
15. Find the general solution of the differential equation $x(1+x^2)y'' - 2x^2y' + 2xy = 1 - x^2$ taking $u(x) = x$ as a solution of the homogeneous equation.
16. Find the general solution of the differential equation $y'' + y = \sec x$ taking $u(x) = \cos x$ as a solution of the homogeneous equation.
17. Show that $y(x) = x^m$ is a solution of the Cauchy–Euler equation $x^2y'' + axy' + by = 0$ if and only if $m^2 + (a-1)m + b = 0$ where a and b are constants.

    HINT: Let $y(x) = x^m$. It is easy to verify that $y'(x) = mx^{m-1}$ and $y''(x) = m(m-1)x^{m-2}$. Hence one can have

$$\begin{aligned} 0 &= x^2y'' + axy' + by \\ &= x^2(m(m-1)x^{m-2}) + axmx^{m-1} + bx^m \\ &= x^m(m(m-1) + am + b) \end{aligned}$$

    In other words, one should have

$$m^2 + (a-1)m + b = 0 \tag{5.36}$$

    Hence the required solution may take one of the form

    (a) $y(x) = c_1x^\alpha + c_2x^\beta$ where α and β are two distinct real roots of the characteristic equation corresponding to the differential equation given in Eq. (5.36)
    (b) $y(x) = c_1x^\alpha + c_2\ln(x)\,x^\alpha$ where α is the only real root of the characteristic equation corresponding to the differential equation given in Eq. (5.36)
    (c) $y(x) = x^\alpha\left(c_1\cos(\beta\ln(x)) + c_2\sin(\beta\ln(x))\right)$ where $\alpha \pm i\beta$ are complex roots of the characteristic equation corresponding to the differential equation given in Eq. (5.36)

5.4.2 Removal of First Derivative

In general, it is not possible to guess one of the homogeneous solution on the differential equation $y'' + p(x)y' + q(x)y = f(x)$. Let $y(x) = u(x)v(x)$ be the solution of the differential equation $y'' + p(x)y' + q(x)y = f(x)$, where both $u(x)$ and $v(x)$ are unknown. Again, we have

$$\begin{aligned} y' &= uv' + u'v \\ y'' &= u''v + 2u'v' + uv'' \end{aligned}$$

Substituting $y = uv$, y' and y'' in the given differential equation, we have

$$\begin{aligned} f(x) &= y'' + p(x)y' + q(x)y \\ &= u''v + 2u'v' + uv'' + p(x)(uv' + u'v) + q(x)uv \\ &= uv'' + (2u' + p(x)u)v' + (u'' + p(x)u' + q(x)u)v \end{aligned} \tag{5.37}$$

Let $2u' + p(x)u = 0$. Hence Eq.(5.37) reduces to

$$uv'' + (u'' + p(x)u' + q(x)u)v = f(x) \tag{5.38}$$

Again, the solution of the first order differential equation $2u' + p(x)u = 0$ is

$$u(x) = e^{-\frac{1}{2}\int p(x)\,dx} \tag{5.39}$$

by the formula given in Eq.(2.2). Accordingly, Eq.(5.38) reduces to

$$\begin{aligned}\frac{f(x)}{u} &= v'' + \frac{1}{u}(u'' + p(x)u' + q(x)u)v \\ &= v'' + \frac{1}{u}\left(-\frac{1}{2}\,p'(x)u + \frac{1}{4}\,p^2(x)u - \frac{1}{2}\,p^2(x)u + q(x)u\right)v \\ &= v'' + \frac{1}{u} - \left(\frac{1}{2}p'(x) - \frac{1}{4}p^2(x) + q(x)\right)uv \\ &= v'' + \left(-\frac{1}{2}p'(x) - \frac{1}{4}p^2(x) + q(x)\right)v \end{aligned} \tag{5.40}$$

Again, the second order differential equation given in Eq.(5.40) is sometimes called the *normal form* of the differential equation $y'' + p(x)y' + q(x)y = f(x)$. This method is preferably applicable to a set of problems for which the normal form becomes a constant coefficient equation.

EXAMPLE 5.59 Find the general solution of the differential equation $y'' + 2xy' + (x^2+5)y = xe^{-\frac{1}{2}x^2}$ by reducing to normal form.

Solution: Clearly, $p(x) = 2x$, $q(x) = x^2 + 5$ and $f(x) = xe^{-\frac{1}{2}x^2}$. According to Eq.(5.39), we have

$$\begin{aligned}u(x) &= e^{-\frac{1}{2}\int p(x)\,dx} \\ &= e^{-\int x\,dx} \\ &= e^{-\frac{x^2}{2}}\end{aligned}$$

Again, Eq.(5.40) becomes

$$\begin{aligned}x = \frac{f(x)}{u} &\\ &= v'' + \left(-\frac{1}{2}p'(x) - \frac{1}{4}p^2(x) + q(x)\right)v \\ &= v'' + \left(-1 - x^2 + x^2 + 5\right)v \\ &= v'' + 4v\end{aligned} \tag{5.41}$$

Clearly, Eq.(5.41) is a second order constant coefficient nonhomogeneous differential equation. Again, the general solution of Eq.(5.41) is

$$v(x) = c_1 \cos 2x + c_2 \sin 2x + \frac{x}{4}$$

Hence the general solution of the given differential is

$$\begin{aligned} y(x) &= u(x)v(x) \\ &= e^{-\frac{x^2}{2}} \left(c_1 \cos 2x + c_2 \sin 2x + \frac{x}{4} \right) \end{aligned}$$

EXAMPLE 5.60 Find the general solution of the differential equation $x^2y'' - 2xy' + (a^2x^2 + 2)y = 0$.

Solution: Clearly, $p(x) = -\frac{2}{x}$, $q(x) = a^2 + \frac{2}{x^2}$ and $f(x) = 0$. According to Eq.(5.39), we have

$$\begin{aligned} u(x) &= e^{-\frac{1}{2}\int p(x)\,\mathrm{d}x} \\ &= e^{\int \frac{1}{x}\,\mathrm{d}x} \\ &= e^{\ln x} \\ &= x \end{aligned}$$

Again, Eq.(5.40) becomes

$$\begin{aligned} 0 &= v'' + \left(-\left(\frac{1}{2}\right) p'(x) - \left(\frac{1}{4}\right) p^2(x) + q(x) \right) v \\ &= v'' + \left(-\left(\frac{1}{2}\right) \left(\frac{2}{x^2}\right) - \left(\frac{1}{4}\right) \left(\frac{4}{x^2}\right) + a^2 + \frac{2}{x^2} \right) v \\ &= v'' + a^2 v \end{aligned} \tag{5.42}$$

Clearly, Eq.(5.42) is a second order constant coefficient homogeneous differential equation. Again, the general solution of Eq.(5.42) is

$$v(x) = c_1 \cos ax + c_2 \sin ax$$

Hence the general solution of the given differential is

$$\begin{aligned} y(x) &= u(x)v(x) \\ &= x\,(c_1 \cos ax + c_2 \sin ax) \end{aligned}$$

EXAMPLE 5.61 Find the general solution of the differential equation $x^2y'' - 2xy' + (2 - a^2x^2)y = 0$.

Solution: Clearly, $p(x) = -\frac{2}{x}$, $q(x) = \frac{2}{x^2} - a^2$ and $f(x) = 0$. According to Eq.(5.39), we have

$$\begin{aligned} u(x) &= e^{-\frac{1}{2}\int p(x)\,\mathrm{d}x} \\ &= e^{\int \frac{1}{x}\,\mathrm{d}x} \\ &= e^{\ln x} \\ &= x \end{aligned}$$

Again, Eq.(5.40) becomes

$$\begin{aligned} 0 &= v'' + \left(-\left(\frac{1}{2}\right)p'(x) - \left(\frac{1}{4}\right)p^2(x) + q(x)\right)v \\ &= v'' + \left(-\left(\frac{1}{2}\right)\left(\frac{2}{x^2}\right) - \left(\frac{1}{4}\right)\left(\frac{4}{x^2}\right) - a^2 + \frac{2}{x^2}\right)v \\ &= v'' - a^2 v \end{aligned} \tag{5.43}$$

Clearly, Eq.(5.43) is a second order constant coefficient homogeneous differential equation. Again, the general solution of Eq.(5.43) is

$$v(x) = c_1\, e^{ax} + c_2\, e^{-ax}$$

Hence the general solution of the given differential is

$$\begin{aligned} y(x) &= u(x)v(x) \\ &= x\left(c_1\, e^{ax} + c_2\, e^{-ax}\right) \end{aligned}$$

Exercises 5.4.2

1. Find the general solution of the differential equation $y'' - 4xy' + (4x^2 - 1)y = -3\sin(2x)e^{x^2}$ by reducing into normal form.
2. Find the general solution of the differential equation $xy'' + 2y' + xy = \sin 2x$ by reducing into normal form.
3. Find the general solution of the differential equation $y'' + 2xy' + (x^2 + 1)y = x(x^2 + 3)$ by reducing into normal form.
4. Find the general solution of the differential equation $x^2y'' - 2x(3x - 2)y' + 3x(3x - 4)y = 2^{3x}$ by reducing into normal form.
5. Find the general solution of the differential equation $y'' + 2\tan(x)\, y' + (1 + \tan^2 x)y = \sec x \tan x$ by reducing into normal form.
6. Find the general solution of the differential equation $xy'' + 2y' + \alpha^2 y = 0$ by reducing into normal form.
7. Find the general solution of the differential equation $y'' + 4xy' + 4x^2y = 0$ by reducing into normal form.
8. Find the general solution of the differential equation $4x^2y'' + 4xy' + (4x^2 - 1)y = 0$ by reducing into normal form.
9. Find the general solution of the differential equation $4x^2y'' + 4xy' - (4x^2 + 1)y = 0$ by reducing into normal form.
10. Find the solution of the differential equation $y'' - 2\cot(x)y' + (1 + 2\cot^2 x)y = 0$ by reducing into normal form.

5.4.3 Change of Independent Variable

It is sometimes useful to change the independent variable x of some variable coefficient differential equation $y'' + p(x)y' + q(x)y = f(x)$, so as to transfer the given differential equation into a constant coefficient differential equation. Let $u = g(x)$ be the required change of the independent variable x into the new variable u. Again, we have

$$y' = g'(x)\left(\frac{\mathrm{d}y}{\mathrm{d}u}\right)$$

and

$$\begin{aligned}
y'' &= \frac{\mathrm{d}}{\mathrm{d}x}\left(\frac{\mathrm{d}y}{\mathrm{d}x}\right) \\
&= \frac{\mathrm{d}}{\mathrm{d}x}\left(g'(x)\left(\frac{\mathrm{d}y}{\mathrm{d}u}\right)\right) \\
&= g''(x)\left(\frac{\mathrm{d}y}{\mathrm{d}u}\right) + g'(x)\left(\frac{\mathrm{d}}{\mathrm{d}x}\left(\frac{\mathrm{d}y}{\mathrm{d}u}\right)\right) \\
&= g''(x)\left(\frac{\mathrm{d}y}{\mathrm{d}u}\right) + g'(x)\left(\frac{\mathrm{d}}{\mathrm{d}u}\left(\frac{\mathrm{d}y}{\mathrm{d}u}\right)\right)\left(\frac{\mathrm{d}u}{\mathrm{d}x}\right) \\
&= g''(x)\left(\frac{\mathrm{d}y}{\mathrm{d}u}\right) + (g'(x))^2\left(\frac{\mathrm{d}^2y}{\mathrm{d}u^2}\right)
\end{aligned}$$

Substituting y' and y'' by the appropriate expression in the given differential equation $y'' + p(x)y' + q(x)y = f(x)$, we have

$$\begin{aligned}
f(x) &= y'' + p(x)y' + q(x)y \\
&= g''(x)\left(\frac{\mathrm{d}y}{\mathrm{d}u}\right) + (g'(x))^2\left(\frac{\mathrm{d}^2y}{\mathrm{d}u^2}\right) + p(x)g'(x)\left(\frac{\mathrm{d}y}{\mathrm{d}u}\right) + q(x)y \\
&= (g'(x))^2\left(\frac{\mathrm{d}^2y}{\mathrm{d}u^2}\right) + g''(x)\left(\frac{\mathrm{d}y}{\mathrm{d}u}\right) + p(x)g'(x)\left(\frac{\mathrm{d}y}{\mathrm{d}u}\right) + q(x)y \\
&= (g'(x))^2\left(\frac{\mathrm{d}^2y}{\mathrm{d}u^2}\right) + (g''(x) + p(x)g'(x))\left(\frac{\mathrm{d}y}{\mathrm{d}u}\right) + q(x)y \qquad (5.44)
\end{aligned}$$

Again, all the expressions $g''(x)$, $g'(x)$, $p(x)$ and $q(x)$ has to be transferred into u by the transform $x = g^{-1}(u)$. One has to first find the transform $u = g(x)$ which can transfer the given differential equation into a constant coefficient differential equations which is given in Eq.(5.44). The type of transform which can reduce Eq.(5.44) to an equation with constant coefficient is given in Exercise 1 of this section.

EXAMPLE 5.62 Find the solution of the differential equation $y'' - \cot(x)\, y' + \sin^2(x)\, y = 0$.

Solution: Clearly, $p(x) = -\cot x$ and $q(x) = \sin^2 x$. It is easy to verify that $\frac{q'(x)+2p(x)q(x)}{q^{\frac{3}{2}}(x)}$ is a constant. According to Exercise 1 of this section, $u = g(x) = \int q^{\frac{1}{2}}(x)\, \mathrm{d}x = -\cos x$ is the required transformation of the independent variable. The corresponding transferred differential equation is given by

$$\begin{aligned} 0 &= (g'(x))^2 \left(\frac{\mathrm{d}^2 y}{\mathrm{d}u^2}\right) + (g''(x) + p(x)g'(x)) \left(\frac{\mathrm{d}y}{\mathrm{d}u}\right) + q(x)y \\ &= \sin^2 x \left(\frac{\mathrm{d}^2 y}{\mathrm{d}u^2}\right) + (\cos x - \cot x \sin x) \left(\frac{\mathrm{d}y}{\mathrm{d}u}\right) + y \sin^2 x \\ &= \frac{\mathrm{d}^2 y}{\mathrm{d}u^2} + y \end{aligned} \tag{5.45}$$

Again, the solution of the differential equation given in Eq. (5.45) is given by

$$y(u) = c_1 \cos u + c_2 \sin u$$

In other words, the solution of the given equation is

$$\begin{aligned} y(x) &= c_1 \cos(-\cos x) + c_2 \sin(-\cos x) \\ &= \alpha \cos(\cos x) + \beta \sin(\cos x) \end{aligned}$$

EXAMPLE 5.63 Find the general solution of Cauchy–Euler differential equation $x^2 y'' + axy' + by = 0$.

Solution: Clearly, $p(x) = \frac{a}{x}$ and $q(x) = \frac{b}{x^2}$. Again, we have

$$\frac{q'(x) + 2p(x)q(x)}{q^{\frac{3}{2}}(x)} = \frac{2a - 2}{b^{\frac{1}{2}}}$$

as a constant. Hence according to Exercise 1 of this section

$$\begin{aligned} u &= g(x) \\ &= \int q^{\frac{1}{2}}(x)\, \mathrm{d}x \\ &= \int \frac{\sqrt{b}}{x}\, \mathrm{d}x \\ &= \sqrt{b} \ln x \end{aligned}$$

Accordingly, Eq. (5.44) reduces to

$$\begin{aligned} 0 &= (g'(x))^2 \left(\frac{\mathrm{d}^2 y}{\mathrm{d}u^2}\right) + (g''(x) + p(x)g'(x)) \left(\frac{\mathrm{d}y}{\mathrm{d}u}\right) + q(x)y \\ &= \frac{\sqrt{b}}{x^2} \left(\sqrt{b} \left(\frac{\mathrm{d}^2 y}{\mathrm{d}u^2}\right) + (a-1) \left(\frac{\mathrm{d}y}{\mathrm{d}u}\right) + \sqrt{b}\, y \right) \end{aligned}$$

The corresponding constant coefficient differential equation is

$$\sqrt{b}\left(\frac{\mathrm{d}^2y}{\mathrm{d}u^2}\right) + (a-1)\left(\frac{\mathrm{d}y}{\mathrm{d}u}\right) + \sqrt{b}\,y = 0 \tag{5.46}$$

Again, Eq.(5.46) is a second order constant coefficient homogeneous differential equation. Hence the required solution may take one of the following forms:

1. $y(x) = c_1 x^{\alpha\sqrt{b}} + c_2 x^{\beta\sqrt{b}}$, where α and β are two distinct real roots of the characteristic equation corresponding to the differential equation given in Eq.(5.46).
2. $y(x) = c_1 x^{\alpha\sqrt{b}} + c_2\, x^{\alpha\sqrt{b}} \ln x$, where α is the only real root of the characteristic equation corresponding to the differential equation given in Eq.(5.46).
3. $y(x) = x^{\alpha\sqrt{b}}\left(c_1 \cos\beta\sqrt{b}\ln x + c_2 \sin\beta\sqrt{b}\ln x\right)$, where $\alpha \pm i\beta$ are complex roots of the characteristic equation corresponding to the differential equation given in Eq.(5.46).

EXAMPLE 5.64 Find the general solution of the differential equation $x^2y'' - 5xy' + 8y = 0$.

Solution: Clearly, $a = -5$ and $b = 8$. Hence the required transform is $u = 2\sqrt{2}\ \ln x$, and Eq.(5.46) of Example 5.63 reduces to

$$2\sqrt{2}\left(\frac{\mathrm{d}^2y}{\mathrm{d}u^2}\right) - 6\left(\frac{\mathrm{d}y}{\mathrm{d}u}\right) + 2\sqrt{2}y = 0 \tag{5.47}$$

Again, $\alpha = \frac{2}{\sqrt{2}}$ and $\beta = \frac{1}{\sqrt{2}}$ are distinct real roots of the characteristic equation associated with Eq.(5.47). Hence the required solution is given by

$$y(x) = c_1x^4 + c_2x^2$$

EXAMPLE 5.65 Find the general solution of the differential equation $x^2y'' - 3xy' + 4y = 0$.

Solution: Clearly, $a = -3$ and $b = 4$. Hence the required transform is $u = 2\ \ln x$, and Eq.(5.46) of Example 5.63 reduces to

$$2\left(\frac{\mathrm{d}^2y}{\mathrm{d}u^2}\right) - 4\left(\frac{\mathrm{d}y}{\mathrm{d}u}\right) + 2y = 0 \tag{5.48}$$

Again, $\alpha = 1$ is the only real root of the characteristic equation associated with Eq.(5.48). Hence the required solution is given by

$$y(x) = c_1x^2 + c_2\,x^2\ln x$$

EXAMPLE 5.66 Find the general solution of the differential equation $x^2y'' + 2xy' + y = 0$.

Solution: Clearly, $a = 2$ and $b = 1$. Hence the required transform is $u = \ln x$, and Eq.(5.46) of Example 5.63 reduces to

$$\frac{\mathrm{d}^2y}{\mathrm{d}u^2} + \frac{\mathrm{d}y}{\mathrm{d}u} + y = 0 \tag{5.49}$$

Again, $-\frac{1}{2} \pm i\left(\frac{3^{\frac{1}{2}}}{2}\right)$ are the complex roots of the characteristic equation associated with Eq.(5.49). Hence the required solution is given by

$$y(x) = x^{-\frac{1}{2}}\left(c_1 \sin\frac{3^{\frac{1}{2}}}{2} \ln x + c_2 \cos\frac{3^{\frac{1}{2}}}{2} \ln x\right)$$

EXAMPLE 5.67 Find the general solution of the differential equation $x^2y'' - xy' + y = \ln x$.

Solution: Clearly, $a = -1$ and $b = 1$. Hence the required transform is $u = \ln x$, and Eq. (5.46) of Example 5.63 reduces to

$$\frac{\mathrm{d}^2y}{\mathrm{d}u^2} - 2\left(\frac{\mathrm{d}y}{\mathrm{d}u}\right) + y = 0 \tag{5.50}$$

Again, $\alpha = 1$ is the only root of the characteristic equation associated with Eq.(5.50). Hence $y_1(x) = x$ and $y_2(x) = x \ln x$ are two independent solutions of the homogeneous equation. If $h(x) = y_1(x)v_1(x) + y_2(x)v_2(x)$ is the particular integral of the given equation, then we have

$$\begin{aligned} v_1(x) &= -\int \frac{f(x)y_2(x)}{x^2 w(y_1, y_2)}\,\mathrm{d}x \\ &= -\int \frac{x\ln^2 x}{x^3}\,\mathrm{d}x \\ &= -\int \frac{\ln^2 x}{x^2}\,\mathrm{d}x \\ &= -\left(-\frac{\ln^2 x}{x} - \frac{2\ln x}{x} - \frac{2}{x}\right) \\ &= \frac{\ln^2 x}{x} + \frac{2\ln x}{x} + \frac{2}{x} \end{aligned}$$

and

$$\begin{aligned} v_2(x) &= \int \frac{f(x)y_1(x)}{x^2 w(y_1, y_2)}\,\mathrm{d}x \\ &= \int \frac{x\ln x}{x^3}\,\mathrm{d}x \\ &= \int \frac{\ln x}{x^2}\,\mathrm{d}x \\ &= -\frac{\ln x}{x} - \frac{1}{x} \end{aligned}$$

where $w(y_1, y_2) = x$ and $r(x) = x^2$ according to Eq.(5.13). Hence the required solution is given by

$$\begin{aligned} y(x) &= c_1 y_1(x) + c_2 y_2(x) + h(x) \\ &= c_1\, x + c_2\, x \ln x + x\left(\frac{\ln^2 x}{x} + \frac{2\ln x}{x} + \frac{2}{x}\right) + x\ln x\left(-\frac{\ln x}{x} - \frac{1}{x}\right) \\ &= c_1\, x + c_2\, x\ln x + 2 + \ln x \end{aligned}$$

EXAMPLE 5.68 Find the general solution of the differential equation $x^2y'' - 3xy' + 4y = \ln x$.

Solution: Clearly, $a = -3$ and $b = 4$. Hence the required transform is $u = \sqrt{2}\ln x$, and Eq.(5.46) of Example 5.63 reduces to

$$2\left(\frac{\mathrm{d}^2 y}{\mathrm{d}u^2}\right) - 4\left(\frac{\mathrm{d}y}{\mathrm{d}u}\right) + 2y = 0 \tag{5.51}$$

Again, $\alpha = 1$ is the only root of the characteristic equation associated with Eq.(5.51). Hence $y_1(x) = x^2$ and $y_2(x) = 2\,x^2 \ln x$ are two independent solutions of the homogeneous equation. If $h(x) = y_1(x)v_1(x) + y_2(x)v_2(x)$ is the particular integral of the given equation, then we have

$$\begin{aligned} v_1(x) &= -\int \frac{f(x)y_2(x)}{r(x)w(y_1,\, y_2)}\,\mathrm{d}x \\ &= -\int \frac{2\,x^2 \ln^2 x}{2x^5}\,\mathrm{d}x \\ &= -\int \frac{\ln^2 x}{x^3}\,\mathrm{d}x \\ &= \frac{1}{2}\frac{\ln^2 x}{x^2} + \frac{1}{2}\frac{\ln x}{x^2} + \frac{1}{4}\frac{1}{x^2} \end{aligned}$$

and

$$\begin{aligned} v_2(x) &= \int \frac{f(x)y_1(x)}{r(x)w(y_1,\, y_2)}\mathrm{d}x \\ &= \int \frac{x^2 \ln x}{2x^5}\mathrm{d}x \\ &= \frac{1}{2}\int \frac{\ln x}{x^3}\mathrm{d}x \\ &= -\frac{1}{4}\frac{\ln x}{x^2} - \frac{1}{8}\frac{1}{x^2} \end{aligned}$$

where $w(y_1, y_2) = 2x^3$ and $r(x) = x^2$ according to Eq.(5.13). Hence the required solution is given by

$$\begin{aligned} y(x) &= c_1 y_1(x) + c_2 y_2(x) + h(x) \\ &= c_1\, x^2 + c_2\, x^2 \ln x + x^2 \left(\left(\frac{1}{2}\right)\frac{\ln^2 x}{x^2} + \left(\frac{1}{2}\right)\frac{\ln x}{x^2} + \left(\frac{1}{4}\right)\frac{1}{x^2}\right) \\ &+ 2x^2 \ln x \left(-\left(\frac{1}{4}\right)\frac{\ln x}{x^2} - \left(\frac{1}{8}\right)\frac{1}{x^2}\right) \\ &= c_1\, x^2 + c_2\, x^2 \ln x + \frac{1}{4}\ln x + \frac{1}{4} \end{aligned}$$

EXAMPLE 5.69 Find the general solution of the differential equation $x^2y'' - 2xy' + 2y = x^3 \sin x$.

Solution: Clearly, $a = -2$ and $b = 2$. Hence the required transform is $u = \sqrt{2}\ln x$, and Eq.(5.46) of Example 5.63 reduces to

$$\sqrt{2}\left(\frac{\mathrm{d}^2 y}{\mathrm{d}u^2}\right) - 3\left(\frac{\mathrm{d}y}{\mathrm{d}u}\right) + \sqrt{2}y = 0 \tag{5.52}$$

Again, $\alpha = \frac{2}{\sqrt{2}}$ and $\beta = \frac{1}{\sqrt{2}}$ are the roots of the characteristic equation associated with Eq.(5.52). Hence $y_1(x) = x$ and $y_2(x) = x^2$ are two independent solutions of the homogeneous equation. If $h(x) = y_1(x)v_1(x) + y_2(x)v_2(x)$ is the particular integral of the given equation, then we have

$$\begin{aligned} v_1(x) &= -\int \frac{f(x)y_2(x)}{r(x)w(y_1,\, y_2)}\,\mathrm{d}x \\ &= -\int \frac{x^5 \sin x}{x^4}\,\mathrm{d}x \\ &= -\int x \sin x \mathrm{d}x \\ &= x\cos x - \sin x \end{aligned}$$

and

$$\begin{aligned} v_2(x) &= \int \frac{f(x)y_1(x)}{r(x)w(y_1,\, y_2)}\,\mathrm{d}x \\ &= \int \frac{x^4 \sin x}{x^4}\,\mathrm{d}x \\ &= \int \sin x\,\mathrm{d}x \\ &= -\cos x \end{aligned}$$

where $w(y_1, y_2) = x^2$ and $r(x) = x^2$ according to Eq.(5.13). Hence the required solution is given by

$$\begin{aligned} y(x) &= c_1 y_1(x) + c_2 y_2(x) + h(x) \\ &= c_1 x + c_2 x^2 + x\,(x \cos x - \sin x) + x^2\,(-\cos x) \\ &= c_1 x + c_2 x^2 - x \sin x \end{aligned}$$

Exercises 5.4.3

1. Show that $u = g(x) = \int q^{\frac{1}{2}}(x)\,dx$ changes the differential equation $y''+p(x)y'+q(x)y = 0$ into an equation with constant coefficient if and only if

$$\frac{q'(x) + 2p(x)q(x)}{q^{\frac{3}{2}}(x)}$$

is constant.

HINT: According to Eq.(5.44), the transferred equation will have constant coefficient if and only if

$$\frac{g''(x) + p(x)g'(x)}{(g'(x))^2} \tag{5.53}$$

and

$$\frac{q(x)}{(g'(x))^2} \tag{5.54}$$

are constants. According to Eq.(5.54), we have $g(x) = c \int q^{\frac{1}{2}}(x)\,dx$ where c is any constant. Again, Eq.(5.53) gives

$$\frac{q'(x) + 2p(x)q(x)}{q^{\frac{3}{2}}(x)}$$

as constant.

2. Find the transform $u = g(x)$, such that the differential equation $y''+\tan(x)y'+\cos^2(x)y = 0$ will be transferred into a constant coefficient differential equation.
3. Find the transform $u = g(x)$ such that the differential equation $xy'' - 3y' + 16x^7 y = 0$ will be transferred into a constant coefficient differential equation.
4. Find the transform $u = g(x)$ such that the differential equation $x^4y''+x^2(2x-3)y'+2y = 0$ will be transferred into a constant coefficient differential equation.
5. Find the transform $u = g(x)$ such that the differential equation $2xy''+(5x^2-2)y'+2x^3y = 0$ will be transferred into a constant coefficient differential equation.
6. Find the transform $u = g(x)$ such that the differential equation $xy''+(8x^2-1)y'+20x^3y = 0$ will be transferred into a constant coefficient differential equation.

7. Show that the variable $u = \log x$ transforms the Cauchy–Euler equation $x^2y'' + axy' + by = 0$ into the constant coefficient differential equation $y'' + (a-1)y' + by = 0$.

   HINTS: It is easy to verify that

$$\begin{aligned}\frac{\mathrm{d}y}{\mathrm{d}x} &= \frac{\mathrm{d}y}{\mathrm{d}u}\frac{\mathrm{d}u}{\mathrm{d}x}\\ &= \frac{1}{x}\frac{\mathrm{d}y}{\mathrm{d}u}\end{aligned}$$

   and

$$\begin{aligned}\frac{\mathrm{d}^2y}{\mathrm{d}x^2} &= \frac{\mathrm{d}}{\mathrm{d}x}\left(\frac{\mathrm{d}y}{\mathrm{d}x}\right)\\ &= \frac{\mathrm{d}}{\mathrm{d}x}\left(\frac{1}{x}\frac{\mathrm{d}y}{\mathrm{d}u}\right)\\ &= \frac{1}{x}\frac{\mathrm{d}}{\mathrm{d}x}\left(\frac{\mathrm{d}y}{\mathrm{d}u}\right) + \frac{\mathrm{d}}{\mathrm{d}x}\left(\frac{1}{x}\right)\left(\frac{\mathrm{d}y}{\mathrm{d}u}\right)\\ &= \frac{1}{x}\frac{\mathrm{d}}{\mathrm{d}u}\left(\frac{\mathrm{d}y}{\mathrm{d}u}\right)\left(\frac{\mathrm{d}u}{\mathrm{d}x}\right) - \left(\frac{1}{x^2}\right)\left(\frac{\mathrm{d}y}{\mathrm{d}u}\right)\\ &= \frac{1}{x^2}\frac{\mathrm{d}^2y}{\mathrm{d}u^2} - \frac{1}{x^2}\frac{\mathrm{d}y}{\mathrm{d}u}\end{aligned}$$

   Again, one can have

$$\begin{aligned}0 &= x^2y'' + axy' + by\\ &= x^2\left(\left(\frac{1}{x^2}\right)\frac{\mathrm{d}^2y}{\mathrm{d}u^2} - \left(\frac{1}{x^2}\right)\frac{\mathrm{d}y}{\mathrm{d}u}\right) + ax\left(\left(\frac{1}{x}\right)\frac{\mathrm{d}y}{\mathrm{d}u}\right) + by\\ &= \frac{\mathrm{d}^2y}{\mathrm{d}u^2} + (a-1)\frac{\mathrm{d}y}{\mathrm{d}u} + by \qquad (5.55)\end{aligned}$$

   which is a differential equation with constant coefficient. Hence the required solution may take one of the following forms:

   (a) $y(x) = c_1x^\alpha + c_2x^\beta$ where α and β are two distinct real roots of the characteristic equation corresponding to the differential equation given in Eq.(5.55)

   (b) $y(x) = c_1x^\alpha + c_2\ln(x)\,x^\alpha$ where α is the only real root of the characteristic equation corresponding to the differential equation given in Eq.(5.55)

   (c) $y(x) = x^\alpha\left(c_1\cos(\beta\ln(x)) + c_2\sin(\beta\ln(x))\right)$, where $\alpha \pm i\beta$ are complex roots of the characteristic equation corresponding to the differential equation given in Eq.(5.55)

8. Find the solution of the differential equation $x^2y'' + 9xy' + 25y = \ln x$.
9. Find the solution of the differential equation $x^2y'' - xy' + 2y = x\ln x$.
10. Find the solution of the differential equation $x^2y'' + 3xy' + y = x\ln x$.
11. Find the solution of the differential equation $x^2y'' - 2xy' - 4y = x^2 + 2\ln x$.

12. Find the solution of the differential equation $x^2y'' + xy' - y = x^2e^{-x}$.
13. Find the solution of the differential equation $x^2y'' + 4xy' + 2y = e^x$.
14. Find the solution of the differential equation $x^2y'' + 4xy' + 2y = \sin x$.
15. Find the solution of the differential equation $x^2y'' + xy' - 4y = x^2$.
16. Find the solution of the differential equation $x^2y'' + xy' - 4y = 0$.
17. Find the solution of the differential equation $4x^2y'' + 24xy' + 25y = 0$.
18. Find the solution of the differential equation $x^2y'' - xy' + y = 0$.
19. Find the solution of the differential equation $x^2y'' + xy' - 4y = x^2 \ln x$.
20. Find the solution of the differential equation $x^2y'' - 2xy' + 2y = x$.
21. Find the solution of the differential equation $x^2y'' - 3xy' + 4y = 2x^2$.
22. Find the solution of the differential equation $x^2y'' - 2xy' + 2y = 2x^3$.
23. Find the solution of the differential equation $(1+2x)^2y''+6(1+2x)y'+16y = 8(1+2x)^2$.
24. Find the solution of the differential equation $(1+x)^2y''+(1+x)y'+y = 4\cos(\ln(1+x))$.

5.5 Initial Value Problem

Every second order differential equation general solution contains two arbitrary constants which appear as integration constants. One can determine them uniquely when two conditions which are satisfied by the solution are given at a fixed point $x = x_0$. In particular, the conditions are:

1. $y(x_0) = y_0$
2. $y'(x_0) = \overline{y}_0$

EXAMPLE 5.70 Solve the differential equation $y'' - 5y' + 6y = 2e^x$ with $y(0) = 1$ and $y'(0) = 1$.

Solution: Clearly, two independent homogeneous solutions are $y_1(x) = e^{2x}$ and $y_2(x) = e^{3x}$. Again, the particular integral is

$$h(x) = e^x$$

Hence the general solution is

$$\begin{aligned} y(x) &= c_1y_1(x) + c_2y_2(x) + h(x) \\ &= c_1\, e^{2x} + c_2\, e^{3x} + e^x \end{aligned}$$

One can use both initial conditions for determining the unknown parameters c_1 and c_2. Hence we have

$$c_1 + c_2 = 0 \tag{5.56}$$

and

$$2c_1 + 3c_2 = 0 \tag{5.57}$$

Again, the solution of Eqs. (5.56) and (5.57) are given by $c_1 = c_2 = 0$. Hence the particular solution is given by $y(x) = e^x$.

EXAMPLE 5.71 Solve the differential equation $y'' - 5y' + 6y = e^{4x}$ with $y(0) = \frac{1}{2}$ and $y'(0) = 2$.

Solution: Clearly, two independent homogeneous solutions are $y_1(x) = e^{2x}$ and $y_2(x) = e^{3x}$. Again, the particular integral is

$$h(x) = \frac{e^{4x}}{2}$$

Hence the general solution is

$$\begin{aligned} y(x) &= c_1 y_1(x) + c_2 y_2(x) + h(x) \\ &= c_2\, e^{2x} + c_2\, e^{3x} + \frac{e^{4x}}{2} \end{aligned}$$

One can use both initial conditions for determining the unknown parameters c_1 and c_2. Hence we have

$$c_1 + c_2 = 0 \tag{5.58}$$

and

$$2c_1 + 3c_2 = 0 \tag{5.59}$$

Again, the solution of Eqs. (5.58) and (5.59) are given by $c_1 = c_2 = 0$. Hence the particular solution is given by $y(x) = \frac{e^{4x}}{2}$.

EXAMPLE 5.72 Solve the differential equation $y'' + 4y = 4\cos 2x$ with $y(0) = 0$ and $y'(0) = 2$.

Solution: Clearly, two independent homogeneous solutions are $y_1(x) = \cos 2x$ and $y_2(x) = \sin 2x$. Again, the particular integral is

$$h(x) = x \sin 2x$$

Hence the general solution is

$$\begin{aligned} y(x) &= c_1\, y_1(x) + c_2\, y_2(x) + h(x) \\ &= c_1 \cos 2x + c_2 \sin 2x + x \sin 2x \end{aligned}$$

One can use both initial conditions for determining the unknown parameter c_1 and c_2. Hence we have

$$c_1 = 0 \tag{5.60}$$

and

$$2c_2 = 2 \tag{5.61}$$

Again, the solution of Eqs.(5.60) and (5.61) are given by $c_1 = 0$ and $c_2 = 1$. Hence the particular solution is given by $y(x) = \sin 2x + x \sin 2x$.

EXAMPLE 5.73 Solve the initial value problem $x^2y'' - 4xy' + 6y = 6\ln x$ with $y(1) = \frac{5}{6}$ and $y'(1) = 1$.

Solution: Clearly, $y_1(x) = x^2$ and $y_2(x) = x^3$ are two independent homogeneous solutions. Again, the particular integral is

$$h(x) = \ln x + \frac{5}{6}$$

Hence the general solution is

$$\begin{aligned} y(x) &= c_1\, y_1(x) + c_2\, y_2(x) + h(x) \\ &= c_1\, x^2 + c_2\, x^3 + \ln x + \frac{5}{6} \end{aligned}$$

One can use the initial conditions. Hence we have

$$c_1 + c_2 = 0 \tag{5.62}$$

and

$$2c_1 + 3c_2 = 0 \tag{5.63}$$

Again, the solution of Eqs.(5.62) and (5.63) are given by $c_1 = 0$ and $c_2 = 0$. Hence the particular solution is given by $y(x) = \ln x + \frac{5}{6}$.

EXAMPLE 5.74 Solve the initial value problem $x^2y'' - 2xy' + 2y = 10\sin(\ln x)$ with $y(1) = 3$ and $y'(1) = 1$.

Solution: Clearly, $y_1(x) = x$ and $y_2(x) = x^2$ are two independent homogeneous solutions. Again, the particular integral is

$$h(x) = 3\cos(\ln x) + \sin(\ln x)$$

Hence the general solution is

$$\begin{aligned} y(x) &= c_1\, y_1(x) + c_2\, y_2(x) + h(x) \\ &= c_1\, x + c_2\, x^2 + 3\cos(\ln x) + \sin(\ln x) \end{aligned}$$

One can use the initial conditions. Hence we have

$$c_1 + c_2 = 0 \tag{5.64}$$

and

$$c_1 + 2c_2 = 0 \tag{5.65}$$

Again, the solution of Eqs.(5.64) and (5.65) are given by $c_1 = 0$ and $c_2 = 0$. Hence the particular solution is given by $y(x) = 3\cos(\ln x) + \sin(\ln x)$.

EXAMPLE 5.75 Find the solution of the differential equation $y''+y'\tan x+y\cos^2 x = \sin x$ with $y(0) = 0$ and $y'(0) = 1$.

Solution: Clearly, $p(x) = \tan x$ and $q(x) = \cos^2 x$. It is easy to verify that $\frac{q'(x)+2p(x)q(x)}{q^{\frac{3}{2}}(x)}$ is a constant. According to Exercise 1 given in Exercises 5.4.3, $u = g(x) = \int q^{\frac{1}{2}}(x)\,\mathrm{d}x = \sin x$ is the required transformation of the independent variable. The corresponding transformed differential equation is given by

$$\begin{aligned} u &= \sin x \\ &= (g'(x))^2\left(\frac{\mathrm{d}^2y}{\mathrm{d}u^2}\right) + (g''(x)+p(x)g'(x))\left(\frac{\mathrm{d}y}{\mathrm{d}u}\right) + q(x)y \\ &= \cos^2 x\left(\frac{\mathrm{d}^2y}{\mathrm{d}u^2}\right) + (\sin x - \tan x\cos x)\left(\frac{\mathrm{d}y}{\mathrm{d}u}\right) + y\cos^2 x \\ &= \frac{\mathrm{d}^2y}{\mathrm{d}u^2} + y \end{aligned} \tag{5.66}$$

Again, the solution of the differential equation given in Eq.(5.66) is given by

$$y(u) = c_1\cos u + c_2\sin u + u$$

In other words, the solution of the given equation is

$$y(x) = c_1\cos(\sin x) + c_2\sin(\sin x) + \sin x$$

Using the initial conditions, we have

$$c_1 = 0$$

and

$$c_2 = 0$$

Hence the required solution is $y(x) = \sin x$.

Exercises 5.5

1. Solve the initial value problem $x^2y'' - 3xy' + 4y = 0$ with $y(1) = 1$ and $y'(1) = 1$.
2. Solve the initial value problem $x^2y'' - 3xy' + 3y = 0$ with $y(1) = 0$ and $y'(1) = -2$.
3. Solve the initial value problem $x^2y'' + 2xy' + 2y = 0$ with $y(1) = 3$ and $y'(1) = 5$.
4. Solve the initial value problem $x^2y'' - 2xy' + 2y = 2x^3$ with $y(1) = 1$ and $y'(1) = 3$.
5. Solve the initial value problem $x^2y'' - 2xy' + 2y = x$ with $y(1) = 4$ and $y'(1) = 0$.
6. Solve the initial value problem $x^2y'' - xy' + y = \ln x$ with $y(1) = 3$ and $y'(1) = 0$.
7. Solve the initial value problem $y'' - 2y' + y = 2x^2 - 8x + 4$ with $y(0) = 3$ and $y'(0) = 3$.
8. Solve the initial value problem $y'' - y' - 2y = 3\,e^{2x}$ with $y(0) = 0$ and $y'(0) = -2$.
9. Solve the initial value problem $y'' + 4y' + 8y = 4\cos x + 7\sin x$ with $y(0) = 1$ and $y'(0) = -1$.
10. Solve the initial value problem $y'' + 2y' + 3y = e^x$ with $y(0) = 0$ and $y'(0) = 1$.
11. Solve the initial value problem $2y'' + y = 2x^2 + 3x = 1$ with $y(0) = 0$ and $y'(0) = 0$.

Chapter 6

Second Order Equation Applications

Linear second order differential equations with constant coefficients have important engineering applications. We shall consider such applications, which are called *fundamental vibrations* in mechanical and electrical systems. Generally speaking, vibrations occur whenever a physical system in stable equilibrium is disturbed, since then it is subject to forces tending to restore its equilibrium. In the present chapter, we shall see how situations of this kind can lead to differential equations of the form

$$y'' + ay' + by = f(x)$$

and also how the study of these equations sheds light on the physical circumstances.

6.1 Freely Falling Body

According to Newton's second law of motion, the acceleration a of a body of mass m is proportional to the total force F acting on it, with $\frac{1}{m}$ as the constant of proportionality. In other words, we have

$$\begin{aligned} ma &= m\left(\frac{1}{m}\right)F \\ &= F \end{aligned} \tag{6.1}$$

Now, a body of mass m falls freely under the influence of gravity alone. In this case, the only force acting on it is mg, where g is the acceleration due to gravity. If y is the distance of the body from some fixed height, then its velocity $v = y'$ is the rate of change of position and its acceleration $a = v' = y''$ is the rate of change of velocity. Accordingly, Eq.(6.1) reduces to

$$y'' = g \tag{6.2}$$

One integration of Eq.(6.2) yields the velocity,

$$\begin{aligned} v &= y' \\ &= gt + c \end{aligned} \tag{6.3}$$

Clearly, $c = v_0$ is the initial velocity at $t = 0$ of the freely falling body. In other words, Eq.(6.3) becomes

$$\begin{aligned} y' &= v \\ &= gt + v_0 \end{aligned} \tag{6.4}$$

On integrating Eq.(6.4) again, we have

$$y = \frac{1}{2}gt^2 + v_0t + d \tag{6.5}$$

Clearly, $d = y_0$ is the initial position at $t = 0$ of the freely falling body. In other words, Eq.(6.5) becomes

$$y = \frac{1}{2}gt^2 + v_0t + y_0 \tag{6.6}$$

If the body falls from rest starting at $y_0 = 0$, then Eqs.(6.3) and (6.6) get reduced to $v = gt$ and $y = \frac{1}{2}gt^2$ as $v_0 = 0$. On eliminating t, we have the useful relation

$$v = (2gy)^{\frac{1}{2}}$$

for the velocity attained in terms of the distance covered.

EXAMPLE 6.1 If a body of mass 10 pounds is dropped vertically from a height of 256 feet from the ground with initial velocity $v_0 = 0$, then find the distance covered by the body at $t = 4$ seconds.

Solution: Clearly, $v_0 = 0 = y_0$ and $g = 32$ feet per second square. According to Eq.(6.6), the displacement $y(t)$ is given by

$$\begin{aligned} y(t) &= \frac{1}{2}gt^2 + v_0t + y_0 \\ &= \frac{1}{2}gt^2 \\ &= 16t^2 \end{aligned}$$

Hence the distance covered at $t = 4$ is given by $y(4) = 256$ feet. In other words, the freely falling body will be grounded at $t = 4$ seconds.

EXAMPLE 6.2 If a body is dropped vertically from a height of 68 feet with initial velocity $v_0 = 2$ feet per second, then find the velocity of the body when it is grounded.

Solution: Clearly, $v_0 = 2$, $y_0 = 0$ and $g = 32$ feet per second square. According to Eq.(6.6), the displacement $y(t)$ is given by

$$\begin{aligned} y(t) &= \frac{1}{2}gt^2 + v_0 t + y_0 \\ &= \frac{1}{2}gt^2 + 2t \\ &= 16t^2 + 2t \end{aligned}$$

Hence the time required to cover 68 feet is given by $t = 2$ seconds. Again, the velocity at $t = 2$ is given by

$$\begin{aligned} v(2) &= gt + v_0 \\ &= (32 \times 2) + 2 \\ &= 66 \text{ feet per second} \end{aligned}$$

Exercises 6.1

1. If a body is dropped vertically from a height of 150 feet from the ground with initial velocity $v_0 = 0$, then find the velocity of the body when it is grounded.
2. If a body is dropped vertically from a height of 175 feet with initial velocity $v_0 = 0$, then how long will it take to achieve the velocity of 15 feet per second?
3. How far from the ground should a body be dropped vertically with initial velocity $v_0 = 0$ so that its velocity is 250 feet per second when it is grounded?
4. How far from the ground should a body be dropped vertically with initial velocity $v_0 = 2$ feet per second, so that its velocity is 150 feet per second when it is grounded?
5. If a body is dropped vertically from a height of 250 feet and its velocity when it is grounded is 100 feet per second, then find its initial velocity v_0.

6.2 Retarded Falling Body

If we assume that the air exerts a resisting force proportional to the velocity of a free falling body, then the corresponding differential equation formed by modifying Eq.(6.2) is

$$y'' = g - ky' \tag{6.7}$$

which is a second order linear nonhomogeneous differential equation, where $k > 0$. If y' is replaced by v, then Eq.(6.7) reduces to

$$v' + kv = g \tag{6.8}$$

Clearly, Eq.(6.8) is a first order linear nonhomogeneous differential equation. Let $p(t) = k$ and $f(t) = g$. According to the formula given in Eq.(2.8),

$$\begin{aligned}
v(t) &= e^{-\int p(t)\,\mathrm{d}t}\left(\int e^{\int p(t)\,\mathrm{d}t} f(t)\,\mathrm{d}t + c\right) \\
&= e^{-kt}\left(\int g e^{kt}\,\mathrm{d}t + c\right) \\
&= e^{-kt}\left(\left(\frac{g}{k}\right)e^{kt} + c\right) \\
&= \frac{g}{k} + ce^{-kt}
\end{aligned} \tag{6.9}$$

The initial condition $v(0) = 0$ gives $c = -\frac{g}{k}$ in Eq.(6.9). Hence we have

$$v(t) = \frac{g}{k}\left(1 - e^{-kt}\right)$$

Since $k > 0$, therefore $\lim\limits_{t\to\infty} v(t) = \frac{g}{k}$. This limiting value of velocity $v(t)$ is called the *terminal velocity*.

EXAMPLE 6.3 A body of mass m falling from rest is subjected to the force of gravity and an air resistance proportional to the square of the velocity. If it falls through a distance x and possesses a velocity v at that instance, then show that $\frac{2kx}{m} = \ln\left(\frac{mg}{mg-kv^2}\right)$.

Solution: Clearly, the net force acting on the body is $mg - kv^2$ and the acceleration at any distance x is $v\left(\frac{\mathrm{d}v}{\mathrm{d}x}\right)$. According to Newton's second law, we have

$$mvv' + kv^2 = mg \tag{6.10}$$

Let $v^2 = u$. Accordingly, Eq.(6.10) reduces to

$$u' + \left(\frac{2k}{m}\right)u = 2g \tag{6.11}$$

Let $p(x) = \frac{2k}{m}$ and $f(x) = 2g$. According to the formula given in Eq.(2.8), we have

$$\begin{aligned}
v^2(x) &= u(x) \\
&= e^{-\int p(x)\,\mathrm{d}x}\left(\int f(x)\,e^{-\int p(x)\,\mathrm{d}x}\,\mathrm{d}x + c\right) \\
&= e^{-\int\left(\frac{2k}{m}\right)\mathrm{d}x}\left(\int 2g\,e^{\int\left(\frac{2k}{m}\right)\mathrm{d}x}\,\mathrm{d}x + c\right) \\
&= e^{-\left(\frac{2k}{m}\right)x}\left(\left(\frac{gm}{k}\right)e^{\left(\frac{2k}{m}\right)x} + c\right) \\
&= \frac{mg}{k} + c\,e^{-\left(\frac{2k}{m}\right)x}
\end{aligned}$$

Using the initial condition $v(0) = 0$, we have $c = -\frac{mg}{k}$. In other words, we have

$$v^2(x) = \frac{mg}{k}\left(1 - e^{-\left(\frac{2k}{m}\right)x}\right) \tag{6.12}$$

One can show that the velocity expression given in Eq.(6.12) can be represented by

$$\left(\frac{2k}{m}\right)x = \ln\left(\frac{mg}{mg - kv^2}\right)$$

Exercises 6.2

1. A torpedo is travelling at a speed of 600 miles per hour at the moment it runs out of fuel. If the water resists its motion with a force proportional to the speed and one mile of travel reduces its speed to 30 miles per hour, then how far will it coast?
2. A rock is thrown upwards from the surface of the earth with initial velocity 128 feet per second. Neglecting air resistance and assuming that the only force acting on the rock is a constant gravitational force, find the maximum height it reaches.
3. A rock is thrown upwards from the surface of the earth with initial velocity 128 feet per second. Neglecting air resistance and assuming that the only force acting on the rock is a constant gravitational force, when does it reach the maximum height?
4. A rock is thrown upwards from the surface of the earth with initial velocity 128 feet per second. Neglecting air resistance and assuming that the only force acting on the rock is a constant gravitational force, when does it reach the ground?

6.3 Simple Harmonic Motion

The motion of a particle is said to be *simple harmonic* if and only if the acceleration of the particle is proportional to its displacement. In other words, we have

$$y'' = -\omega^2 y \tag{6.13}$$

where $y(t)$ is the displacement of the particle from some fixed reference point and ω^2 is the constant of proportionality. The negative sign indicates that the acceleration and the displacement are opposite in direction. Clearly, Eq.(6.13) is a second order linear homogeneous differential equation. According to Section 5.2.1, the solution of Eq.(6.13) is given by

$$y(t) = c_1 \sin\omega t + c_2 \cos\omega t \tag{6.14}$$

On introducing $c_1 = A\cos\phi$ and $c_2 = A\sin\phi$, the solution given in Eq.(6.14) gets reduced to

$$y(t) = A\sin(\omega t + \phi) \tag{6.15}$$

where ω is the *natural frequency* of the pendulum. If the particle starts from a point which is at a distance a from the reference point with zero velocity, then the initial conditions are $y(0) = a$ and $y'(0) = 0$. Accordingly, we have $c_1 = 0$ and $c_2 = a$. Hence the corresponding displacement of the particle which is in harmonic motion is given by

$$y(t) = a\cos\omega t \tag{6.16}$$

The *amplitude* of the simple harmonic motion $y(t)$ given in Eq.(6.16) is a, which is the maximum displacement of the particle from the reference point. Again, the *period* $\mathcal{T}$ required for one complete cycle is $\mathcal{T} = \frac{2\pi}{\omega}$. The *frequency* $\mathcal{F}$, the number of cycles per unit time, of the simple harmonic motion of the pendulum is $\mathcal{F} = \frac{1}{\mathcal{T}} = \frac{\omega}{2\pi}$.

EXAMPLE 6.4 If a particle is in simple harmonic motion with amplitude 20 centimetres and period 4 seconds, then compute the time required by the particle to cross two points which are at distances of 15 centimetres and 5 centimetres from the fixed reference point.

Solution: Clearly, $a = 20$. According to Eq.(6.16), we have

$$y(t) = 20\cos\omega t$$

Again, $\mathcal{T} = 4$. Accordingly, $\omega = \frac{2\pi}{4} = \frac{\pi}{2}$. Hence we have

$$y(t) = 20\cos\frac{\pi}{2}t$$

Let t_1 and t_2 be the respective time periods required by the particle to pass the points situated 5 centimetres and 15 centimetres away from the fixed reference point. Clearly, we have

$$t_1 = \frac{2}{\pi}\cos\frac{1}{4}$$

and

$$t_2 = \frac{2}{\pi}\cos\frac{3}{4}$$

The required time taken by the particle for covering the distance from the position at 5 centimetres to the position at 15 centimetres measured from the fixed reference point is given by

$$\begin{aligned} t &= t_2 - t_1 \\ &= \frac{2}{\pi}\cos\frac{3}{4} - \frac{2}{\pi}\cos\frac{1}{4} \\ &= \frac{2}{\pi}\left(\cos\frac{3}{4} - \cos\frac{1}{4}\right) \end{aligned}$$

EXAMPLE 6.5 A particle of mass m grams moves in a straight line under the action of a force $m\omega^2 y$ which is always directed towards a fixed reference point on the line. If the resistance to the motion is $2\lambda m\omega y'$, then find the displacement of the particle subject to the conditions $y(0) = 0$ and $y'(0) = y_0$ when $0 < \lambda < 1$.

Solution: Clearly, the differential equation describing the motion of the particle is given by

$$m\,y'' = -2\lambda m\omega y' - m\omega^2 y \tag{6.17}$$

Again, Eq.(6.17) is a second order linear homogeneous differential equation. According to Section 5.2.1, the solution of Eq.(6.17) is given by

$$y(t) = e^{-\lambda\omega t}\left(c_1\cos\left(\omega(1-\lambda^2)^{\frac{1}{2}}t\right) + c_2\sin\left(\omega(1-\lambda^2)^{\frac{1}{2}}t\right)\right) \tag{6.18}$$

Using the initial conditions $y(0) = 0$ and $y'(0) = y_0$, we have $c_1 = 0$ and $c_2 = \frac{y_0}{\omega(1-\lambda^2)^{\frac{1}{2}}}$. Hence the corresponding displacement of the particle is given by

$$y(t) = \frac{y_0}{\omega(1-\lambda^2)^{\frac{1}{2}}} e^{-\lambda \omega t} \sin\left(\omega(1-\lambda^2)^{\frac{1}{2}} t\right)$$

Exercises 6.3

1. A pendulum consisting of a bob of mass m at the end of a rod of negligible mass and of length a. If the bob is pulled to one side through an angle ψ and released, then find the differential equation which describes the motion of the pendulum.
2. If a particle in simple harmonic motion satisfies the conditions $y(1) = 2$, $y(2) = 4$ and $y(3) = 7$, then find the displacement of the particle.
3. Determine the equation for the motion of a particle of mass m attached to one end of a stretched elastic vertical string whose other end is fixed.
4. Find the period of a particle of mass m, in simple harmonic motion, attached to the middle point of a elastic string of length $2a$ stretched between two points which are $4a$ apart.

6.4 Free Motion of Spring–Mass System

There are many physical phenomena that are described with second order linear homogeneous differential equations. Airplanes, bridges, ships, machines, cars, etc. are vibrating mechanical systems. We wish to discuss one such phenomenon—the free motion of a spring–mass system. The spring-mass system is the simplest mechanical system which consists of a coil spring of natural length L suspended vertically from a fixed point support such as a ceiling or beam. A constant mass m attached to the lower end of the spring stretches the spring to a length $L + l$ and comes to rest at a point which is known as the *static equilibrium position.* Clearly, $l > 0$ is the static deflection due to the hanging of the mass on the spring. Now, the mass is set in motion either by pushing or pulling the mass from the equilibrium position or by imparting a non-zero velocity to the mass. Since the motion takes place in the vertical direction, we consider the downward direction as positive. In order to determine the displacement $y(t)$ of the mass from the static equilibrium position, we use Newton's second law and Hooke's law.

We shall make the following assumptions:

1. The mass m is constrained to move in the vertical directions only.
2. The gravitational force mg acts downwards.
3. The viscous damping c, in units of mass per second, is proportional to the velocity y'.
4. The force in the spring is $k(y(t) + l)$.
5. The mass of the spring is negligible compared with the mass m attached to it.
6. No external force acts on the system.

Since the force on the mass exerted by the spring must be equal and opposite to the gravitational force on the mass, we have $kl = mg$. According to Newton's second law of motion, which says that the mass times its acceleration equals the total force acting on it, we have

$$\begin{aligned} my'' &= mg - cy' - k(y+l) \\ &= -cy' - ky \end{aligned}$$

In other words, we have

$$y'' + \left(\frac{c}{m}\right)y' + \left(\frac{k}{m}\right)y = 0 \tag{6.19}$$

Clearly, Eq.(6.19) is a second order linear homogeneous differential equation. The motion discussed above is known as *free motion*, because all the forces acting on the system are internal to the system itself.

6.4.1 Undamped Motion

As our next step is in developing this physical problem, we suppose that the medium in which the mass moves has no viscosity effect. In other words, $c = 0$. Hence the differential Eq.(6.19) reduces to

$$y'' + \left(\frac{k}{m}\right)y = 0 \tag{6.20}$$

Again, the solution of differential equation (6.20) is given by

$$y(t) = c_1 \sin\frac{k}{m} + c_2 \cos\frac{k}{m}$$

If the mass is pulled aside to position $y = y_0$ and released without any initial velocity at time $t = 0$, then our initial conditions are

$$y(0) = y_0 \tag{6.21}$$

and

$$y'(0) = 0 \tag{6.22}$$

Using Eqs.(6.21) and (6.22) in Eq.(6.19), we have

$$y(t) = y_0 \cos \omega_0 t \tag{6.23}$$

where $\omega_0 = \left(\frac{k}{m}\right)^{\frac{1}{2}}$ is the *natural frequency* of the mass. The *amplitude* of the simple harmonic motion $y(t)$ of the mass given in Eq.(6.23) is y_0, and the *period* $\mathcal{T}$ required for one complete cycle is $\mathcal{T} = \frac{2\pi}{\omega_0} = 2\pi\left(\frac{k}{m}\right)^{\frac{1}{2}}$. The *frequency* $\mathcal{F}$, the number of cycles per unit time, of the simple harmonic motion of the mass is $\mathcal{F} = \frac{1}{\mathcal{T}} = \frac{\omega_0}{2\pi} = \frac{1}{2\pi}\left(\frac{m}{k}\right)^{\frac{1}{2}}$.

6.4.2 Damped Motion

As our next step is in developing this physical problem, we consider some medium in which the mass moves has viscosity effect. In other words, $c \neq 0$. Hence for the sake of convenience, the differential Eq.(6.19) can be written in the form

$$y'' + 2\left(\frac{c}{2m}\right)y' + \left(\frac{k}{m}\right)y = 0 \tag{6.24}$$

Again, the solution of differential Eq.(6.24) is given by

$$y(t) = e^{-\left(\frac{c}{2m}\right)t}\left(c_1 e^{\left(\frac{\left(c^2-4mk\right)^{\frac{1}{2}}}{2m}\right)t} + c_2 e^{-\left(\frac{\left(c^2-4mk\right)^{\frac{1}{2}}}{2m}\right)t}\right) \tag{6.25}$$

according to Section 5.2.1. The physical problem corresponding to $c^2 - 4mk > 0$ is called *overdamping.* If $c^2 - 4mk < 0$, then solution in Eq.(6.25) gets reduced to

$$y(t) = e^{-\left(\frac{c}{2m}\right)t}\left(c_1 \sin\left(\frac{\left(c^2-4mk\right)^{\frac{1}{2}}}{2m}\right) + c_2 \cos\left(\frac{\left(c^2-4mk\right)^{\frac{1}{2}}}{2m}\right)\right)$$

according to Example 5.12. The physical problem corresponding to $c^2 - 4mk < 0$ is called *underdamping.* According to Example 5.12, the solution of Eq.(6.24) is

$$y(t) = e^{-\left(\frac{c}{2m}\right)t}(c_1 + c_2 t)$$

when $c^2 - 4mk = 0$. The physical problem corresponding to $c^2 - 4mk = 0$ is called *critical damping.*

EXAMPLE 6.6 A 8 lb weight is placed at one end of a spring suspended from the ceiling. The weight is raised to 5 inches above the equilibrium position and left free. Assuming spring constant $k = 12$ lb per feet, find the equation of motion, displacement function $y(t)$, amplitude, period, frequency and maximum velocity.

Solution: Clearly, $k = 12$ lb per feet. Again, the mass is $m = \frac{8}{32} = \frac{1}{4}$ pound. Hence the corresponding differential equation according to Eq.(6.19) is given by

$$y'' + 48y = 0 \tag{6.26}$$

Clearly, Eq.(6.26) is a constant coefficient homogeneous differential equation. According to Section 5.2.1, the solution of Eq.(6.26) is given by

$$y(t) = c_1 \cos 4\sqrt{3}t + c_2 \sin 4\sqrt{3}t$$

Using the initial conditions $y(0) = -5$ and $y'(0) = 0$, we have $c_1 = -\frac{5}{12}$ and $c_2 = 0$. Hence the displacement $y(t)$ is given by

$$\begin{aligned} y(t) &= -\frac{5}{12}\cos 4\sqrt{3}t \\ &= \frac{5}{12}\sin\left(4\sqrt{3}t - \frac{\pi}{2}\right) \end{aligned}$$

and the velocity $y'(t)$ is given by

$$y'(t) = \frac{5}{\sqrt{3}} \sin 4\sqrt{3}t$$

Clearly, amplitude is $\frac{5}{12}$, period is $\frac{\pi\sqrt{3}}{6}$ and frequency is $\frac{6}{\pi\sqrt{3}}$. Again, the maximum velocity is $\frac{5}{\sqrt{3}}$.

EXAMPLE 6.7 A 2 lb weight suspended from one end of a spring stretches it to 6 inches. A velocity of 5 feet per second square upwards is imparted to the weight at its equilibrium position. Assuming the damping constant $0 < c < 1$, determine the position and the velocity of the mass at any time.

Solution: Clearly, $l = 6$ inches $= \frac{1}{2}$ feet. Let k be the spring constant. According to Hooke's law, we have $2 = \frac{k}{2}$. Hence $k = 4$ lb per feet. Again, the mass is $m = \frac{2}{32} = \frac{1}{16}$ pound. Hence the corresponding differential equation according to Eq.(6.19) is given by

$$y'' + 16cy' + 4y = 0 \tag{6.27}$$

Clearly, Eq.(6.27) is a constant coefficient homogeneous differential equation. According to Section 5.2.1, the solution of Eq.(6.27) is given by

$$y(t) = e^{-8ct}\left(c_1 \cos\left(8(1-c^2)^{\frac{1}{2}}t\right) + c_2 \sin\left(8(1-c^2)^{\frac{1}{2}}t\right)\right)$$

Using initial conditions $y(0) = 0$ and $y'(0) = -5$, we have $c_1 = 0$ and $c_2 = -\frac{5}{8(1-c^2)^{\frac{1}{2}}}$. Hence the displacement $y(t)$ is given by

$$y(t) = -\frac{5e^{-8ct}}{8(1-c^2)^{\frac{1}{2}}} \sin\left(8(1-c^2)^{\frac{1}{2}}t\right)$$

and the velocity $y'(t)$ is given by

$$\begin{aligned} y'(t) &= \frac{5e^{-8ct}}{8(1-c^2)^{\frac{1}{2}}}\left(8c \sin\left(8(1-c^2)^{\frac{1}{2}}t\right) - 8(1-c^2)^{\frac{1}{2}} \cos\left(8(1-c^2)^{\frac{1}{2}}t\right)\right) \\ &= \frac{5e^{-8ct}}{(1-c^2)^{\frac{1}{2}}}\left(c \sin\left(8(1-c^2)^{\frac{1}{2}}t\right) - (1-c^2)^{\frac{1}{2}} \cos\left(8(1-c^2)^{\frac{1}{2}}t\right)\right) \end{aligned}$$

Exercises 6.4

1. A weight of 4.5 lb stretches a spring 9.8 centimetres long. If the lower end of the spring attached with the weight 4.5 lb is pulled down by 5.0 cm and given an upward velocity of 30.0 centimetres per second square, then find the position of the weight at all latter times, assuming zero damping.
2. Determine the displacement and the maximum velocity of a body of weight 10 kg attached to a spring, given that a 20 kg weight will stretch the spring to 10 centimetres.
3. A 6 lb weight stretches a spring by 6 inches. If the weight is pulled 4 inches below the equilibrium position, find the displacement, velocity and acceleration of the weight.

4. A 32 lb weight, suspended from a coil spring, stretches the spring to 2 feet. If the weight is pulled down 6 inches from the equilibrium position and released at $t = 0$, find the position $y(t)$ of the weight at time t.
5. A body weighing 10 lb is hung from a spring. A pull of 20 lb weight will stretch the spring to 10 feet. The body is pulled down to 20 feet below the static equilibrium position and released. Find the displacement of the body from its equilibrium position at time t, the maximum velocity and the period of oscillation.

6.5 Forced Motion of Spring–Mass System

There are many physical phenomena that are described with second order linear non-homogeneous differential equations. We wish to discuss one such phenomenon—the free motion of a spring–mass system. In order to determine the displacement $y(t)$ of the mass from the static equilibrium position, we use Newton's second law and Hooke's law.

We shall make the following assumptions:

1. The mass m is constrained to move in the vertical direction only.
2. The gravitational force mg acts downwards.
3. The viscous damping c, in units of mass per second, is proportional to the velocity y'.
4. The force in the spring is $k(y(t) + l)$.
5. The mass of the spring is negligible compared with the mass m attached to it.
6. An external force $f(t)$ acts on the system.

Since the force on the mass exerted by the spring must be equal and opposite to the gravitational force on the mass, we have $kl = mg$. According to Newton's second law of motion, which says that the mass times its acceleration equals the total force acting on it, we have

$$\begin{aligned} my'' &= mg - cy' - k(y + l) + f(t) \\ &= -cy' - ky + f(t) \end{aligned}$$

In other words, we have

$$y'' + \left(\frac{c}{m}\right) y' + \left(\frac{k}{m}\right) y = \left(\frac{1}{m}\right) f(t) \tag{6.28}$$

Clearly, Eq.(6.28) is a second order linear non-homogeneous differential equation. The motion discussed above is known as *forced motion* because all the forces acting on the system are not internal to the system itself. In particular, we take up the sinusoidal forcing function.

Let $f(t) = a\cos\omega t$, where ω is called the *input frequency*. Accordingly, Eq.(6.28) reduces to

$$y'' + \left(\frac{c}{m}\right) y' + \left(\frac{k}{m}\right) y = \left(\frac{a}{m}\right) \cos\omega t \tag{6.29}$$

One can solve the differential Eq.(6.29) by any one of the methods discussed in Section 5.3. According to the undetermined coefficient method discussed in Section 5.3.2, one can assume that the particular integral $h(t)$ is given by

$$h(t) = \alpha \cos \omega t + \beta \sin \omega t \tag{6.30}$$

Substituting all required forms of $h(t)$ given in Eq.(6.30) into Eq.(6.29), we have

$$((k - m\,\omega^2)\,\alpha + \omega\,\beta c)\cos \omega t + ((k - m\,\omega^2)\beta - \omega\,\alpha c)\sin \omega\, t = a\cos \omega t \tag{6.31}$$

Again, we have

$$\alpha = \frac{a(k - m\,\omega^2)}{(k - m\,\omega^2)^2 + \omega^2 c^2}$$

and

$$\beta = \frac{a\omega c}{(k - m\,\omega^2)^2 + \omega^2 c^2}$$

by comparing coefficients of $\cos \omega t$ and $\sin \omega t$ in Eq.(6.31). Substituting expressions for α and β into Eq.(6.30), we have

$$h(t) = \left(\frac{a(k - m\,\omega^2)}{(k - m\,\omega^2)^2 + \omega^2 c^2}\right)\cos \omega t + \left(\frac{a\omega c}{(k - m\,\omega^2)^2 + \omega^2 c^2}\right)\sin \omega t \tag{6.32}$$

The general solution of the differential Eq.(6.29) can be obtained by adding the expression for $h(t)$ in Eq.(6.32) to the solution given in Eq.(6.25). Hence we have

$$y(t) = e^{-\left(\frac{c}{2m}\right)t}\left(c_1\, e^{\left(\frac{\left(c^2-4mk\right)^{\frac{1}{2}}}{2m}\right)t} + c_2\, e^{-\left(\frac{\left(c^2-4mk\right)^{\frac{1}{2}}}{2m}\right)t}\right)$$
$$+ \left(\frac{a(k - m\,\omega^2)}{(k - m\,\omega^2)^2 + \omega^2 c^2}\right)\cos \omega t + \left(\frac{a\omega c}{(k - m\,\omega^2)^2 + \omega^2 c^2}\right)\sin \omega t \tag{6.33}$$

Let $f(t) = a\sin \omega t$, where ω is called the *input frequency*. Accordingly, Eq.(6.28) gets reduced to

$$y'' + \left(\frac{c}{m}\right)y' + \left(\frac{k}{m}\right)y = \left(\frac{a}{m}\right)\sin \omega t \tag{6.34}$$

One can solve the differential Eq.(6.34) by any one of the methods discussed in Section 5.3. According to the undetermined coefficient method discussed in Section 5.3.2, one can denote the particular integral $h(t)$ by

$$h(t) = \alpha \cos \omega t + \beta \sin \omega t \tag{6.35}$$

Substituting all required forms of $h(t)$ given in Eq.(6.35) into Eq.(6.34), we have

$$((k - m\,\omega^2)\alpha + \omega\,\beta c)\cos \omega t + ((k - m\,\omega^2)\beta - \omega\,\alpha c)\sin \omega\, t = a\cos \omega t \tag{6.36}$$

Again, we have

$$\alpha = \frac{a\omega c}{(k - m\,\omega^2)^2 + \omega^2 c^2}$$

and

$$\beta = \frac{a(k - m\,\omega^2)}{(k - m\,\omega^2)^2 + \omega^2 c^2}$$

by comparing coefficients of $\cos\omega t$ and $\sin\omega t$ in Eq.(6.36). Substituting expressions for α and β into Eq.(6.35), we have

$$h(t) = \left(\frac{a\omega c}{(k - m\,\omega^2)^2 + \omega^2 c^2}\right)\cos\omega t + \left(\frac{a(k - m\,\omega^2)}{(k - m\,\omega^2)^2 + \omega^2 c^2}\right)\sin\omega t \tag{6.37}$$

The general solution of the differential Eq.(6.34) can be obtained by adding the expression for $h(t)$ in Eq.(6.37) to the solution given in Eq.(6.25). Hence we have

$$y(t) = e^{-\left(\frac{c}{2m}\right)t}\left(c_1\, e^{\left(\frac{(c^2-4mk)^{\frac{1}{2}}}{2m}\right)t} + c_2\, e^{-\left(\frac{(c^2-4mk)^{\frac{1}{2}}}{2m}\right)t}\right)$$
$$+\left(\frac{a\omega c}{(k - m\,\omega^2)^2 + \omega^2 c^2}\right)\cos\omega t + \left(\frac{a(k - m\,\omega^2)}{(k - m\,\omega^2)^2 + \omega^2 c^2}\right)\sin\omega t \tag{6.38}$$

6.5.1 Resonance

An interesting and very important phenomenon is observed in the solution given in Eq.(6.33) when the damping coefficient c is zero. The general solution given in Eq.(6.33) with $c = 0$ gets reduced to

$$\begin{aligned}
y(t) &= c_1\, e^{\left(\frac{(-4mk)^{\frac{1}{2}}}{2m}\right)t} + c_2\, e^{-\left(\frac{(-4mk)^{\frac{1}{2}}}{2m}\right)t} + \left(\frac{a}{k - m\,\omega^2}\right)\cos\omega t \\
&= c_1\, e^{\left(\frac{(-mk)^{\frac{1}{2}}}{m}\right)t} + c_2\, e^{-\left(\frac{(-mk)^{\frac{1}{2}}}{m}\right)t} + \left(\frac{a}{k - m\,\omega^2}\right)\cos\omega t \\
&= b_1 \cos\left(\left(\frac{k}{m}\right)^{\frac{1}{2}} t\right) + b_2 \cos\left(\left(\frac{k}{m}\right)^{\frac{1}{2}} t\right) + \left(\frac{a}{k - m\,\omega^2}\right)\cos\omega t \\
&= b_1 \cos\omega_0 t + b_2 \cos\omega_0 t + \left(\frac{a}{k - m\,\omega^2}\right)\cos\omega t \\
&= b_1 \cos\omega_0 t + b_2 \cos\omega_0 t + \left(\frac{a}{m\left(\left(\frac{k}{m}\right) - \omega^2\right)}\right)\cos\omega t \\
&= b_1 \cos\omega_0 t + b_2 \cos\omega_0 t + \left(\frac{a}{m\,(\omega_0^2 - \omega^2)}\right)\cos\omega t
\end{aligned} \tag{6.39}$$

where $\omega_0 = \left(\frac{k}{m}\right)^{\frac{1}{2}}$. Now, consider the case where $\omega = \omega_0$. The physical phenomenon in which the natural frequency is equal to the input frequency is called *resonance.* In other words, the input frequency is equal to the natural frequency. Clearly, the expression for $y(t)$ given in Eq.(6.39) is not a solution of the differential Eq.(6.29) when $\omega = \omega_0$ and $c = 0$. The corresponding differential equation is

$$\begin{aligned} y'' + \omega_0^2 y &= y'' + \left(\frac{k}{m}\right) y \\ &= \frac{a}{m} \cos \omega_0 t \end{aligned} \tag{6.40}$$

One can solve the differential Eq.(6.40) by any one of the methods discussed in Section 5.3. According to the undetermined coefficient method discussed in Section 5.3.2, one can denote the particular integral $h(t)$ by

$$h(t) = t\,(\alpha \cos \omega_0 t + \beta \sin \omega_0 t) \tag{6.41}$$

substituting the expression for $h(t)$ given in Eq.(6.41) into Eq.(6.40), and comparing the coefficients of $\sin \omega_0 t$ and $\cos \omega_0 t$, we have

$$\alpha = 0$$

and

$$\beta = \frac{a}{2m\omega_0}$$

In other words, the particular integral $h(t)$ becomes

$$h(t) = \frac{at}{2m\omega_0} \sin \omega_0 t \tag{6.42}$$

It is clear from Eq.(6.42) that as the time period becomes large, the amplitude becomes large. Hence the general solution of the differential equation given in Eq.(6.40) is given by

$$y(t) = c_1 \cos \omega_0 t + c_2 \sin \omega_0 t + \frac{at}{2m\omega_0} \sin \omega_0 t$$

6.5.2 Near Resonance

An interesting and very important phenomenon is observed in the solution given in Eq.(6.33) when the damping coefficient c is zero, natural frequency ω_0 is not equal to input frequency ω, but $\omega_0 - \omega$, is small. According to Eq.(6.39),

$$y(t) = b_1 \cos \omega_0 t + b_2 \cos \omega_0 t + \left(\frac{a}{m\,(\omega_0^2 - \omega^2)}\right) \cos \omega t \tag{6.43}$$

is a solution of the differential Eq.(6.29) when $\omega \neq \omega_0$ and $c = 0$. Now, consider the initial condition $y'(0) = 0$ and $y(0) = 0$. Clearly, the parameters b_1 and b_2 present in Eq.(6.43) become

$$b_1 = -\frac{a}{m(\omega_0^2 - \omega^2)}$$

and

$$b_2 = 0$$

Hence the corresponding solution is

$$\begin{aligned} y(t) &= -\frac{a}{m(\omega_0^2 - \omega^2)} \cos \omega_0 t + \frac{a}{m\,(\omega_0^2 - \omega^2)} \cos \omega t \\ &= \frac{a}{m(\omega_0^2 - \omega^2)} (\cos \omega t - \cos \omega_0 t) \\ &= \frac{2a}{m(\omega_0^2 - \omega^2)} \sin\left((\omega_0 + \omega)\frac{t}{2}\right) \sin\left((\omega_0 - \omega)\frac{t}{2}\right) \end{aligned} \tag{6.44}$$

Let $\frac{\omega_0 - \omega}{2} = \epsilon$. Hence $\frac{\omega_0 + \omega}{2} \simeq \omega$. Clearly, under near resonance condition, the solution given in Eq.(6.44) reduces to

$$y(t) = \frac{2a \sin \epsilon t}{m(\omega_0^2 - \omega^2)} \sin \omega t \tag{6.45}$$

It is clear from Eq.(6.45) that the amplitude varies slowly and is bounded by $\frac{2a}{m(\omega_0^2 - \omega^2)}$ as t varies.

6.5.3 Forced Motion with Damping

According to Eq.(6.33),

$$\begin{aligned} y(t) &= e^{-\left(\frac{c}{2m}\right)t} \left(c_1\, e^{\left(\frac{\left(c^2 - 4mk\right)^{\frac{1}{2}}}{2m}\right)t} + c_2\, e^{-\left(\frac{\left(c^2 - 4mk\right)^{\frac{1}{2}}}{2m}\right)t} \right) \\ &\quad + \frac{a(k - m\omega^2)}{(k - m\,\omega^2)^2 + \omega^2 c^2} \cos \omega t + \frac{a\omega c}{(k - m\,\omega^2)^2 + \omega^2 c^2} \sin \omega t \end{aligned} \tag{6.46}$$

is the general solution of the differential Eq.(6.29) when $c \neq 0$. It is clear from Eq.(6.46) that for a large time period t, the homogeneous solution vanishes. In other words, the solution given in Eq.(6.46) becomes *steady-state* for a large time period t. For a short time period t, the homogeneous solution must be included in the general solution and the corresponding solution is called the *transient solution.*

Now, we are interested in amplitude of the solution for large time t. In other words, the amplitude of particular integral for the input $f(t) = a \cos \omega t$. Let $\tan \phi = \frac{\omega c}{k - \omega^2 m}$. Hence we have

$$\begin{aligned} \cos \phi &= \frac{1}{\sec \phi} \\ &= \left(\frac{1}{1 + \tan^2 \phi}\right)^{\frac{1}{2}} \end{aligned}$$

$$= \left(\frac{(k - m\omega^2)^2}{(k - m\omega^2)^2 + \omega^2 c^2} \right)^{\frac{1}{2}}$$

$$= \frac{k - m\omega^2}{((k - m\omega^2)^2 + \omega^2 c^2)^{\frac{1}{2}}}$$

and

$$\sin\phi = \left(1 + \cos^2\phi\right)^{\frac{1}{2}}$$

$$= \frac{\omega c}{((k - m\omega^2)^2 + \omega^2 c^2)^{\frac{1}{2}}}$$

Clearly, we have

$$y(t) = \left(\frac{a(k - m\omega^2)}{(k - m\omega^2)^2 + \omega^2 c^2} \right) \cos\omega t + \left(\frac{a\omega c}{(k - m\omega^2)^2 + \omega^2 c^2} \right) \sin\omega t$$

$$= \left(\frac{a}{(k - m\omega^2)^2 + \omega^2 c^2} \right) ((k - m\omega^2)\cos\omega t + \omega c \sin\omega t)$$

$$= \left(\left(\frac{k - m\omega^2}{((k - m\omega^2)^2 + \omega^2 c^2)^{\frac{1}{2}}} \right) \cos\omega t + \left(\frac{\omega c}{((k - m\omega^2)^2 + \omega^2 c^2)^{\frac{1}{2}}} \right) \sin\omega t \right)$$

$$\left(\frac{a}{((k - m\omega^2)^2 + \omega^2 c^2)^{\frac{1}{2}}} \right)$$

$$= \left(\frac{a}{((k - m\omega^2)^2 + \omega^2 c^2)^{\frac{1}{2}}} \right) (\cos\phi \cos\omega t + \sin\phi \sin\omega t)$$

$$= \left(\frac{a}{((k - m\omega^2)^2 + \omega^2 c^2)^{\frac{1}{2}}} \right) \cos(\omega t - \phi)$$

$$= \left(\frac{a}{(m^2(\omega_0^2 - \omega^2)^2 + \omega^2 c^2)^{\frac{1}{2}}} \right) \cos(\omega t - \phi)$$

where $\omega_0 = \left(\frac{k}{m}\right)^{\frac{1}{2}}$ and the angle ϕ is called the *phase angle*. In other words, the amplitude of the motion is

$$\delta(\omega) = \frac{a}{(m^2(\omega_0^2 - \omega^2)^2 + \omega^2 c^2)^{\frac{1}{2}}} \tag{6.47}$$

it is easy to verify that the amplitude is maximum when

$$\omega^2 = \omega_0^2 - \frac{c^2}{2m^2} \tag{6.48}$$

One can obtain the maximum amplitude of the motion by substituting the expression for ω^2 given in Eq.(6.48) into Eq.(6.47) by

$$\delta_{\max} = \frac{2ma}{c\,(4m^2\omega_0^2 - c^2)^{\frac{1}{2}}}$$

EXAMPLE 6.8 A 16 lb weight is suspended from a spring having spring constant $k = 5$ lb per feet in a medium with damping constant $c = 4$. Assuming that an external force $f(t) = 24\sin 10\,t$ acts on the mass and the weight is at rest at its equilibrium position, find the position of the weight at any time.

Solution: Clearly, the suspended mass m is given by $m = \frac{16}{32} = \frac{1}{2}$ pound. According to Eq.(6.28), the corresponding differential equation is given by

$$y'' + 8y' + 10y = 48\sin 10\,t \tag{6.49}$$

Clearly, Eq.(6.49) is a second order nonhomogeneous differential equation. According to Eq.(6.38), the solution of Eq.(6.49) is given by

$$y(t) = c_1 e^{-(4+\sqrt{6})t} + c_2 e^{-(4-\sqrt{6})t} - \frac{24}{725}\,(8\cos 10t + 9\sin 10t) \tag{6.50}$$

Using the initial conditions $y(0) = 0$ and $y'(0) = 0$ in Eq.(6.50), we have $c_1 = \frac{3}{145}\left(27 + \frac{8(4-\sqrt{6})}{5}\right)$ and $c_2 = \frac{3}{145}\left(27 - \frac{8(4+\sqrt{6})}{5}\right)$. Hence the required displacement of the suspended mass is

$$y(t) = \frac{3}{145}\left(27 + \frac{8(4-\sqrt{6})}{5}\right)e^{-(4+\sqrt{6})t} + \frac{3}{145}\left(27 - \frac{8(4+\sqrt{6})}{5}\right)e^{-(4-\sqrt{6})t}$$
$$- \frac{24}{725}\,(8\cos 10t + 9\sin 10t)$$

EXAMPLE 6.9 A 32 lb weight is suspended from a spring having spring constant $k = 4$ lb per feet in a medium without damping. If an external force $16\sin 2t$ is applied and initially the weight is at rest in the equilibrium position, show that the motion is one of resonance.

Solution: Clearly, the suspended mass m is given by $m = \frac{32}{32} = 1$ pound. According to Eq.(6.28), the corresponding differential equation is given by

$$y'' + 4y = 16\sin 2t \tag{6.51}$$

Clearly, Eq.(6.49) is a second order non-homogeneous differential equation. According to Section 5.3.2, the solution of Eq.(6.51) is given by

$$y(t) = c_1\cos 2t + c_2\sin 2t - 4t\cos 10t \tag{6.52}$$

Using the initial conditions $y(0) = 0$ and $y'(0) = 0$ in Eq.(6.52), we have $c_1 = 0$ and $c_2 = 2$. Hence the required displacement of the suspended mass is

$$y(t) = 2\sin 2t - 4t\sin 2t$$

and the velocity is

$$y(t) = 8t\sin 2t$$

Since the frequency of the external force is equal to the natural frequency, therefore, resonance occurs.

Exercises 6.5

1. A cart of weight 128 pounds is attached to a wall by a spring with spring constant $k = 64$ lb per feet. The cart is pulled 6 inches in the direction away from the wall and released with no initial velocity; simultaneously a periodic external force $F(t) = 32\sin 4t$ is applied to the cart. If there is no air resistance, then find the position $y(t)$ of the cart at time t.
2. A cart of weight 128 pounds is attached to a wall by a spring with spring constant $k = 64$ lb per feet. The cart is pulled 6 inches in the direction away from the wall and released with initial velocity 30.0 centimetres per second; simultaneously, a periodic external force $F(t) = 32\sin 4t$ is applied to the cart. If there is no air resistance, then find the position $y(t)$ of the cart at time t.
3. A 32 lb weight suspended from a coil spring stretches the spring to 2 feet. If the weight is pulled down 6 inches from the equilibrium position and released at $t = 0$, find the position $y(t)$ of the cart at time t.
4. A spring of negligible weight which stretches 1 inch under a tension of 2 lb is fixed at one end and is attached to a weight of w lb at the other end. It is found that resonance occurs when an axial periodic force $f(t) = 2\cos 2t$ lb acts on the weight. Show that when the free vibrations have died out the forced vibrations are given by $y(t) = ct\sin 2t$, and find the value of w and c.
5. A spring which stretches to an extent under a force $m\lambda^2 l$ is suspended from a support and a mass m at the lower end. Initially the mass is at rest in its equilibrium position. A vertical oscillation is now given to the support such that at any time t, its displacement below the initial position is $a\sin nt$. Show that the displacement $y(t)$ of the mass below the equilibrium satisfies the differential equation $y'' + \lambda^2 y = \lambda^2 a\sin nt$.

6.6 Electrical Circuit

One can transfer an electrical circuit containing a resistance R, inductance L, capacitance C and source of electromotive force $E(t)$ in series to a second order nonhomogeneous differential equation using Kirchhoff's law. Precisely, we have the following:

1. A source of electromotive force $E(t)$ which drives charge and produces a current $i(t)$.
2. A resistance R which opposes the current by producing a drop in electromotive force of magnitude Ri.
3. A inductor of inductance L which opposes any change in the current by producing a drop in electromotive force of magnitude Li'.

4. A condenser of capacitance C which stores the charge $q(t)$. The drop in electromotive force for resisting the inflow of additional charge is given by $\frac{q}{C} = \int_0^t \left(\frac{i(u)}{C}\right) du$.

Let $i(t)$ be the current in the circuit at time t. Let a known electromotive force $E(t)$ be impressed across the circuit. Our problem is to find the current in the system as a function of time when the switch is put on. According to Kirchhoff's voltage law, the voltage impressed on a circuit is equal to the sum of the voltage drops in the rest of the circuit. Hence we have

$$L\,i' + R\,i + \int_0^t \left(\frac{i(u)}{C}\right) du = E(t) \tag{6.53}$$

One can differentiate Eq.(6.53) to get

$$L\,i'' + R\,i' + \frac{i}{C} = E'(t) \tag{6.54}$$

and replace $i(t) = q'$ in Eq.(6.53) to get

$$L\,q'' + R\,q' + \frac{q}{C} = E(t) \tag{6.55}$$

Equations (6.54) and (6.55) are nonhomogeneous second order linear differential equations in current $i(t)$ and charge $q(t)$ respectively. One can solve either Eq.(6.54) or Eq.(6.55) according to the requirement.

Let $E(t) = E_0 \cos\omega\, t$. According to Section 5.3.2, the solution of Eq.(6.54) is given by

$$i(t) = e^{-\left(\frac{R}{2L}\right)t}\left(c_1\, e^{\left(\frac{\left(R^2-4\left(\frac{L}{C}\right)\right)^{\frac{1}{2}}}{2L}\right)t} + c_2\, e^{-\left(\frac{\left(R^2-4\left(\frac{L}{C}\right)\right)^{\frac{1}{2}}}{2L}\right)t}\right)$$
$$+\left(\frac{E_0\omega^2 R}{\left(\frac{1}{C} - L\omega^2\right)^2 + \omega^2 R^2}\right)\cos\omega t - \left(\frac{E_0\omega\left(\frac{1}{C} - L\omega^2\right)}{\left(\frac{1}{C} - L\omega^2\right)^2 + \omega^2 R^2}\right)\sin\omega t \tag{6.56}$$

Let $E(t) = E_0 \sin\omega\, t$. According to Section 5.3.2, the solution of Eq.(6.54) is given by

$$i(t) = e^{-\left(\frac{R}{2L}\right)t}\left(c_1\, e^{\left(\frac{\left(R^2-4\left(\frac{L}{C}\right)\right)^{\frac{1}{2}}}{2L}\right)t} + c_2\, e^{-\left(\frac{\left(R^2-4\left(\frac{L}{C}\right)\right)^{\frac{1}{2}}}{2L}\right)t}\right)$$
$$+\left(\frac{E_0\omega\left(\frac{1}{C} - L\omega^2\right)}{\left(\frac{1}{C} - L\omega^2\right)^2 + \omega^2 R^2}\right)\cos\omega t + \left(\frac{E_0\omega^2 R}{\left(\frac{1}{C} - L\omega^2\right)^2 + \omega^2 R^2}\right)\sin\omega t \tag{6.57}$$

Using initial current i_0 and i_0' in the circuit at $t = 0$, one can determine the constants c_1 and c_2 in the solution given in Eqs.(6.56) and (6.57).

Let $E(t) = E_0 \sin \omega t$. According to Section 5.3.2, the solution of Eq.(6.55) is given by

$$q(t) = e^{-\left(\frac{R}{2L}\right)t}\left(c_1\, e^{\left(\frac{\left(R^2-4\left(\frac{L}{C}\right)\right)^{\frac{1}{2}}}{2L}\right)t} + c_2\, e^{-\left(\frac{\left(R^2-4\left(\frac{L}{C}\right)\right)^{\frac{1}{2}}}{2L}\right)t}\right)$$

$$+\left(\frac{E_0\left(\frac{1}{C}-L\,\omega^2\right)}{\left(\frac{1}{C}-L\,\omega^2\right)^2+R^2\,\omega^2}\right)\sin \omega t - \left(\frac{E_0 R\,\omega}{\left(\frac{1}{C}-L\right)^2+R^2\,\omega^2}\right)\cos \omega t \tag{6.58}$$

Let $E(t) = E_0 \cos \omega t$. According to Section 5.3.2, the solution of Eq.(6.55) is given by

$$q(t) = e^{-\left(\frac{R}{2L}\right)t}\left(c_1\, e^{\left(\frac{\left(R^2-4\left(\frac{L}{C}\right)\right)^{\frac{1}{2}}}{2L}\right)t} + c_2\, e^{-\left(\frac{\left(R^2-4\left(\frac{L}{C}\right)\right)^{\frac{1}{2}}}{2L}\right)t}\right)$$

$$+\left(\frac{E_0\left(\frac{1}{C}-L\,\omega^2\right)}{\left(\frac{1}{C}-L\,\omega^2\right)^2+R^2\,\omega^2}\right)\cos \omega t + \left(\frac{E_0 R\,\omega}{\left(\frac{1}{C}-L\right)^2+R^2\,\omega^2}\right)\sin \omega t \tag{6.59}$$

Using initial charge q_0 and q_0' in the circuit at $t = 0$, one can determine the constants c_1 and c_2 in the solution given in Eqs.(6.58) and (6.59).

EXAMPLE 6.10 If a voltage of $100 \cos 10\,t$ V is impressed on a circuit containing elements with resistance $40\,\Omega$, inductance 1 H and capacitance 16×10^{-4} F in series, then find the charge at time t in the circuit.

Solution: Clearly, $R = 40$, $L = 1$, $C = 16 \times 10^{-4}$ and $E_0 = 100$. According to Eq.(6.55), we have

$$q'' + 40\,q' + 625\,q = 100 \cos 10t \tag{6.60}$$

Again, the solution of Eq.(6.60) according to Eq.(6.59) is given by

$$q(t) = e^{-20t}\left(c_1 \cos 15t + c_2 \sin 15t\right) + \frac{4}{697}\left(21 \cos 10\,t + 16 \sin 10t\right)$$

Imposing the initial conditions $q(0) = 0$ and $q'(0) = 0$, we have $c_1 = -\frac{84}{697}$ and $c_2 = -\frac{464}{2091}$. In other words, we have

$$q(t) = -\frac{4e^{-20t}}{697}\left(21 \cos 15t + \frac{116}{7} \sin 15t\right) + \frac{4}{697}\left(21 \cos 10t + 16 \sin 10t\right)$$

EXAMPLE 6.11 If a voltage of $10 \cos 10\,t$ V is impressed on a circuit containing elements with resistance $40\,\Omega$, inductance 1 H and capacitance 16×10^{-4} F in series, then find the current at time t in the circuit.

Solution: Clearly, $R = 40$, $L = 1$, $C = 16 \times 10^{-4}$ and $E_0 = 10$. According to Eq.(6.54), we have

$$i'' + 40\,i' + 625\,i = -100 \sin 10t \tag{6.61}$$

Again, the solution of Eq.(6.61) according to Eq.(6.56) is given by

$$i(t) = e^{-20t}\,(c_1 \cos 15t + c_2 \sin 15t) + \frac{4}{697}\,(16 \cos 10t - 21 \sin 10t)$$

Imposing the initial conditions $i(0) = 0$ and $i'(0) = 0$, we have $c_1 = -\frac{84}{697}$ and $c_2 = -\frac{464}{2091}$. In other words, we have

$$i(t) = -\frac{8e^{-20t}}{697}\left(8 \cos 15\,t + \frac{11}{3} \sin 15\,t\right) + \frac{4}{697}(16 \cos 10t - 21 \sin 10t)$$

Exercises 6.6

1. If a constant voltage of 12 V is impressed on a circuit containing elements with resistance $30\,\Omega$, inductance 10^{-4} H and capacitance 10^{-6} F in series, then determine the expressions for charge $q(t)$ on the capacitor and the current $i(t)$ in the circuit when $q(0) = i(0) = 0$.
2. If a voltage of $120 \cos \pi\, t$ V is impressed on a circuit containing elements with resistance $60\,\Omega$, inductance 10^{-3} H and capacitance 10^{-5} F in series at $t = 0$, then determine the expressions for steady-state current in the circuit.
3. A circuit is composed of resistance $80\,\Omega$, inductance 10^{-4} H and capacitance 10^{-6} F in series. If the capacitor has an initial charge of 10^{-4} C and there is no current flowing through the circuit at $t = 0$, then what is the current flowing through the circuit at $t = 10^{-4}$ second?
4. A circuit is composed of resistance $80\,\Omega$, inductance 10^{-4} H and capacitance 10^{-6} F in series. If a voltage of $120 \cos \pi\, t$ V is impressed on the circuit, then what is the current flowing through the circuit at $t = 10^{-4}$ second, then when the capacitor has an initial charge of 10^{-4} C and there is no current flowing through the circuit at $t = 0$?

Chapter 7

Series Solutions

There are certain variable differential equations which cannot be transformed into constant coefficient differential equations. Hence, the closed form solution of the corresponding differential equation is not possible. Alternatively, one can expect a solution in a series form called the *series solution.* In this chapter, we have illustrated the condition under which a series solution exists, and its region of validity. In other words, the *radius of convergence* R of the series solution. If $y(x) = \sum_{n=0}^{\infty} a_n(x - x_0)^n$ is any series solution of some differential equation, then the radius of convergence is the interval $|x - x_0| < \rho$, where $\rho = \lim_{n\to\infty} \left|\frac{a_n}{a_{n+1}}\right|$, called the *ratio test*, or $\rho = \lim_{n\to\infty} \left|\frac{1}{a_n}\right|^{\frac{1}{n}}$, called the *root test.* Again, $|x - x_0| < \rho$ is equivalent to $-\rho + x_0 < x < \rho + x_0$. If a_n and a_{n+k} are two consecutive terms of any series solution, then the radius of convergence is the interval $|x - x_0| < \rho^{\frac{1}{k}}$.

7.1 First Order Equations

Let $(x - x_0)y' + p(x)y = 0$ be any arbitrary first order linear differential equation. In order to have a series solution, the coefficient function $p(x)$ has to satisfy certain conditions. Let $x = x_0$ be any point, and we expect to have a series solution of the form $y(x) = \sum_{n=0}^{\infty} a_n(x - x_0)^{n+r}$. For this purpose, one has to verify the nature of the point $x = x_0$. If $\frac{p(x)}{x-x_0}$ can be expanded in terms of power series about the point $x = x_0$, then the point $x = x_0$ is said to be an *ordinary point*, and the form of the solution is $y(x) = \sum_{n=0}^{\infty} a_n(x - x_0)^n$. If $\frac{p(x)}{x-x_0}$ cannot be expanded in terms of power series about the point $x = x_0$ but $p(x)$ can be expanded in terms of power series about the point $x = x_0$, then the point $x = x_0$ is said to be the *regular singular point*, and the form of the solution is $y(x) = \sum_{n=0}^{\infty} a_n(x - x_0)^{n+r}$.

EXAMPLE 7.1 Find the series solution of the differential equation $xy' - y = x$ about $x = 1$.

Solution: It is easy to find that $\frac{1}{x} = \sum_{n=0}^{\infty} (-1)^n (x-1)^n$ for $0 < x < 2$ is the valid expansion. Hence $x = 1$ is an ordinary point. Again, $y(x) = \sum_{n=0}^{\infty} a_n (x-1)^n$ is the expected series solution. Clearly, we have

$$
\begin{aligned}
1 + (x-1) &= x \\
&= xy' - y \\
&= x\left(\sum_{n=1}^{\infty} n a_n (x-1)^{n-1}\right) - \sum_{n=0}^{\infty} a_n (x-1)^n \\
&= (1 + (x-1))\left(\sum_{n=1}^{\infty} n a_n (x-1)^{n-1}\right) - \sum_{n=0}^{\infty} a_n (x-1)^n \\
&= \sum_{n=1}^{\infty} n a_n (x-1)^{n-1} + \sum_{n=1}^{\infty} n a_n (x-1)^n - \sum_{n=0}^{\infty} a_n (x-1)^n \\
&= \sum_{n=0}^{\infty} (n+1) a_{n+1} (x-1)^n + \sum_{n=1}^{\infty} n a_n (x-1)^n - \sum_{n=0}^{\infty} a_n (x-1)^n \\
&= \sum_{n=1}^{\infty} ((n+1) a_{n+1} + n a_n - a_n)(x-1)^n + a_1 - a_0 \\
&= \sum_{n=2}^{\infty} ((n+1) a_{n+1} + (n-1) a_n)(x-1)^n + 2a_2(x-1) + a_1 - a_0 \qquad (7.1)
\end{aligned}
$$

By comparing both sides of Eq.(7.1), one can conclude that

$$a_1 - a_0 = 1$$

$$2a_2 = 1$$

and

$$(n+1)a_{n+1} + (n-1)a_n = 0 \qquad n \geq 2$$

In general, one can conclude that

$$a_1 = 1 + a_0$$

$$a_2 = \frac{1}{2}$$

and

$$a_{n+1} = -\frac{(n-1)a_n}{n+1}$$

$$
\begin{aligned}
&= (-1)^{n-1}\frac{1}{n(n+1)} \\
&= (-1)^{n-1}\left(\frac{1}{n} - \frac{1}{n+1}\right) \qquad n \geq 2
\end{aligned}
$$

The general series solution of the given equation is presented by

$$
\begin{aligned}
y(x) &= \sum_{n=0}^{\infty} a_n(x-1)^n \\
&= a_0 + (1+a_0)(x-1) + \frac{1}{2}(x-1)^2 + \sum_{n=2}^{\infty}(-1)^{n+1}\frac{1}{n(n+1)}(x-1)^{n+1} \\
&= a_0 + (1+a_0)(x-1) + \frac{1}{2}(x-1)^2 + \sum_{n=2}^{\infty}(-1)^{n+1}\left(\frac{1}{n} - \frac{1}{n+1}\right)(x-1)^{n+1} \\
&= a_0 + (1+a_0)(x-1) + (x-1)^2 - \frac{1}{2}(x-1)^2 + \sum_{n=2}^{\infty}(-1)^{n+1}\frac{1}{n}(x-1)^{n+1} \\
&\quad - \sum_{n=2}^{\infty}(-1)^{n+1}\frac{1}{n+1}(x-1)^{n+1} \\
&= a_0 + a_0(x-1) + \sum_{n=1}^{\infty}(-1)^{n+1}\frac{(x-1)^n}{n} + \sum_{n=1}^{\infty}(-1)^{n+1}\frac{(x-1)^{n+1}}{n} \\
&= a_0 x + \sum_{n=1}^{\infty}(-1)^{n+1}\frac{(x-1)^n}{n} + (x-1)\sum_{n=1}^{\infty}(-1)^{n+1}\frac{(x-1)^n}{n} \\
&= a_0 x + \ln x + (x-1)\ln x \\
&= a_0 x + x\ln x
\end{aligned}
$$

EXAMPLE 7.2 Find the series solution of the differential equation $y' - xy = 0$ about the point $x = 0$.

Solution: Clearly, $x = 0$ is an ordinary point. Hence $y(x) = \sum\limits_{n=0}^{\infty} a_n x^n$ is the expected series solution. Clearly, we have

$$
\begin{aligned}
0 &= y' - xy \\
&= \sum_{n=1}^{\infty} n a_n x^{n-1} - x\left(\sum_{n=0}^{\infty} a_n x^n\right) \\
&= \sum_{n=1}^{\infty} n a_n x^{n-1} - \sum_{n=0}^{\infty} a_n x^{n+1}
\end{aligned}
$$

$$= \sum_{n=1}^{\infty} n a_n x^{n-1} - \sum_{n=2}^{\infty} a_{n-2} x^{n-1}$$

$$= a_1 + \sum_{n=2}^{\infty} n a_n x^{n-1} - \sum_{n=2}^{\infty} a_{n-2} x^{n-1}$$

$$= a_1 + \sum_{n=2}^{\infty} (n a_n - a_{n-2}) x^{n-1} \tag{7.2}$$

By comparing both sides of Eq.(7.2), one can conclude that

$$a_1 = 0$$

and

$$n a_n - a_{n-2} = 0 \qquad n \geq 2$$

In general, one can conclude that

$$\begin{aligned} a_{2n+1} &= 0 \\ a_{2n} &= \frac{a_{2n-2}}{2n} \\ &= \frac{a_0}{2^n n!} \qquad n \geq 0 \end{aligned}$$

The general series solution of the given equation is presented by

$$\begin{aligned} y(x) &= \sum_{n=0}^{\infty} a_n x^n \\ &= \sum_{n=0}^{\infty} a_{2n} x^{2n} \\ &= \sum_{n=0}^{\infty} \left(\frac{a_0}{2^n n!} \right) x^{2n} \\ &= \sum_{n=0}^{\infty} \left(\frac{a_0}{n!} \right) \left(\frac{x^2}{2} \right)^n \\ &= \sum_{n=0}^{\infty} a_0 \frac{\left(\frac{x^2}{2} \right)^n}{n!} \\ &= a_0 \left(\sum_{n=0}^{\infty} \frac{\left(\frac{x^2}{2} \right)^n}{n!} \right) \\ &= a_0 e^{\frac{x^2}{2}} \end{aligned}$$

EXAMPLE 7.3 Find the series solution of the differential equation $y' - xy = x$ about the point $x = 0$.

Solution: Clearly, $x = 0$ is an ordinary point. Hence $y(x) = \sum_{n=0}^{\infty} a_n x^n$ is the expected series solution. Clearly, we have

$$\begin{aligned}
x &= y' - xy \\
&= \sum_{n=1}^{\infty} n a_n x^{n-1} - x\left(\sum_{n=0}^{\infty} a_n x^n\right) \\
&= \sum_{n=1}^{\infty} n a_n x^{n-1} - \sum_{n=0}^{\infty} a_n x^{n+1} \\
&= \sum_{n=1}^{\infty} n a_n x^{n-1} - \sum_{n=2}^{\infty} a_{n-2} x^{n-1} \\
&= a_1 + 2a_2 x + \sum_{n=3}^{\infty} n a_n x^{n-1} - a_0 x - \sum_{n=3}^{\infty} a_{n-2} x^{n-1} \\
&= a_1 + (2a_2 - a_0)x + \sum_{n=3}^{\infty} (n a_n - a_{n-2})\, x^{n-1} \qquad (7.3)
\end{aligned}$$

By comparing both sides of Eq.(7.3), one can conclude that

$$a_1 = 0$$

$$2a_2 - a_0 = 1$$

and

$$n a_n - a_{n-2} = 0 \qquad n \geq 3$$

In general, one can conclude that

$$\begin{aligned}
a_{2n+1} &= 0 \qquad n \geq 0 \\
a_{2n} &= \frac{a_{2n-2}}{2n} \\
&= \frac{1 + a_0}{2^n n!} \qquad n \geq 1
\end{aligned}$$

The general series solution of the given equation is given by

$$y(x) = \sum_{n=0}^{\infty} a_n x^n$$

$$= a_0 + \sum_{n=1}^{\infty} a_{2n}x^{2n}$$

$$= a_0 + \sum_{n=1}^{\infty} \left(\frac{1+a_0}{2^n n!}\right) x^{2n}$$

$$= a_0 + \sum_{n=1}^{\infty} \left(\frac{1+a_0}{n!}\right) \left(\frac{x^2}{2}\right)^n$$

$$= -1 + \sum_{n=0}^{\infty} (1+a_0) \frac{\left(\frac{x^2}{2}\right)^n}{n!}$$

$$= -1 + (1+a_0) \left(\sum_{n=0}^{\infty} \frac{\left(\frac{x^2}{2}\right)^n}{n!}\right)$$

$$= (1+a_0)e^{\frac{x^2}{2}} - 1$$

EXAMPLE 7.4 Find the series solution of the differential equation $xy' - y = 0$ about the point $x = 0$.

Solution: Clearly, $x = 0$ is a regular singular point. Hence the expected form of the solution is $y(x) = \sum_{n=0}^{\infty} a_n x^{n+r}$. Clearly, we have $y' = \sum_{n=0}^{\infty} (n+r)a_n x^{n+r-1}$. Again, we have

$$0 = xy' - y$$

$$= x \left(\sum_{n=0}^{\infty} (n+r)a_n x^{n+r-1}\right) - \sum_{n=0}^{\infty} a_n x^{n+r}$$

$$= \sum_{n=0}^{\infty} (n+r)a_n x^{n+r} - \sum_{n=0}^{\infty} a_n x^{n+r}$$

$$= \sum_{n=0}^{\infty} \left((n+r)a_n - a_n\right) x^{n+r}$$

$$= \sum_{n=0}^{\infty} (n+r-1)\, a_n x^{n+r}$$

Equating the lowest power of x to zero, we have $r = 1$. Again, $a_n = 0$ for $n \geq 1$. Hence the required solution is given by

$$y(x) = \sum_{n=0}^{\infty} a_n x^{n+1}$$

$$= a_0 x$$

EXAMPLE 7.5 Find the series solution of the differential equation $xy' - y = x^2$ about the point $x = 0$.

Solution: Clearly, $x = 0$ is a regular singular point. Hence the expected form of the solution is $y(x) = \sum_{n=0}^{\infty} a_n x^{n+r}$. Clearly, we have $y' = \sum_{n=0}^{\infty} (n+r) a_n x^{n+r-1}$. Again, we have

$$\begin{aligned}
x^2 &= xy' - y \\
&= x\left(\sum_{n=0}^{\infty}(n+r)a_n x^{n+r-1}\right) - \sum_{n=0}^{\infty} a_n x^{n+r} \\
&= \sum_{n=0}^{\infty}(n+r)a_n x^{n+r} - \sum_{n=0}^{\infty} a_n x^{n+r} \\
&= \sum_{n=0}^{\infty}\left((n+r)a_n - a_n\right) x^{n+r} \\
&= \sum_{n=0}^{\infty}(n+r-1)\, a_n x^{n+r}
\end{aligned}$$

Equating the lowest power of x to zero, we have $r = 1$. Again, $a_n = 0$ for $n \geq 2$ and $a_1 = 1$. Hence the required solution is given by

$$\begin{aligned}
y(x) &= \sum_{n=0}^{\infty} a_n x^{n+1} \\
&= a_0 x + x^2
\end{aligned}$$

EXAMPLE 7.6 Find the series solution of the differential equation $xy' - 3y = k$ about the point $x = 0$.

Solution: Clearly, $x = 0$ is a regular singular point. According to Exercise 2 given in Exercises 7.1, the expected form of the solution is $y(x) = \sum_{n=0}^{\infty} a_n\, x^n$. Clearly, we have $y' = \sum_{n=0}^{\infty} n\, a_n\, x^{n-1}$. Again, we have

$$\begin{aligned}
k &= xy' - y \\
&= x\left(\sum_{i=1}^{\infty} n\, a_n\, x^{n-1}\right) - 3\sum_{i=0}^{\infty} a_n\, x^n \\
&= \sum_{i=1}^{\infty} n\, a_n\, x^n - 3\sum_{i=0}^{\infty} a_n\, x^n \\
&= -3\, a_0 + \sum_{i=1}^{\infty}(n-3)\, a_n\, x^n
\end{aligned}$$

Equating the coefficients of different powers of x, we have $a_0 = -\frac{k}{3}$, $a_1 = 0$, $a_2 = 0$ and $a_n = 0$ for $n \geq 4$. Hence the required solution is given by

$$y(x) = \sum_{n=0}^{\infty} a_n x^n$$
$$= a_3 x^3 - \frac{k}{3}$$

Exercises 7.1

1. Show that the differential equation $(x - x_0)y' + p(x)y = f(x)$ has a series solution of the form $y(x) = \sum\limits_{n=0}^{\infty} a_n(x - x_0)^n$ if and only if $\frac{p(x)}{x-x_0}$ and $\frac{f(x)}{x-x_0}$ can be expanded in series about the point $x = x_0$.
2. Show that the differential equation $(x - x_0)y' + p(x)y = f(x)$ has a series solution of the form $y(x) = \sum\limits_{n=0}^{\infty} a_n(x - x_0)^{n+r}$ if and only if $\frac{p(x)}{x-x_0}$ and $\frac{f(x)}{x-x_0}$ can be expanded in series about the point $x = x_0$, and the non-vanishing lowest degree term of $f(x)$ is greater than r.
3. Find the two independent series solutions of the differential equation $y' + y = 0$ about the point $x = 0$.
4. Find the two independent series solutions of the differential equation $y' + xy = 0$ about the point $x = 0$.
5. Find the two independent series solutions of the differential equation $y' + (x - 1)y = 0$ about the point $x = 0$.
6. Find the two independent series solutions of the differential equation $y' + (1 + x)y = 0$ about the point $x = 0$.
7. Find the two independent series solutions of the differential equation $y' + y \sin x = 0$ about the point $x = 0$.
8. Find the two independent series solutions of the differential equation $y' + y \cos x = 0$ about the point $x = 0$.
9. Find the two independent series solutions of the differential equation $y' + e^x\, y = 0$ about the point $x = 0$.
10. Find the two independent series solutions of the differential equation $y' + e^{-x}\, y = 0$ about the point $x = 0$.

7.2 Second Order Equations

Consider $(x-x_0)^2 y'' + (x-x_0)p(x)y' + q(x)y = 0$ be any arbitrary second order linear differential equation. In order to have a series solution, the coefficient functions $p(x)$ and $q(x)$ have to satisfy certain conditions. Let $x = x_0$ be any point, and we expect to have a series solution of the form $y(x) = \sum\limits_{n=0}^{\infty} a_n(x - x_0)^{n+r}$. For this purpose, one has to verify the nature of the point $x = x_0$. If $\frac{p(x)}{x-x_0}$ and $\frac{q(x)}{(x-x_0)^2}$ can be expanded in terms of power series about the point $x = x_0$, then the point $x = x_0$ is said to be an *ordinary point*, and the form of the solution is $y(x) = \sum_{n=0}^{\infty} a_n(x-x_0)^n$. If at least one or both $\frac{p(x)}{x-x_0}$ and $\frac{q(x)}{(x-x_0)^2}$ cannot be expanded in terms of power series about the point $x = x_0$, but $p(x)$ and $q(x)$ can be expanded in terms of power series about the point $x = x_0$, then the point $x = x_0$ is said to be a *regular singular point*, and the form of the solution is $y(x) = \sum_{n=0}^{\infty} a_n(x-x_0)^{n+r}$.

EXAMPLE 7.7 Let $y'' + p(x)y' + y = 0$ be any differential equation. If one assumes $y(x) = \sum_{n=0}^{\infty} a_n(x-x_0)^n$ as the series solution of the given differential equation, then what should be the unknown function $p(x)$.

Solution: Obviously, the question is very tricky, and the answer is very simple. One can say that both $p(x)$ and $(x-x_0)p(x)$ can be expanded in series about $x = x_0$. In other words, one can conclude that $p(x) = \sum_{n=0}^{\infty} c_n(x-x_0)^n$.

EXAMPLE 7.8 Let $y'' + p(x)y' + y = 0$ be any differential equation. If one assumes $y(x) = \sum_{n=0}^{\infty} a_n(x-x_0)^{n+r}$ as the series solution of the given differential equation, then what should be the unknown function $p(x)$.

Solution: Obviously, the question is very tricky, and the answer is very simple. One can say that $(x-x_0)p(x)$ can be expanded in series about $x = x_0$, but $p(x)$ can not be expanded in series about $x = x_0$. In other words, one can conclude that $p(x) = \frac{b}{x-x_0} + \sum_{n=0}^{\infty} c_n(x-x_0)^n$ where $b \neq 0$.

7.2.1 Ordinary Point Solution

EXAMPLE 7.9 Find the general series solution of the differential equation $y'' + x^2y' + 2xy = 0$ about the point $x = 0$.

Solution: Clearly, $x = 0$ is an ordinary point. Hence the expected form of the solution is $y(x) = \sum_{n=0}^{\infty} a_n x^n$. Hence we have

$$\begin{aligned} 0 &= \sum_{n=2}^{\infty} n(n-1)a_n x^{n-2} + x^2\left(\sum_{n=1}^{\infty} n a_n x^{n-1}\right) + 2x\left(\sum_{n=0}^{\infty} a_n x^n\right) \\ &= \sum_{n=2}^{\infty} n(n-1)a_n x^{n-2} + \sum_{n=1}^{\infty} a_n n x^{n+1} + \sum_{n=0}^{\infty} 2a_n x^{n+1} \\ &= \sum_{n=-1}^{\infty} (n+3)(n+2)a_{n+3} x^{n+1} + \sum_{n=1}^{\infty} a_n n x^{n+1} + \sum_{n=0}^{\infty} 2a_n x^{n+1} \end{aligned}$$

$$= 2a_2 + (6a_3 + 2a_0)x + \sum_{n=1}^{\infty}(n+3)(n+2)a_{n+3}x^{n+1} + \sum_{n=1}^{\infty} nx^{n+1} + \sum_{n=1}^{\infty} 2a_n x^{n+1}$$

$$= 2a_2 + (6a_3 + 2a_0)x + \sum_{n=1}^{\infty}\left((n+3)(n+2)a_{n+3} + na_n + 2a_n\right)x^{n+1}$$

$$= 2a_2 + 2(3a_3 + a_0)x + \sum_{n=1}^{\infty}(n+2)\left((n+3)a_{n+3} + a_n\right)x^{n+1} \tag{7.4}$$

Equating the coefficient of each power of x in Eq.(7.4) to zero, we have

$$a_2 = 0$$

$$3a_3 + a_0 = 0$$

$$(n+3)a_{n+3} + a_n = 0 \qquad n \geq 1$$

In other words, one can conclude that

$$a_2 = 0$$

$$a_3 = -\frac{a_0}{3}$$

$$a_{n+3} = -\frac{a_n}{n+3} \qquad n \geq 1$$

Hence the required solution is given by

$$y(x) = a_0\left(1 - \frac{x^3}{3} + \frac{x^6}{18} + \cdots\right) + a_1\left(x - \frac{x^4}{4} + \frac{x^7}{28} + \cdots\right)$$

EXAMPLE 7.10 Find the general series solution of the differential equation $y'' + x^2y' - 2xy = 0$ about the point $x = 0$.

Solution: Clearly, $x = 0$ is an ordinary point. Hence the expected form of the solution is $y(x) = \sum_{n=0}^{\infty} a_n x^n$. Hence we have

$$0 = \sum_{n=2}^{\infty} n(n-1)a_n x^{n-2} + x^2\left(\sum_{n=1}^{\infty} na_n x^{n-1}\right) - 2x\left(\sum_{n=0}^{\infty} a_n x^n\right)$$

$$= \sum_{n=2}^{\infty} n(n-1)a_n x^{n-2} + \sum_{n=1}^{\infty} a_n n x^{n+1} - \sum_{n=0}^{\infty} 2a_n x^{n+1}$$

$$= \sum_{n=-1}^{\infty}(n+3)(n+2)a_{n+3}x^{n+1} + \sum_{n=1}^{\infty} a_n n x^{n+1} - \sum_{n=0}^{\infty} 2a_n x^{n+1}$$

$$= 2a_2 + (6a_3 - 2a_0)x + \sum_{n=1}^{\infty}(n+3)(n+2)a_{n+3}x^{n+1} + \sum_{n=1}^{\infty} nx^{n+1} - \sum_{n=1}^{\infty} 2a_n x^{n+1}$$

$$= 2a_2 + (6a_3 - 2a_0)x + \sum_{n=1}^{\infty} \left((n+3)(n+2)a_{n+3} + na_n - 2a_n\right) x^{n+1}$$

$$= 2a_2 + 2(3a_3 - a_0)x + \sum_{n=1}^{\infty} \left((n+3)(n+2)a_{n+3} + (n-2)a_n\right) x^{n+1} \tag{7.5}$$

Equating the coefficient of each power of x in Eq.(7.5) to zero, we have

$$a_2 = 0$$

$$3a_3 - a_0 = 0$$

$$(n+3)(n+2)a_{n+3} + (n-2)a_n = 0 \qquad n \geq 1$$

In other words, one can conclude that

$$a_2 = 0$$

$$a_3 = \frac{a_0}{3}$$

$$a_{n+3} = -\left(\frac{n-2}{(n+3)(n+2)}\right) a_n \qquad n \geq 1$$

Hence the required solution is given by

$$y(x) = a_0 \left(1 + \frac{x^3}{3} - \frac{x^6}{90} + \cdots\right) + a_1 \left(x + \frac{x^4}{12} - \frac{x^7}{252} + \cdots\right)$$

EXAMPLE 7.11 Find the general series solution of the differential equation $y'' + x^2y' - xy = 0$ about the point $x = 0$.

Solution: Clearly, $x = 0$ is an ordinary point. Hence the expected form of the solution is $y(x) = \sum\limits_{n=0}^{\infty} a_n x^n$. Hence we have

$$0 = \sum_{n=2}^{\infty} n(n-1)a_n x^{n-2} + x^2 \left(\sum_{n=1}^{\infty} na_n x^{n-1}\right) - x \left(\sum_{n=0}^{\infty} a_n x^n\right)$$

$$= \sum_{n=2}^{\infty} n(n-1)a_n x^{n-2} + \sum_{n=1}^{\infty} a_n n x^{n+1} - \sum_{n=0}^{\infty} a_n x^{n+1}$$

$$= \sum_{n=-1}^{\infty} (n+3)(n+2)a_{n+3} x^{n+1} + \sum_{n=1}^{\infty} a_n n x^{n+1} - \sum_{n=0}^{\infty} a_n x^{n+1}$$

$$= 2a_2 + (6a_3 - a_0)x + \sum_{n=1}^{\infty} (n+3)(n+2)a_{n+3} x^{n+1} + \sum_{n=1}^{\infty} n x^{n+1} - \sum_{n=1}^{\infty} a_n x^{n+1}$$

$$= 2a_2 + (6a_3 - a_0)x + \sum_{n=1}^{\infty} ((n+3)(n+2)a_{n+3} + na_n - a_n)\, x^{n+1}$$

$$= 2a_2 + (6a_3 - a_0)x + \sum_{n=1}^{\infty} ((n+3)(n+2)a_{n+3} + (n-1)a_n)\, x^{n+1} \tag{7.6}$$

Equating the coefficient of each power of x in Eq.(7.6) to zero, we have

$$a_2 = 0$$

$$6a_3 - a_0 = 0$$

$$(n+3)(n+2)a_{n+3} + (n-1)a_n = 0 \qquad n \geq 1$$

In other words, one can conclude that

$$a_2 = 0$$

$$a_3 = \frac{a_0}{6}$$

$$a_{n+3} = -\left(\frac{n-1}{(n+3)(n+2)}\right) a_n \qquad n \geq 1$$

Hence the required solution is given by

$$y(x) = a_0 \left(1 + \frac{x^3}{6} - \frac{x^6}{90} + \cdots\right) + a_1 x$$

EXAMPLE 7.12 Find the general series solution of the differential equation $y'' - 9y = 0$ about the point $x = 0$.

Solution: Clearly, $x = 0$ is an ordinary point. Hence the expected type of solution is $y(x) = \sum_{n=0}^{\infty} a_n x^n$. Again, we have

$$0 = y'' - 9y$$

$$= \sum_{n=2}^{\infty} n(n-1)a_n x^{n-2} - 9\left(\sum_{n=0}^{\infty} a_n x^n\right)$$

$$= \sum_{n=0}^{\infty} (n+2)(n+1)a_{n+2}x^n - \sum_{n=0}^{\infty} 9a_n x^n$$

$$= \sum_{n=0}^{\infty} ((n+2)(n+1)a_{n+2} - 9a_n)\, x^n \tag{7.7}$$

Equating the coefficient of each power of x in Eq.(7.7) to zero, we have

$$(n+2)(n+1)a_{n+2} - 9a_n = 0 \qquad n \geq 0$$

In other words, one can conclude that

$$\begin{aligned} a_{2n} &= \left(\frac{9}{2n(2n-1)}\right) a_{2(n-1)} \\ &= \left(\frac{3^{2n}}{(2n)!}\right) a_0 \qquad n \geq 0 \end{aligned}$$

and

$$\begin{aligned} a_{2n+1} &= \left(\frac{9}{2n(2n+1)}\right) a_{2(n-1)} \\ &= \left(\frac{3^{2n}}{(2n+1)!}\right) a_1 \qquad n \geq 0 \end{aligned}$$

Hence the required solution is given by

$$\begin{aligned} y(x) &= \sum_{n=0}^{\infty} a_n x^n \\ &= \sum_{n=1}^{\infty} a_{2n} x^{2n} + \sum_{n=0}^{\infty} a_{2n+1} x^{2n+1} \\ &= \sum_{n=1}^{\infty} \left(\frac{3^{2n}}{(2n)!}\right) a_0 x^{2n} + \sum_{n=0}^{\infty} \left(\frac{3^{2n}}{(2n+1)!}\right) a_1 x^{2n+1} \\ &= a_0 \left(\sum_{n=1}^{\infty} \left(\frac{(3x)^{2n}}{(2n)!}\right)\right) + \frac{a_1}{3} \left(\sum_{n=0}^{\infty} \left(\frac{(3x)^{2n+1}}{(2n+1)!}\right)\right) \\ &= (\alpha+\beta) \left(\sum_{n=1}^{\infty} \left(\frac{(3x)^{2n}}{(2n)!}\right)\right) + (\alpha-\beta) \left(\sum_{n=0}^{\infty} \left(\frac{(3x)^{2n+1}}{(2n+1)!}\right)\right) \\ &= \alpha \left(\sum_{n=0}^{\infty} \frac{(3x)^n}{n!}\right) + \beta \left(\sum_{n=0}^{\infty} (-1)^n \frac{(3x)^n}{n!}\right) \\ &= \alpha e^{3x} + \beta e^{-3x} \end{aligned}$$

EXAMPLE 7.13 Find the general series solution of the differential equation $y'' + 9y = 0$ about the point $x = 0$.

Solution: Clearly, $x = 0$ is an ordinary point. Hence the expected type of solution is $y(x) = \sum_{n=0}^{\infty} a_n x^n$. Again, we have

$$0 = y'' + 9y$$

$$
\begin{aligned}
&= \sum_{n=2}^{\infty} n(n-1)a_n x^{n-2} + 9\left(\sum_{n=0}^{\infty} a_n x^n\right) \\
&= \sum_{n=0}^{\infty} (n+2)(n+1)a_{n+2}x^n + \sum_{n=0}^{\infty} 9a_n x^n \\
&= \sum_{n=0}^{\infty} \left((n+2)(n+1)a_{n+2} + 9a_n\right) x^n \qquad (7.8)
\end{aligned}
$$

Equating the coefficient of each power of x in Eq.(7.8) to zero, we have

$$(n+2)(n+1)a_{n+2} + 9a_n = 0 \qquad n \geq 0$$

In other words, one can conclude that

$$
\begin{aligned}
a_{2n} &= -\left(\frac{9}{2n(2n-1)}\right) a_{2(n-1)} \\
&= (-1)^n \left(\frac{3^{2n}}{(2n)!}\right) a_0 \qquad n \geq 0
\end{aligned}
$$

and

$$
\begin{aligned}
a_{2n+1} &= -\left(\frac{9}{2n(2n+1)}\right) a_{2n-1} \\
&= (-1)^{n+1} \left(\frac{3^{2n}}{(2n+1)!}\right) a_1 \qquad n \geq 0
\end{aligned}
$$

Hence the required solution is given by

$$
\begin{aligned}
y(x) &= \sum_{n=0}^{\infty} a_n x^n \\
&= \sum_{n=0}^{\infty} a_{2n} x^{2n} + \sum_{n=0}^{\infty} a_{2n+1} x^{2n+1} \\
&= \sum_{n=0}^{\infty} (-1)^n \left(\frac{3^{2n}}{(2n)!}\right) a_0 x^{2n} + \sum_{n=0}^{\infty} (-1)^{n+1} \left(\frac{3^{2n}}{(2n+1)!}\right) a_1 x^{2n+1} \\
&= a_0 \left(\sum_{n=0}^{\infty} (-1)^n \left(\frac{(3x)^{2n}}{(2n)!}\right)\right) + \left(\frac{a_1}{3}\right)\left(\sum_{n=0}^{\infty} (-1)^n \left(\frac{(3x)^{2n+1}}{(2n+1)!}\right)\right) \\
&= \alpha \left(\sum_{n=0}^{\infty} (-1)^n \frac{(3x)^{2n}}{n!}\right) + \beta \left(\sum_{n=0}^{\infty} (-1)^n \frac{(3x)^{2n+1}}{(2n+1)!}\right) \\
&= \alpha \cos 3x + \beta \sin 3x
\end{aligned}
$$

EXAMPLE 7.14 Find the general series solution of the differential equation $y'' - 2x^2y' + 4xy = x^2 + 2x + 4$ about the point $x = 0$.

Solution: Clearly, $x = 0$ is an ordinary point. Hence the expected type of solution is $y(x) = \sum_{n=0}^{\infty} a_n x^n$. Again, we have

$$\begin{aligned}
x^2 + 2x + 4 &= y'' - 2x^2y' + 4xy \\
&= \sum_{n=2}^{\infty} n(n-1)a_n x^{n-2} - 2x^2 \left(\sum_{n=1}^{\infty} na_n x^{n-1}\right) + 4x\left(\sum_{n=0}^{\infty} a_n x^n\right) \\
&= \sum_{n=2}^{\infty} n(n-1)a_n x^{n-2} - 2\left(\sum_{n=1}^{\infty} na_n x^{n+1}\right) + 4\left(\sum_{n=0}^{\infty} a_n x^{n+1}\right) \\
&= \sum_{n=-1}^{\infty} (n+3)(n+2)a_{n+3} x^{n+1} - 2\left(\sum_{n=1}^{\infty} na_n x^{n+1}\right) + 4\left(\sum_{n=0}^{\infty} a_n x^{n+1}\right) \\
&= \sum_{n=2}^{\infty} \left((n+3)(n+2)a_{n+3} - 2(n-2)a_n\right) x^{n+1} + 2a_2 + (6a_3 + 4a_0)x \\
&\quad + (12a_4 + 4a_1)x^2 \qquad (7.9)
\end{aligned}$$

Comparing the coefficients of each power of x in Eq.(7.9), we have

$$a_2 = 2$$

$$3a_3 + 2a_0 = 1$$

$$12a_4 + 4a_1 = 1$$

and

$$a_{n+3} = \left(\frac{2(n-2)}{(n+3)(n+2)}\right) a_n \qquad n \geq 2$$

In other words, we have

$$a_2 = 2$$

$$a_3 = \frac{1}{3} - \frac{2}{3}a_0$$

$$a_4 = \frac{1}{12} - \frac{1}{3}a_1$$

and

$$a_{n+3} = \left(\frac{2(n-2)}{(n+3)(n+2)}\right) a_n \qquad n \geq 2 \qquad (7.10)$$

One can conclude by Eq.(7.10) that

$$a_{3n+2} = 0 \qquad n \geq 1$$

$$a_{3n} = \left(\frac{2}{3}\right)^{n-1} \left(\prod_{i=1}^{n-1} \frac{3i-2}{(i+1)(3i+2)}\right) \left(\frac{1}{3} - \frac{2}{3}a_0\right) \qquad n \geq 2$$

and

$$a_{3n+1} = \left(\frac{2}{3}\right)^{n-1} \left(\prod_{i=1}^{n-1} \frac{3i-1}{(i+1)(3i+4)}\right) \left(\frac{1}{12} - \frac{1}{3}a_1\right) \qquad n \geq 2$$

Hence the general solution is given by

$$y(x) = 2x^2 + \left(\frac{1}{3} - \frac{2}{3}a_0\right) \left(x^3 + \sum_{n=2}^{\infty} \left(\frac{2}{3}\right)^{n-1} \left(\prod_{i=1}^{n-1} \frac{3i-2}{(i+1)(3i+2)}\right) x^{3n}\right)$$

$$+ \left(\frac{1}{12} - \frac{1}{3}a_1\right) \left(x^4 + \sum_{n=2}^{\infty} \left(\frac{2}{3}\right)^{n-1} \left(\prod_{i=1}^{n-1} \frac{3i-1}{(i+1)(3i+4)}\right) x^{3n+1}\right) + a_0 + a_1 x$$

Exercises 7.2.1

1. The differential equation $(x-x_0)^2y''+(x-x_0)p(x)y'+q(x)y = f(x)$ has a series solution of the form $y(x) = \sum_{n=0}^{\infty} a_n(x-x_0)^n$ if and only if $\frac{p(x)}{x-x_0}$, $\frac{q(x)}{(x-x_0)^2}$ and $\frac{f(x)}{(x-x_0)^2}$ can be expanded in series about the point $x = x_0$.
2. Find the power series solution of the differential equation $(x^2+2)y'' - xy' - 3y = 0$ about $x = 0$.
3. Find the power series solution of the differential equation $x(2-x)y'' - 6(x-1)y' - 4y = 0$ about $x = 1$.
4. Find the power series solution of the differential equation $(x^2+1)y'' + xy' + xy = 0$ about $x = 0$.
5. Find the power series solution of the differential equation $(x-1)y'' - (3x-2)y' + 2xy = 0$ about $x = 0$.
6. Find the power series solution of the differential equation $(x+3)y'' + (x+2)y' + y = 0$ about $x = 0$.
7. Find the power series solution of the differential equation $y'' - x^3y = 0$ about $x = 0$.
8. Find the power series solution of the differential equation $y''+8xy'-4y = 0$ about $x = 0$.

7.2.2 Regular Singular Point Solution

Theorem 7.2.1 If both $p(x)$ and $q(x)$ are analytic at $x = 0$ with $p(x) = \sum_{n=0}^{\infty} c_n x^n$ and $q(x) = \sum_{n=0}^{n} b_n x^n$ but at least $\frac{p(x)}{x}$ or $\frac{q(x)}{x^2}$ is not analytic at $x = 0$, then the differential equation

$x^2y'' + xp(x)y' + q(x)y = 0$ admits a solution of the form $y(x) = \sum_{n=0}^{\infty} a_n x^{n+r}$ where $a_0 \neq 0$ and r is a root of the equation $r^2 + (c_0 - 1)r + b_0 = 0$.

Proof: The given differential equation admits a solution of the type $y(x) = \sum_{n=0}^{\infty} a_n x^{n+r}$ where $a_0 \neq 0$ provides all the unknown parameters a_n for $0 \leq n < \infty$ are well defined and r is determined. It is easy to verify that

$$\begin{aligned}
x^2y'' + xp(x)y' + q(x)y &= \sum_{n=0}^{\infty}(n+r)(n+r-1)a_n x^{n+r} + p(x)\sum_{n=0}^{\infty}(n+r)a_n x^{n+r} \\
&\quad + q(x)\sum_{n=0}^{\infty} a_n x^{n+r} \\
&= \sum_{n=0}^{\infty}\left(\sum_{k=0}^{n}(n+r-k)c_k a_{n-k}\right)x^{n+r} + \sum_{n=0}^{\infty}\left(\sum_{k=0}^{n} b_k a_{n-k}\right)x^{n+r} \\
&\quad + \sum_{n=0}^{\infty}(n+r)(n+r-1)a_n x^{n+r} \\
&= \sum_{n=0}^{\infty}\left(\sum_{k=0}^{n}\left((n+r-k)c_k + b_k\right)a_{n-k}\right)x^{n+r} \\
&\quad + \sum_{n=0}^{\infty}(n+r)(n+r-1)a_n x^{n+r} \\
&= \sum_{n=1}^{\infty}\Bigg(((n+r)(n+r-1) + (n+r)c_0 + b_0)a_n \\
&\quad + \sum_{k=1}^{n}\left((n+r-k)c_k + b_k\right)a_{n-k}\Bigg)x^{n+r} + (r(r-1) + rc_0 + b_0)a_0 x^r \\
&= F(r)a_0 x^r + \sum_{n=1}^{\infty}\left(F(r+n)a_n + \sum_{k=1}^{n} G(n,\, r,\, k)a_{n-k}\right)x^{n+r}
\end{aligned}$$

where

$$F(n+r) = (n+r)(n+r-1) + (n+r)c_0 + b_0$$

and

$$G(n,\, r,\, k) = (n+r-k)c_k + b_k$$

In order to make $y(x)$ as a solution with $a_0 \neq 0$, we have to have

$$\begin{aligned}
F(r) &= r(r-1) + rc_0 + b_0 \\
&= 0
\end{aligned}$$

and for $n \geq 1$,

$$\begin{aligned} 0 &= F(n+r)a_n + \sum_{k=1}^{n} G(n, r, k)a_{n-k} \\ &= ((n+r)(n+r-1) + (n+r)c_0 + b_0)a_n + \sum_{k=1}^{n} ((n+r-k)c_k + b_k)\, a_{n-k} \end{aligned}$$

One can make

$$((n+r)(n+r-1) + (n+r)c_0 + b_0)a_n + \sum_{k=1}^{n} ((n+r-k)c_k + b_k)\, a_{n-k} = 0$$

by the condition

$$a_n = -\frac{\sum\limits_{k=1}^{n} G(n, r, k)a_{n-k}}{F(n+r)} \qquad n \geq 1 \tag{7.11}$$

whereas

$$F(r) = r(r-1) + rc_0 + b_0 = r^2 + (c_0 - 1)r + b_0 = 0$$

by suitable r given by

$$r = \frac{1 - c_0 \pm \left((c_0 - 1)^2 - 4b_0\right)^{\frac{1}{2}}}{2}$$

In other words, we have

$$F(n+r) = (n+r-r_1)(n+r-r_2)$$

Let $r_1 - r_2 = s \geq 0$. It is easy to conclude that

$$\begin{aligned} F(n+r_1) &= n(n+r_1-r_2) \\ &= n(n+s) \end{aligned}$$

and

$$\begin{aligned} F(n+r_2) &= n(n+r_2-r_1) \\ &= n(n-s) \end{aligned}$$

Since $F(n+r_1) \neq 0$ for all $n \geq 1$; therefore, all the unknown parameters a_n for $0 \leq n < \infty$ are well defined by Eq.(7.11). Hence the given differential equation admits a solution of the type $y(x) = \sum\limits_{n=0}^{\infty} a_n x^{n+r}$ where $a_0 \neq 0$.

Theorem 7.2.2 If both $p(x)$ and $q(x)$ are analytic at $x = 0$ with $p(x) = \sum\limits_{n=0}^{\infty} c_n x^n$ and $q(x) = \sum\limits_{n=0}^{n} b_n x^n$ but at least $\frac{p(x)}{x}$ or $\frac{q(x)}{x^2}$ is not analytic at $x = 0$, then the differential equation

$x^2y''+xp(x)y'+q(x)y=0$ admits a solution of the form $y(x)=y_1(x)\ln x+\sum_{n=0}^{\infty}A_nx^{n+r}$ where $y_1(x)=\sum_{n=0}^{\infty}a_nx^{n+r}$ and r is the only root of the equation $r^2+(c_0-1)r+b_0=0$.

Proof: We know that $y_1(x)=\sum_{n=0}^{\infty}a_nx^{n+r}$ is a solution of the given differential equation by Theorem 7.2.1. Again, r is the only root of the equation $r^2+(c_0-1)r+b_0=0$. Hence $r=\frac{1-c_0}{2}$. In other words, we have $2r+c_0=1$. Let the second solution be $y_2(x)=u(x)y_1(x)$. Again, we have

$$\begin{aligned}
y_2' &= uy_1'+u'y_1\\
y_2'' &= u''y_1+2u'y_1'+uy_1''
\end{aligned}$$

Substituting y_2, y_2' and y_2'' in the given differential equation, we have

$$\begin{aligned}
0 &= x^2y''+xp(x)y'+q(x)y\\
&= x^2(u''y_1+2u'y_1'+uy_1'')+xp(x)(uy_1'+u'y_1)+q(x)uy_1\\
&= (x^2y_1''+xp(x)y_1'+q(x)y_1)u+x^2(u''y_1+2u'y_1')+xp(x)u'y_1\\
&= x^2y_1u''+(2y_1'x^2+xp(x)y_1)u'
\end{aligned}$$

In other words, we have

$$\begin{aligned}
0 &= u''+\left(2\left(\frac{y_1'}{y_1}\right)+\frac{p(x)}{x}\right)u'\\
&= u''+\left(2\left(\frac{y_1'}{y_1}\right)+\sum_{n=0}^{\infty}c_nx^{n-1}\right)u'\\
&= u''+\left(2\left(\frac{x^{r-1}\sum_{n=0}^{\infty}(r+n)a_nx^n}{x^r\sum_{n=0}^{\infty}a_nx^n}\right)+\sum_{n=0}^{\infty}c_nx^{n-1}\right)u'\\
&= u''+\left(\frac{2}{x}\left(\frac{\sum_{n=0}^{\infty}(r+n)a_nx^n}{\sum_{n=0}^{\infty}a_nx^n}\right)+\sum_{n=0}^{\infty}c_nx^{n-1}\right)u'\\
&= u''+\left(\frac{2}{x}\left(r+\sum_{n=1}^{\infty}B_nx^n\right)+\sum_{n=0}^{\infty}c_nx^{n-1}\right)u'\\
&= u''+\left(\frac{2r}{x}+\sum_{n=1}^{\infty}B_nx^{n-1}+\sum_{n=0}^{\infty}c_nx^{n-1}\right)u'
\end{aligned}$$

$$= u'' + \left(\frac{2r}{x} + \sum_{n=1}^{\infty} B_n x^{n-1} + \frac{c_0}{x} + \sum_{n=1}^{\infty} c_n x^{n-1}\right) u'$$

$$= u'' + \left(\frac{2r}{x} + \frac{c_0}{x} + \sum_{n=1}^{\infty} B_n x^{n-1} + \sum_{n=1}^{\infty} c_n x^{n-1}\right) u'$$

$$= u'' + \left(\frac{2r + c_0}{x} + \sum_{n=1}^{\infty} B_n x^{n-1} + \sum_{n=1}^{\infty} c_n x^{n-1}\right) u'$$

$$= u'' + \left(\frac{2r + c_0}{x} + \sum_{n=1}^{\infty} (B_n + c_n) x^{n-1}\right) u'$$

$$= u'' + \left(\frac{1}{x} + \sum_{n=1}^{\infty} (B_n + c_n) x^{n-1}\right) u' \tag{7.12}$$

Integration of Eq.(7.12) gives

$$u' = \frac{1}{x} e^{-\int \sum_{n=1}^{\infty} (B_n + c_n x^{n-1})}$$

$$= \frac{1}{x}\left(1 + \sum_{n=1}^{\infty} C_n x^n\right)$$

$$= \frac{1}{x} + \sum_{n=1}^{\infty} C_n x^{n-1} \tag{7.13}$$

Again, integration of Eq.(7.13) gives

$$u(x) = \ln x + \sum_{n=1}^{\infty} \left(\frac{C_n}{n}\right) x^n$$

In other words, we have

$$y_2(x) = u(x) y_1(x)$$

$$= y_1(x) \ln x + \sum_{n=0}^{\infty} A_n x^{n+r}$$

Theorem 7.2.3 If both $p(x)$ and $q(x)$ are analytic at $x = 0$ with $p(x) = \sum_{n=0}^{\infty} c_n x^n$ and $q(x) = \sum_{n=0}^{n} b_n x^n$ but at least $\frac{p(x)}{x}$ or $\frac{q(x)}{x^2}$ is not analytic at $x = 0$, then the differential equation $x^2 y'' + xp(x)y' + q(x)y = 0$ admits a solution of the form $y(x) = C_s y_1(x) \ln x + \sum_{n=0}^{\infty} A_n x^{n+r_2}$ where $y_1(x) = \sum_{n=0}^{\infty} a_n x^{n+r}$ such that the equation $r^2 + (c_0 - 1)r + b_0 = 0$ has two roots r and r_2 with $s = r - r_2 \in \mathbb{I}$.

Proof: We know that $y_1(x) = \sum\limits_{n=0}^{\infty} a_n x^{n+r}$ is a solution of the given differential equation by Theorem 7.2.1. Again, $r_1 = r$ and r_2 are the roots of the equation $r^2 + (c_0 - 1)r + b_0 = 0$. Hence $c_0 - 1 = -r - r_2 = r - 2r - r_2 = r - r_2 - 2r = s - 2r$. In other words, we have $2r + c_0 = s + 1$. Let the second solution be $y_2(x) = u(x)y_1(x)$. Again, we have

$$\begin{aligned} y_2' &= uy_1' + u'y_1 \\ y_2'' &= u''y_1 + 2u'y_1' + uy_1'' \end{aligned}$$

Substituting y_2, y_2' and y_2'', in the given differential equation, we have

$$\begin{aligned} 0 &= x^2y'' + xp(x)y' + q(x)y \\ &= x^2(u''y_1 + 2u'y_1' + uy_1'') + xp(x)(uy_1' + u'y_1) + q(x)uy_1 \\ &= (x^2y_1'' + xp(x)y_1' + q(x)y_1)u + x^2(u''y_1 + 2u'y_1') + xp(x)u'y_1 \\ &= x^2y_1u'' + (2y_1'x^2 + xp(x)y_1)u' \end{aligned}$$

In other words, we have

$$\begin{aligned} 0 &= u'' + \left(2\left(\frac{y_1'}{y_1}\right) + \frac{p(x)}{x}\right)u' \\ &= u'' + \left(2\left(\frac{y_1'}{y_1}\right) + \sum_{n=0}^{\infty} c_n x^{n-1}\right)u' \\ &= u'' + \left(2\left(\frac{x^{r-1}\sum\limits_{n=0}^{\infty}(r+n)a_n x^n}{x^r \sum\limits_{n=0}^{\infty} a_n x^n}\right) + \sum_{n=0}^{\infty} c_n x^{n-1}\right)u' \\ &= u'' + \left(\frac{2}{x}\left(\frac{\sum\limits_{n=0}^{\infty}(r+n)a_n x^n}{\sum\limits_{n=0}^{\infty} a_n x^n}\right) + \sum_{n=0}^{\infty} c_n x^{n-1}\right)u' \\ &= u'' + \left(\frac{2}{x}\left(r + \sum_{n=1}^{\infty} B_n x^n\right) + \sum_{n=0}^{\infty} c_n x^{n-1}\right)u' \\ &= u'' + \left(\frac{2r}{x} + \sum_{n=1}^{\infty} B_n x^{n-1} + \sum_{n=0}^{\infty} c_n x^{n-1}\right)u' \\ &= u'' + \left(\frac{2r}{x} + \sum_{n=1}^{\infty} B_n x^{n-1} + \frac{c_0}{x} + \sum_{n=1}^{\infty} c_n x^{n-1}\right)u' \end{aligned}$$

$$= u'' + \left(\frac{2r}{x} + \frac{c_0}{x} + \sum_{n=1}^{\infty} B_n x^{n-1} + \sum_{n=1}^{\infty} c_n x^{n-1}\right) u'$$

$$= u'' + \left(\frac{2r + c_0}{x} + \sum_{n=1}^{\infty} B_n x^{n-1} + \sum_{n=1}^{\infty} c_n x^{n-1}\right) u'$$

$$= u'' + \left(\frac{2r + c_0}{x} + \sum_{n=1}^{\infty} (B_n + c_n)\, x^{n-1}\right) u'$$

$$= u'' + \left(\frac{s+1}{x} \sum_{n=1}^{\infty} (B_n + c_n)\, x^{n-1}\right) u' \tag{7.14}$$

Integration of Eq.(7.14) gives

$$u' = \frac{1}{x^{s+1}} e^{-\int \sum_{n=1}^{\infty} (B_n + c_n) x^{n-1}}$$

$$= \frac{1}{x^{s+1}} \left(1 + \sum_{n=1}^{\infty} C_n x^n\right)$$

$$= \frac{1}{x^{s+1}} + \sum_{n=1}^{\infty} C_n x^{n-s-1} \tag{7.15}$$

Again, integration of Eq.(7.15) gives

$$u(x) = -\frac{1}{s x^s} + \sum_{n=1}^{s-1} \left(\frac{C_n}{s-n}\right) x^{n-s} + C_s \ln x + \sum_{n=s+1}^{\infty} \left(\frac{C_n}{n-s}\right) x^{n-s}$$

In other words, we have

$$y_2(x) = u(x) y_1(x)$$

$$= C_s y_1(x) \ln x + \sum_{n=0}^{\infty} A_n x^{n+r_2}$$

where $C_s = 0$ or $C_s \neq 0$. One can easily identify the situation when $C_s = 0$ and $C_s \neq 0$.

There are certain precise procedures for determining the two independent solutions of the differential equation $x^2 y'' + x p(x) y' + q(x) y = 0$ when both $p(x)$ and $q(x)$ are analytic at $x = 0$ with $p(x) = \sum_{n=0}^{\infty} c_n x^n$ and $q(x) = \sum_{n=0}^{n} b_n x^n$, but at least $\frac{p(x)}{x}$ or $\frac{q(x)}{x^2}$ is not analytic at $x = 0$. According to Theorem 7.2.1, $y(r, x) = \sum_{n=0}^{\infty} a_n x^{n+r}$ is one of the solutions where $a_0 \neq 0$.

In other words, we have

$$
\begin{aligned}
x^2y'' + xp(x)y' + q(x)y &= \sum_{n=0}^{\infty}(n+r)(n+r-1)a_n x^{n+r} + p(x)\sum_{n=0}^{\infty}(n+r)a_n x^{n+r} \\
&\quad + q(x)\sum_{n=0}^{\infty} a_n x^{n+r} \\
&= \sum_{n=0}^{\infty}\left(\sum_{k=0}^{n}(n+r-k)c_k a_{n-k}\right)x^{n+r} + \sum_{n=0}^{\infty}\left(\sum_{k=0}^{n} b_k a_{n-k}\right)x^{n+r} \\
&\quad + \sum_{n=0}^{\infty}(n+r)(n+r-1)a_n x^{n+r} \\
&= \sum_{n=0}^{\infty}\left(\sum_{k=0}^{n}((n+r-k)c_k + b_k)\,a_{n-k}\right)x^{n+r} \\
&\quad + \sum_{n=0}^{\infty}(n+r)(n+r-1)a_n x^{n+r} \\
&= \sum_{n=1}^{\infty}(((n+r)(n+r-1) + (n+r)c_0 + b_0)a_n \\
&\quad + \sum_{k=1}^{n}((n+r-k)c_k + b_k)\,a_{n-k})x^{n+r} + (r(r-1) + rc_0 + b_0)a_0 x^r \\
&= F(r)a_0 x^r + \sum_{n=1}^{\infty}\left(F(r+n)a_n + \sum_{k=1}^{n} G(n,\, r,\, k)a_{n-k}\right)x^{n+r}
\end{aligned}
$$

where

$$F(n+r) = (n+r)(n+r-1) + (n+r)c_0 + b_0$$

and

$$G(n,\, r,\, k) = (n+r-k)c_k + b_k$$

In order to make $y(r,\, x)$ as a solution with $a_0 \neq 0$, we have to have

$$
\begin{aligned}
F(r) &= r(r-1) + rc_0 + b_0 \\
&= 0
\end{aligned}
$$

and for $n \geq 1$,

$$
\begin{aligned}
0 &= F(n+r)a_n + \sum_{k=1}^{n} G(n,\, r,\, k)a_{n-k} \\
&= ((n+r)(n+r-1) + (n+r)c_0 + b_0)a_n + \sum_{k=1}^{n}((n+r-k)c_k + b_k)\,a_{n-k}
\end{aligned}
$$

One can make

$$((n+r)(n+r-1)+(n+r)c_0+b_0)a_n+\sum_{k=1}^{n}((n+r-k)c_k+b_k)\,a_{n-k}=0$$

by the condition

$$a_n=-\frac{\sum\limits_{k=1}^{n}G(n,r,k)a_{n-k}}{F(n+r)}\qquad n\geq 1 \tag{7.16}$$

whereas

$$F(r)=r(r-1)+rc_0+b_0=r^2+(c_0-1)r+b_0=0$$

by suitable r given by

$$r=\frac{1-c_0\pm\left((c_0-1)^2-4b_0\right)^{\frac{1}{2}}}{2}$$

In other words, we have

$$F(n+r)=(n+r-r_1)(n+r-r_2)$$

Let $r_1-r_2=s\geq 0$. It is easy to conclude that

$$\begin{aligned}F(n+r_1)&=n(n+r_1-r_2)\\&=n(n+s)\end{aligned}$$

and

$$\begin{aligned}F(n+r_2)&=n(n+r_2-r_1)\\&=n(n-s)\end{aligned}$$

It is easy to verify that when $s=0$, $F(n+r)\neq 0$ for all $n\geq 1$. Hence all a_n are well defined. Again, $F(r_1)=0$ and $F'(r)|_{r=r_1}=0$. It is easy to verify that when $s\in\mathbb{I}$, $F(n+r_2)=0$. Hence some a_n are not well defined. Again, $F(r)|_{r=r_1}=0$ and $\frac{\mathrm{d}}{\mathrm{d}r}\left((r-r_2)F(r)\right)|_{r=r_1}=0$. Hence we have the following conclusions:

1. Let $y(r,x)=\sum\limits_{n=0}^{\infty}a_nx^{n+r}$. If r_1 is the only root of $r^2+(c_0-1)r+b_0=0$, then $y_1(x)=y(r,x)|_{r=r_1}$ and $y_2(x)=\frac{\partial}{\partial r}\left(y(r,x)\right)|_{r=r_1}$ are two independent solutions of the differential equation $x^2y''+xp(x)y'+q(x)y=0$.

2. Let $y(r,x)=\sum\limits_{n=0}^{\infty}a_nx^{n+r}$. If r_1 and r_2 are the roots of $r^2+(c_0-1)r+b_0=0$ such that $r_1-r_2=s\in\mathbb{I}$ and $F(n+r_2)=0$, then $y_1(x)=(r-r_2)y(r,x)|_{r=r_1}$ and $y_2(x)=\frac{\partial}{\partial r}\left((r-r_2)y(r,x)\right)|_{r=r_2}$ are two independent solutions of the differential equation $x^2y''+xp(x)y'+q(x)y=0$. In other words, $C_s\neq 0$.

3. Let $y(r, x) = \sum\limits_{n=0}^{\infty} a_n x^{n+r}$. If r_1 and r_2 are the roots of $r^2 + (c_0 - 1)r + b_0 = 0$ such that $r_1 - r_2 = s \in \mathbb{I}$ but $F(n + r_2) \neq 0$, then $y_1(x) = y(r, x)|_{r=r_1}$ and $y_2(x) = y(r, x)|_{r=r_2}$ are two independent solutions of the differential equation $x^2y'' + xp(x)y' + q(x)y = 0$. In other words, $C_s = 0$.

EXAMPLE 7.15 Find two independent series solutions of the differential equation $y'' - y = 0$.

Solution: One can rewrite the given differential equation in the standard form $x^2y'' - x^2y = 0$. Hence $p(x) = 0$ and $q(x) = -x^2$. Clearly, we have $c_n = 0$ for all $0 \leq n < \infty$, $b_2 = -1$ and $b_n = 0$ for $0 \leq n \neq 2 < \infty$. Accordingly, Eq.(7.16) gets reduced to

$$a_1 = 0$$

and

$$\begin{aligned} a_n &= -\left(\frac{-1}{F(n+r)}\right) a_{n-2} \\ &= \left(\frac{1}{F(n+r)}\right) a_{n-2} \qquad n \geq 2 \end{aligned}$$

In other words, we have

$$\begin{aligned} F(r) &= r(r-1) + rc_0 + b_0 \\ &= r^2 - r \\ &= r(r-1) \end{aligned}$$

Hence $r_1 = 1$ and $r_2 = 0$. Again, we have

$$\begin{aligned} F(n + r_1) &= (n + r_1)(n + r_1 - 1) \\ &= n(n+1) \end{aligned}$$

and

$$\begin{aligned} F(n + r_2) &= (n + r_2)(n + r_2 - 1) \\ &= n(n-1) \end{aligned}$$

Clearly, all unknowns a_n are well defined. Let $a_0 = 1$. Hence we have

$$a_{2k} = \left(\frac{1}{(r+1)(r+2)\cdots(r+2k)}\right) \qquad k \geq 1$$

The general solution $y(r, x)$ is given by

$$y(r, x) = x^r \left(1 + \sum_{k=1}^{\infty} \frac{x^{2k}}{(r+1)(r+2)\cdots(r+2k)}\right)$$

Clearly, $r_1 - r_2 \in \mathbb{I}$ but $F(n+r_2) \neq 0$ for $n \geq 2$. Hence $y_1(x) = y(r_1, x)$ and $y_2(x) = y(r_2, x)$. In other words, we have

$$\begin{aligned} y_1(x) &= y(1, x) \\ &= x + \sum_{k=1}^{\infty} \frac{x^{2k+1}}{(2k+1)!} \end{aligned}$$

and

$$\begin{aligned} y_2(x) &= y(0, x) \\ &= 1 + \sum_{k=1}^{\infty} \frac{x^{2k}}{(2k)!} \end{aligned}$$

EXAMPLE 7.16 Find two independent series solutions of the differential equation $xy'' + 2y' + xy = 0$.

Solution: Clearly, the given differential equation is not in standard form. The corresponding differential equation in standard form is $x^2y'' + 2xy' + x^2y = 0$. Hence $p(x) = 2$ and $q(x) = x^2$. Clearly, we have $c_0 = 2$, $c_n = 0$ for all $1 \leq n < \infty$, $b_2 = 1$ and $b_n = 0$ for $0 \leq n \neq 2 < \infty$. Accordingly, Eq.(7.16) gets reduced to

$$a_1 = 0$$

and

$$\begin{aligned} a_n &= -\left(\frac{1}{F(n+r)}\right) a_{n-2} \\ &= -\left(\frac{1}{(n+r)(n+r+1)}\right) a_{n-2} \qquad n \geq 2 \end{aligned}$$

In other words, we have

$$\begin{aligned} F(r) &= r(r-1) + rc_0 + b_0 \\ &= r^2 + r \\ &= r(r+1) \end{aligned}$$

Hence $r_1 = 0$ and $r_2 = -1$. Again, we have

$$F(n + r_1) = n(n+1)$$

and

$$F(n + r_2) = n(n-1)$$

Clearly, all unknowns a_n are well defined. Let $a_0 = 1$. Hence we have

$$a_{2k} = \frac{(-1)^k}{\prod_{i=1}^{k} (2i+1+r)(2i+r)} \qquad k \geq 1$$

The general solution $y(r, x)$ is given by

$$y(r, x) = x^r + \sum_{k=1}^{\infty} \left(\frac{(-1)^k}{\prod_{i=1}^{k} (2i+1+r)(2i+r)} \right) x^{2k+r}$$

Clearly, $r_1 - r_2 \in \mathbb{I}$ but $F(n+r_2) \neq 0$ for $n \geq 2$. Hence $y_1(x) = y(r_1, x)$ and $y_2(x) = y(r_2, x)$. In other words, we have

$$y_1(x) = 1 + \sum_{k=1}^{\infty} \left(\frac{(-1)^k}{\prod_{i=1}^{k} (2i+1)2i} \right) x^{2k}$$

$$= 1 + \sum_{k=1}^{\infty} \left(\frac{(-1)^k}{(2k+1)!} \right) x^{2k}$$

and

$$y_2(x) = \frac{1}{x} + \sum_{k=1}^{\infty} \left(\frac{(-1)^k}{\prod_{i=1}^{k} 2i(2i-1)} \right) x^{2k-1}$$

$$= \frac{1}{x} + \sum_{k=1}^{\infty} \left(\frac{(-1)^k}{(2k)!} \right) x^{2k-1}$$

EXAMPLE 7.17 Find two independent series solutions of the differential equation $x^2y'' + xy' + \left(x^2 - \frac{1}{4}\right) y = 0$.

Solution: Clearly, the given differential equation is in standard form. Hence $p(x) = 1$ and $q(x) = x^2 - \frac{1}{4}$. Clearly, we have $c_0 = 1$, $c_n = 0$ for all $1 \leq n < \infty$, $b_0 = -\frac{1}{4}$, $b_2 = 1$ and $b_n = 0$ for $1 \leq n \neq 2 < \infty$. Accordingly, Eq.(7.16) gets reduced to

$$a_1 = 0$$

and

$$a_n = -\frac{\sum_{k=1}^{n} G(n, r, k)a_{n-k}}{F(n+r)}$$

$$= -\left(\frac{1}{(n+r-1)(n+r) + (n+r)c_0 + b_0} \right) a_{n-2}$$

$$= -\left(\frac{1}{(n+r)^2 - \frac{1}{4}}\right) a_{n-2}$$

$$= -\left(\frac{1}{\left(n+r+\frac{1}{2}\right)\left(n+r-\frac{1}{2}\right)}\right) a_{n-2}$$

$$= -\left(\frac{2^2}{(2n+2r+1)(2n+2r-1)}\right) a_{n-2} \qquad n \geq 2$$

Again, we have

$$\begin{aligned} F(r) &= r(r-1) + rc_0 + b_0 \\ &= r^2 - \frac{1}{4} \\ &= \left(r+\frac{1}{2}\right)\left(r-\frac{1}{2}\right) \end{aligned}$$

Hence $r_1 = \frac{1}{2}$ and $r_2 = -\frac{1}{2}$. Again, we have

$$\begin{aligned} F(n+r) &= (n+r-1)(n+r) + (n+r)c_0 + b_0 \\ &= (n+r)^2 - \frac{1}{4} \\ &= \left(n+r+\frac{1}{2}\right)\left(n+r-\frac{1}{2}\right) \end{aligned}$$

It is clear that $r_1 - r_2 = 1 \in \mathbb{I}$ but $F(n+r) \neq 0$ for $r = r_1$ and $r = r_2$ when $n \geq 2$. Hence all the unknowns a_n for $n \geq 2$ are well defined. In general, the unknown a_n is given by

$$a_{2n} = (-1)^n \left(\frac{2^{2n}}{\prod_{i=1}^{n}(4i+2r+1)(4i+2r-1)}\right) a_0 \qquad n \geq 1$$

Let $a_0 = 1$. In other words, the general solution is given by

$$y(r,\, x) = \left(1 + \sum_{n=1}^{\infty}(-1)^n \left(\frac{2^{2n}}{\prod_{i=1}^{n}(4i+2r+1)(4i+2r-1)}\right)\right) x^{2n+r}$$

Again, two independent solutions are given by

$$
\begin{aligned}
y_1(x) &= y(r,\, x)|_{r=r_1} \\
&= \left.\left(1+\sum_{n=1}^{\infty}(-1)^n\left(\frac{2^{2n}}{\prod_{i=1}^{n}(4i+2r+1)(4i+2r-1)}\right)\right)x^{2n+r}\right|_{r=\frac{1}{2}} \\
&= \left(1+\sum_{n=1}^{\infty}(-1)^n\left(\frac{2^{2n}}{\prod_{i=1}^{n}(4i+2)4i}\right)\right)x^{2n+\frac{1}{2}} \\
&= \left(1+\sum_{n=1}^{\infty}(-1)^n\left(\frac{2^{2n}}{\prod_{i=1}^{n}2(2i+1)4i}\right)\right)x^{2n+\frac{1}{2}} \\
&= \left(1+\sum_{n=1}^{\infty}(-1)^n\left(\frac{2^{2n}}{\prod_{i=1}^{n}2^3(2i+1)i}\right)\right)x^{2n+\frac{1}{2}} \\
&= \left(1+\sum_{n=1}^{\infty}(-1)^n\left(\frac{1}{\prod_{i=1}^{n}2(2i+1)i}\right)\right)x^{2n+\frac{1}{2}} \\
&= x^{\frac{1}{2}}\left(1+\sum_{n=1}^{\infty}\left(\frac{(-1)^n}{(2n+1)!}\right)x^{2n}\right)
\end{aligned}
$$

and

$$
\begin{aligned}
y_2(x) &= y(r,\, x)|_{r=r_2} \\
&= \left.\left(1+\sum_{n=1}^{\infty}(-1)^n\left(\frac{2^{2n}}{\prod_{i=1}^{n}(4i+2r+1)(4i+2r-1)}\right)\right)x^{2n+r}\right|_{r=-\frac{1}{2}} \\
&= \left(1+\sum_{n=1}^{\infty}(-1)^n\left(\frac{2^{2n}}{\prod_{i=1}^{n}(4i-2)4i}\right)\right)x^{2n-\frac{1}{2}}
\end{aligned}
$$

$$= \left(1+\sum_{n=1}^{\infty}(-1)^n\left(\frac{2^{2n}}{\prod_{i=1}^{n}2(2i-1)4i}\right)\right)x^{2n-\frac{1}{2}}$$

$$= \left(1+\sum_{n=1}^{\infty}(-1)^n\left(\frac{2^{2n}}{\prod_{i=1}^{n}2^3(2i-1)i}\right)\right)x^{2n-\frac{1}{2}}$$

$$= \left(1+\sum_{n=1}^{\infty}(-1)^n\left(\frac{1}{\prod_{i=1}^{n}2(2i-1)i}\right)\right)x^{2n-\frac{1}{2}}$$

$$= x^{-\frac{1}{2}}\left(1+\sum_{n=1}^{\infty}\left(\frac{(-1)^n}{(2n)!}\right)x^{2n}\right)$$

EXAMPLE 7.18 Find the two independent series solutions of the differential equation $x^2y'' + xy' - \frac{(1+x^2)}{4}y = 0$.

Solution: Clearly, the given differential equation is not in standard form. The corresponding standard form is $x^2y''+xy'+\left(-\frac{1}{4}-\frac{x}{4}\right)y = 0$. Hence $p(x) = 1$ and $q(x) = -\frac{1}{4}-\frac{x^2}{4}$. Clearly, we have $c_0 = 1$ and $c_n = 0$ for all $1 \le n < \infty$, $b_0 = -\frac{1}{4}$, $b_2 = -\frac{1}{4}$ and $b_n = 0$ for $1 \le n \ne 2 < \infty$. Accordingly, Eq.(7.16) gets reduced to

$$a_1 = 0$$

and

$$a_n = -\frac{\sum_{k=1}^{n} G(n,\, r,\, k)a_{n-k}}{F(n+r)}$$

$$= -\left(\frac{-\frac{1}{4}}{\left(n+r-\frac{1}{2}\right)\left(n+r+\frac{1}{2}\right)}\right)a_{n-2}$$

$$= \left(\frac{1}{(2r+2n+1)(2r+2n-1)}\right)a_{n-2} \qquad n \ge 2$$

Again, we have

$$F(n+r) = (n+r)(n+r-1) + (n+r)c_0 + b_0$$

$$= (n+r)(n+r-1) + (n+r) - \frac{1}{4}$$

$$= \left(n+r-\frac{1}{2}\right)\left(n+r+\frac{1}{2}\right)$$

In general, we have

$$a_{2n} = \left(\frac{1}{\prod_{i=1}^{n} (2r + 4i + 1)(2r + 4i - 1)} \right) a_0 \qquad n \geq 1$$

Again, we have

$$\begin{aligned} F(r) &= r(r-1) + rc_0 + b_0 \\ &= \left(r - \frac{1}{2}\right)\left(r + \frac{1}{2}\right) \end{aligned}$$

Hence $r_1 = \frac{1}{2}$ and $r_2 = -\frac{1}{2}$. Again, $r_1 - r_2 = 1 \in \mathbb{I}$ but all unknowns a_n are well defined. Let $a_0 = 1$. Hence the general solution is given

$$y(r, x) = \left(1 + \sum_{n=1}^{\infty} \left(\frac{1}{\prod_{i=1}^{n} (2r + 4i + 1)(2r + 4i - 1)} \right) \right) x^{2n+r}$$

In other words, two independent solutions are

$$\begin{aligned} y_1(x) &= y(r, x)|_{r=r_1} \\ &= \left. \left(1 + \sum_{n=1}^{\infty} \left(\frac{1}{\prod_{i=1}^{n} (2r + 4i + 1)(2r + 4i - 1)} \right) \right) x^{2n+r} \right|_{r=\frac{1}{2}} \\ &= \left(1 + \sum_{n=1}^{\infty} \left(\frac{1}{\prod_{i=1}^{n} 4i(4i + 2)} \right) \right) x^{2n+\frac{1}{2}} \\ &= x^{\frac{1}{2}} \left(1 + \sum_{n=1}^{\infty} \left(\frac{1}{\prod_{i=1}^{n} 2^3 i(2i + 1)} \right) x^{2n} \right) \\ &= x^{\frac{1}{2}} \left(1 + \sum_{n=1}^{\infty} \left(\frac{x^{2n}}{2^{2n}(2n + 1)!} \right) \right) \end{aligned}$$

and

$$
\begin{aligned}
y_2(x) &= y(r, x)|_{r=r_2} \\
&= \left. \left(1 + \sum_{n=1}^{\infty} \left(\frac{1}{\prod_{i=1}^{n} (2r+4i+1)(2r+4i-1)} \right) \right) x^{2n+r} \right|_{r=-\frac{1}{2}} \\
&= \left(1 + \sum_{n=1}^{\infty} \left(\frac{1}{\prod_{i=1}^{n} 4i(4i-2)} \right) \right) x^{2n-\frac{1}{2}} \\
&= x^{-\frac{1}{2}} \left(1 + \sum_{n=1}^{\infty} \left(\frac{1}{\prod_{i=1}^{n} 2^3 i(2i-1)} \right) x^{2n} \right) \\
&= x^{-\frac{1}{2}} \left(1 + \sum_{n=1}^{\infty} \left(\frac{x^{2n}}{2^{2n}(2n)!} \right) \right)
\end{aligned}
$$

EXAMPLE 7.19 Find two independent series solutions of the differential equation $xy'' + y' - y = 0$.

Solution: One can rewrite the given differential equation in the standard form $x^2y'' + xy' - xy = 0$. Hence $p(x) = 1$ and $q(x) = -x$. Clearly, we have $c_0 = 1$, $c_n = 0$ for all $1 \le n < \infty$, $b_1 = -1$ and $b_n = 0$ for $0 \le n \ne 1 < \infty$. Accordingly, Eq.(7.16) gets reduced to

$$
\begin{aligned}
a_n &= -\left(\frac{-1}{F(n+r)} \right) a_{n-1} \\
&= \left(\frac{1}{F(n+r)} \right) a_{n-1} \qquad n \ge 1
\end{aligned}
$$

In other words, we have

$$
\begin{aligned}
F(r) &= r(r-1) + rc_0 + b_0 \\
&= r^2
\end{aligned}
$$

Hence $r = 0$ is the only root. Again, we have

$$F(n+r) = (n+r)^2$$

Clearly, all unknowns a_n are well defined. Let $a_0 = 1$. Hence we have

$$a_n = \left(\frac{1}{(r+1)^2(r+2)^2 \cdots (r+n)^2} \right) \qquad n \ge 1$$

The general solution $y(r, x)$ is given by

$$y(r, x) = x^r \left(1 + \sum_{n=1}^{\infty} \left(\frac{x^n}{(r+1)^2(r+2)^2 \cdots (r+n)^2}\right)\right)$$

Clearly, $r_1 = r_2$. Hence $y_1(x) = y(r_1, x)$ and $y_2(x) = \frac{\partial}{\partial r}(y(r, x))\big|_{r=r_1}$. In other words, we have

$$\begin{aligned} y_1(x) &= y(0, x) \\ &= 1 + \sum_{n=1}^{\infty} \left(\frac{x^n}{(n!)^2}\right) \end{aligned}$$

and

$$\begin{aligned} y_2(x) &= \frac{\partial}{\partial r}(y(r, x))\,|_{r=0} \\ &= \ln x\, y_1(x) + \sum_{n=1}^{\infty} \frac{\partial}{\partial r}\left(\frac{x^n}{(r+1)^2(r+2)^2 \cdots (r+n)^2}\right)|_{r=0} \\ &= \ln x\, y_1(x) - \sum_{n=1}^{\infty} \left(\left(\frac{2x^n}{(n!)^2}\right)\left(\sum_{k=1}^{n} \frac{1}{k}\right)\right) \end{aligned}$$

EXAMPLE 7.20 Find two independent series solutions of the differential equation $xy'' + y' - xy = 0$.

Solution: Clearly, the given differential equation is not in standard form. The corresponding standard form is $x^2y'' + xy' - x^2y = 0$. Hence $p(x) = 1$ and $q(x) = -x^2$. Clearly, we have $c_0 = 1$, $c_n = 0$ for all $1 \leq n < \infty$, $b_2 = -1$ and $b_n = 0$ for $1 \leq n \neq 2 < \infty$. Accordingly, Eq.(7.16) gets reduced to

$$a_1 = 0$$

and

$$a_n = \left(\frac{1}{F(n+r)}\right) a_{n-2} \qquad n \geq 2$$

Again, we have

$$F(n+r) = (n+r)^2$$

Let $a_0 = 1$. In other words, we have

$$\begin{aligned} F(r) &= r(r-1) + rc_0 + b_0 \\ &= r^2 \end{aligned}$$

Hence $r_1 = r_2 = 0$. Again, all the unknowns a_n are well defined for $r = r_1$. In general, we have

$$a_{2k} = \left(\frac{1}{\prod_{i=1}^{k} (r+2i)^2} \right) \qquad k \geq 1$$

The general solution $y(r, x)$ is given by

$$y(r, x) = x^r + \sum_{k=1}^{\infty} \left(\frac{1}{\prod_{i=1}^{k} (r+2i)^2} \right) x^{2k+r}$$

Clearly, $r_1 = r_2$. Hence $y_1(x) = y(r_1, x)$ and $y_2(x) = \frac{\partial}{\partial r}(y(r, x))\big|_{r=r_1}$. In other words, we have

$$\begin{aligned} y_1(x) &= y(r, x)|_{r=0} \\ &= 1 + \sum_{k=1}^{\infty} \left(\frac{1}{\prod_{i=1}^{k} (2i)^2} \right) x^{2k} \\ &= 1 + \sum_{k=1}^{\infty} \left(\frac{1}{2^k k!} \right)^2 x^{2k} \end{aligned}$$

and

$$\begin{aligned} y_2(x) &= \frac{\partial}{\partial r} (y(r, x))\bigg|_{r=0} \\ &= \frac{\partial}{\partial r} \left(x^r + \sum_{k=1}^{\infty} \left(\frac{1}{\prod_{i=1}^{k} (r+2i)^2} \right) x^{2k+r} \right)\Bigg|_{r=0} \\ &= \ln x \, y_1 \, x - \sum_{k=1}^{\infty} \left(\frac{1}{2^k k!} \right)^2 \left(\sum_{i=1}^{k} \frac{1}{i} \right) x^{2k} \end{aligned}$$

EXAMPLE 7.21 Find two independent series solutions of the differential equation $x^2y'' - 3xy' + (4+4x)y = 0$.

Solution: Clearly, the given differential equation is in standard form. Hence $p(x) = -3$ and $q(x) = 4 + 4x$. Clearly, we have $c_0 = -3$, $c_n = 0$ for all $1 \leq n < \infty$, $b_0 = 4$, $b_1 = 4$ and $b_n = 0$

for $2 \leq n < \infty$. Accordingly, Eq.(7.16) gets reduced to

$$a_n = -\left(\frac{4}{F(n+r)}\right) a_{n-1} \qquad n \geq 1$$

Again, we have

$$F(n+r) = (n+r-2)^2$$

Let $a_0 = 1$. In other words, we have

$$\begin{aligned} F(r) &= r(r-1) + rc_0 + b_0 \\ &= r^2 - 4r + 4 \end{aligned}$$

Hence $r_1 = r_2 = 2$. Again, all the unknowns a_n are well defined for $r = r_1$. In general, we have

$$a_n = (-1)^n \left(\frac{4^n}{\prod_{i=1}^{n} (r-2+i)^2} \right) \qquad n \geq 1$$

The general solution $y(r, x)$ is given by

$$y(r, x) = x^r + \sum_{n=1}^{\infty} (-1)^n \left(\frac{4^n}{\prod_{i=1}^{n} (r-2+i)^2} \right) x^{n+r}$$

Clearly, $r_1 = r_2$. Hence $y_1(x) = y(r_1, x)$ and $y_2(x) = \frac{\partial}{\partial r}(y(r, x))\big|_{r=r_1}$. In other words, we have

$$\begin{aligned} y_1(x) &= y(r, x)\big|_{r=2} \\ &= x^2 + \sum_{n=1}^{\infty} (-1)^n \left(\frac{4^n}{\prod_{i=1}^{n} i^2} \right) x^{n+2} \\ &= x^2 + \sum_{n=1}^{\infty} (-1)^n \left(\frac{2^n}{n!} \right)^2 x^{2+n} \end{aligned}$$

and

$$\begin{aligned} y_2(x) &= \frac{\partial}{\partial r} (y(r, x)) \bigg|_{r=2} \\ &= \frac{\partial}{\partial r} \left(x^r + \sum_{n=1}^{\infty} (-1)^n \left(\frac{4^n}{\prod_{i=1}^{n} (r-2+i)^2} \right) x^{n+r} \right) \Bigg|_{r=2} \end{aligned}$$

$$= y_1(x)\ln x - \sum_{n=1}^{\infty}(-1)^n\left(\frac{2^n}{n!}\right)^2\left(\sum_{i=1}^{n}\left(\frac{2}{i}\right)\right)x^{n+2}$$

EXAMPLE 7.22 Find two independent series solutions of the differential equation $x^2y'' + xy' + (x^2 - 1)y = 0$.

Solution: Clearly, the given differential equation is in standard form. Hence $p(x) = 1$ and $q(x) = x^2 - 1$. Clearly, we have $c_0 = 1$, $c_n = 0$ for all $1 \le n < \infty$, $b_0 = -1$, $b_2 = 1$ and $b_n = 0$ for $1 \le n \ne 2 < \infty$. Accordingly, Eq.(7.16) gets reduced to

$$a_1 = 0$$

and

$$a_n = -\left(\frac{1}{F(n+r)}\right)a_{n-2} \qquad n \ge 2$$

In other words, we have

$$\begin{aligned} F(r) &= r(r-1) + rc_0 + b_0 \\ &= r^2 - 1 \\ &= (r+1)(r-1) \end{aligned}$$

Hence $r_1 = 1$ and $r_2 = -1$ are the roots. Again, we have

$$\begin{aligned} F(n+r) &= (n+r)^2 - 1 \\ &= (n+r-1)(n+r+1) \end{aligned}$$

Clearly, $s = r_1 - r_2 = 2$. Hence we have

$$\begin{aligned} F(n+r_1) &= (n+r_1-1)(n+r_1+1) \\ &= n(n+2) \end{aligned}$$

and

$$\begin{aligned} F(n+r_2) &= (n+r_2-1)(n+r_2+1) \\ &= n(n-2) \end{aligned}$$

Clearly, some of the unknowns a_n are not well defined for $r = r_2$. Let $a_0 = 1$. Hence we have

$$a_{2k} = \left(\frac{(-1)^k}{(r+1)(r+3)^2\cdots(r+2k-1)^2(r+2k+1)}\right) \qquad k \ge 2$$

The general solution $y(r, x)$ is given by

$$y(r, x) = x^r\left(1 + \sum_{k=1}^{\infty}\left(\frac{(-1)^k x^{2k}}{(r+1)(r+3)^2\cdots(r+2k-1)^2(r+2k+1)}\right)\right)$$

In other words, we have

$$\begin{aligned}
y_1(x) &= (r - r_2)y(r, x)\big|_{r=r_1} \\
&= (r + 1)y(r, x)\big|_{r=1} \\
&= 2y(1, x) \\
&= x\left(2 + \sum_{k=1}^{\infty}\left(\frac{(-1)^k x^{2k}}{2^{2k-1}k!(k+1)!}\right)\right)
\end{aligned}$$

and

$$\begin{aligned}
y_2(x) &= \frac{\partial}{\partial r}\left((r - r_2)y(r, x)\right)|_{r=r_2} \\
&= \frac{\partial}{\partial r}\left((r + 1)y(r, x)\right)|_{r=-1} \\
&= \frac{\ln x}{x}\left(\sum_{k=1}^{\infty}\left(\frac{(-1)^k x^{2k}}{2^{2k-1}k!(k-1)!}\right)\right) \\
&\quad + \frac{1}{x}\left(1 + \sum_{k=1}^{\infty}\frac{\partial}{\partial r}\left(\frac{(-1)^k x^{2k}(r + 2k + 1)}{\prod_{i=1}^{k}(r + 2i + 1)^2}\right)\Bigg|_{r=-1}\right) \\
&= \frac{\ln x}{x}\left(\sum_{k=1}^{\infty}\left(\frac{(-1)^k x^{2k}}{2^{2k-1}k!(k-1)!}\right)\right) \\
&\quad - \frac{1}{x}\left(1 - \sum_{k=1}^{\infty}\left(\left(\frac{(-1)^k x^{2k}}{2^{2k-1}k!(k-1)!}\right)\left(\sum_{i=1}^{k-1}\frac{1}{i} + \frac{1}{2k}\right)\right)\right)
\end{aligned}$$

EXAMPLE 7.23 Find two independent series solutions of the differential equation $x^2y'' + xy' - y = 0$.

Solution: Clearly, the given differential equation is in standard form. Hence $p(x) = 1$ and $q(x) = -1$. Clearly, we have $c_0 = 1$, $c_n = 0$ for all $1 \leq n < \infty$, $b_0 = -1$ and $b_n = 0$ for $1 \leq n < \infty$. Accordingly, Eq.(7.16) gets reduced to

$$a_n = -\left(\frac{1}{F(n + r)}\right)a_{n-1} \qquad n \geq 1$$

Again, we have

$$F(n + r) = (n + r - 1)(n + r + 1)$$

Let $a_0 = 1$. In other words, we have

$$\begin{aligned} F(r) &= r(r-1) + rc_0 + b_0 \\ &= r^2 - 1 \\ &= (r+1)(r-1) \end{aligned}$$

Hence $r_1 = 1$ and $r_2 = -1$. Again, all the unknowns a_n are well defined for $r = r_1$ but not $r = r_2$. The general solution $y(r, x)$ is given by

$$y(r, x) = x^r$$

In other words, we have

$$\begin{aligned} y_1(x) &= \left.(r - r_2)y(r, x)\right|_{r=r_1} \\ &= \left.(r+1)x^r\right|_{r=1} \\ &= 2x \end{aligned}$$

and

$$\begin{aligned} y_2(x) &= \left.\frac{\partial}{\partial r}\left((r - r_2)y(r, x)\right)\right|_{r=r_2} \\ &= \left.\frac{\partial}{\partial r}\left((r+1)y(r, x)\right)\right|_{r=-1} \\ &= \left.\frac{\partial}{\partial r}\left((r+1)x^r\right)\right|_{r=-1} \\ &= \left.(x^r + (r+1)x^r \ln x)\right|_{r=-1} \\ &= \frac{1}{x} \end{aligned}$$

EXAMPLE 7.24 Find the two independent series solutions of the differential equation $4x^2y'' + 2x(2-x)y' - (1+3x)y = 0$.

Solution: Clearly, the given differential equation is not in standard form. The corresponding standard form is $x^2y'' + x\left(1 - \frac{x}{2}\right)y' + \left(-\frac{1}{4} - \frac{3x}{4}\right)y = 0$. Hence $p(x) = 1 - \frac{x}{2}$ and $q(x) = -\frac{1}{4} - \frac{3x}{4}$. Clearly, we have $c_0 = 1$, and $c_1 = -\frac{1}{2}$ and $c_n = 0$ for all $2 \le n < \infty$, $b_0 = -\frac{1}{4}$, $b_1 = -\frac{3}{4}$ and $b_n = 0$ for $2 \le n < \infty$. Accordingly, Eq.(7.16) gets reduced to

$$\begin{aligned} a_n &= -\frac{\sum\limits_{k=1}^{n} G(n, r, k)a_{n-k}}{F(n+r)} \\ &= -\left(\frac{-\left(\frac{1}{4}\right)(2n + 2r - 1)}{\left(n + r - \frac{1}{2}\right)\left(n + r + \frac{1}{2}\right)}\right)a_{n-1} \\ &= \left(\frac{1}{2r + 2n - 1}\right)a_{n-1} \qquad n \ge 1 \end{aligned}$$

Again, we have

$$\begin{aligned}F(n+r) &= (n+r)(n+r-1)+(n+r)c_0+b_0\\ &= (n+r)(n+r-1)+(n+r)-\frac{1}{4}\\ &= \left(n+r-\frac{1}{2}\right)\left(n+r+\frac{1}{2}\right)\end{aligned}$$

In general, we have

$$a_n = \left(\frac{1}{\prod_{i=1}^{n}(2r+2i-1)}\right)a_0 \qquad n \geq 1$$

Again, we have

$$\begin{aligned}F(r) &= r(r-1)+rc_0+b_0\\ &= \left(r-\frac{1}{2}\right)\left(r+\frac{1}{2}\right)\end{aligned}$$

Hence $r_1 = \frac{1}{2}$ and $r_2 = -\frac{1}{2}$. Again, $r_1 - r_2 = 1 \in \mathbb{I}$ and all unknowns a_n are not well defined for $r = r_2$. Let $a_0 = 1$. Hence the general solution is given by

$$y(r,\, x) = \left(1+\sum_{n=1}^{\infty}\left(\frac{1}{\prod_{i=1}^{n}(2r+2i-1)}\right)\right)x^{n+r}$$

In other words, the two independent solutions are

$$\begin{aligned}y_1(x) &= (r-r_2)y(r,\, x)|_{r=r_1}\\ &= \left(r+\frac{1}{2}\right)\left(1+\sum_{n=1}^{\infty}\left(\frac{1}{\prod_{i=1}^{n}(2r+2i-1)}\right)\right)x^{n+r}\Bigg|_{r=\frac{1}{2}}\\ &= 2\left(1+\sum_{n=1}^{\infty}\left(\frac{1}{\prod_{i=1}^{n}2i}\right)\right)x^{n+\frac{1}{2}}\\ &= 2x^{\frac{1}{2}}\left(1+\sum_{n=1}^{\infty}\left(\frac{1}{2^n n!}\right)x^n\right)\\ &= 2x^{\frac{1}{2}}\left(1+\sum_{n=1}^{\infty}\left(\frac{x^n}{2^n n!}\right)\right)\end{aligned}$$

and

$$
\begin{aligned}
y_2(x) &= \frac{\partial}{\partial r}\left((r-r_2)y(r,x)\right)\Big|_{r=r_2} \\
&= \frac{\partial}{\partial r}\left(\left(r+\frac{1}{2}\right)y(r,x)\right)\Big|_{r=-\frac{1}{2}} \\
&= \frac{1}{2}\,\frac{\partial}{\partial r}\left((2r+1)\,y(r,x)\right)\Big|_{r=-\frac{1}{2}} \\
&= \frac{1}{2}\frac{\partial}{\partial r}\left(x^r\left((2r+1)+(2r+1)\sum_{n=1}^{\infty}\left(\frac{1}{\prod_{i=1}^{n}(2r+2i-1)}\right)x^n\right)\right)\Bigg|_{r=-\frac{1}{2}} \\
&= \frac{1}{2}\frac{\partial}{\partial r}\left(x^r\left((2r+1)+\sum_{n=1}^{\infty}\left(\frac{1}{\prod_{i=2}^{n}(2r+2i-1)}\right)x^n\right)\right)\Bigg|_{r=-\frac{1}{2}} \\
&= \frac{1}{2}\ln(x)x^{-\frac{1}{2}}\left(\sum_{n=1}^{\infty}\left(\frac{1}{\prod_{i=2}^{n}(2i-2)}\right)x^n\right)+\left(\frac{1}{2}\right)x^{-\frac{1}{2}} \\
&\quad+\left(2-\sum_{n=1}^{\infty}\left(\frac{1}{\prod_{i=2}^{n}(2i-2)}\right)\left(\sum_{i=2}^{n}\frac{1}{i-1}\right)x^n\right) \\
&= \frac{1}{2}\ln(x)x^{-\frac{1}{2}}\left(\sum_{n=1}^{\infty}\left(\frac{1}{\prod_{i=2}^{n}2(i-1)}\right)x^n\right)+\left(\frac{1}{2}\right)x^{-\frac{1}{2}} \\
&\quad+\left(2-\sum_{n=1}^{\infty}\left(\frac{1}{\prod_{i=2}^{n}2(i-1)}\right)\left(\sum_{i=2}^{n}\frac{1}{i-1}\right)x^n\right) \\
&= \frac{1}{2}\ln(x)x^{-\frac{1}{2}}\left(\sum_{n=1}^{\infty}\left(\frac{1}{2^{n-1}(n-1)!}\right)x^n\right)+\left(\frac{1}{2}\right)x^{-\frac{1}{2}} \\
&\quad+\left(2-\sum_{n=1}^{\infty}\left(\frac{1}{2^{n-1}(n-1)!}\right)\left(\sum_{i=2}^{n}\frac{1}{i-1}\right)x^n\right) \\
&= \left(\frac{\ln(x)}{x^{\frac{1}{2}}}\right)\left(\sum_{n=1}^{\infty}\left(\frac{x^n}{2^n(n-1)!}\right)\right)+\left(\frac{1}{x^{\frac{1}{2}}}\right) \\
&\quad+\left(1-\sum_{n=1}^{\infty}\left(\frac{x^n}{2^n(n-1)!}\right)\left(\sum_{i=2}^{n}\frac{1}{i-1}\right)\right)
\end{aligned}
$$

Exercises 7.2.2

1. The differential equation $(x-x_0)^2y''+(x-x_0)p(x)y'+q(x)y=f(x)$ has a series solution of the form $y(x)=\sum_{n=0}^{\infty}a_n(x-x_0)^{n+r}$ if and only if either $\frac{p(x)}{x-x_0}$ or $\frac{q(x)}{(x-x_0)^2}$ cannot be expanded in power series and $\frac{f(x)}{(x-x_0)^2}$ can be expanded in series about the point $x=x_0$, but both $p(x)$ and $q(x)$ can be expanded in series about $x=x_0$.
2. Find the two independent series solutions of the differential equation $xy''-y=0$ about the point $x=0$.
3. Find the two independent series solutions of the differential equation $x^2y''+xy'+(x^2+1)y=0$ about the point $x=0$.
4. Find the two independent series solutions of the differential equation $x^2y''+x^2y'-2xy=0$ about the point $x=0$.
5. Find the two independent series solutions of the differential equation $xy''+(3+2x)y'+4y=0$ about the point $x=0$.
6. Find the two independent series solutions of the differential equation $xy''+(4+3x)y'+3y=0$ about the point $x=0$.
7. Find the two independent series solutions of the differential equation $xy''-xy'+y=0$ about the point $x=0$.
8. Find the two independent series solutions of the differential equation $y''+(1-x)y=0$ about the point $x=0$.
9. Find the two independent series solutions of the differential equation $x^2y''+y=0$ about the point $x=0$.
10. Find the two independent series solutions of the differential equation corresponding to distinct roots not differing by integer about the point $x=0$.

    (a) $2x^2y''-xy'+(x-5)y=0$

    (b) $2xy''+(1+x)y'-2y=0$

    (c) $2x^2y''+xy'+(x^2-3)y=0$

    (d) $2x(x-1)y''+3(x-1)y'-y=0$

    (e) $2xy''+5(1+2x)y'+5y=0$

    (f) $4xy''+2y'+y=0$

11. Find the two independent series solutions of the differential equation corresponding to distinct roots differing by integer about the point $x=0$.

    (a) $x^2y''-xy'-\left(x^2+\frac{5}{4}\right)y=0$

    (b) $xy''-(4+x)y'+2y=0$

    (c) $x^2y''+2x(x-2)y'+2(2-3x)y=0$

    (d) $(1-x^2)y''+2xy'+y=0$

12. Find the two independent series solutions of the differential equation corresponding to distinct roots differing by integer about the point $x=0$.

(a) $x^2y'' + (x^2 - 3x)y' + 3y = 0$

(b) $x^2y'' + x(1 - x)y' - (1 + 3x)y = 0$

(c) $(x^2 - x)y'' - xy' + y = 0$

(d) $xy'' - 3y' + xy = 0$

(e) $(x - x^2)y'' - (1 + 3x)y' - y = 0$

13. Find the two independent series solutions of the differential equation corresponding to equal roots about the point $x = 0$.

(a) $x(x - 1)y'' + (3x - 1)y' + y = 0$

(b) $x^2y'' + 3xy' + (1 - 2x)y = 0$

(c) $x^2y'' - x(1 + x)y' + y = 0$

(d) $(x - x^2)y'' + (1 - x)y' - y = 0$

(e) $xy'' + y' + xy = 0$

Chapter 8

Special Differential Equations

There are some differential equations whose solutions are very important for finding solutions of certain real life problems when they are modelled in mathematical form. In particular, they are

1. Legendre Differential Equation
2. Hermite Differential Equation
3. Chebyshev Differential Equation
4. Hypergeometric Differential equation
5. Bessel Differential Equation
6. Laguerre Differential Equation.

In this chapter, we study in detail the derivation of solutions to each type and their properties.

8.1 Legendre Differential Equation

The differential equation $(1-x^2)y''-2xy'+m(m+1)y=0$ where $m \in \mathbb{N}$ is called the *Legendre differential equation*. The Legendre differential equation can be rewritten in the form

$$x^2y''+x\left(-\frac{2x^2}{1-x^2}\right)y'+\left(\frac{m(m+1)x^2}{1-x^2}\right)y=0 \tag{8.1}$$

Clearly, $\frac{p(x)}{x}=-\frac{2x}{1-x^2}$ and $\frac{q(x)}{x^2}=\frac{m(m+1)}{1-x^2}$ are analytic at $x=0$. Hence $x=0$ is an ordinary point. In other words, one can expect the series solution in the form $y(x)=\sum\limits_{n=0}^{\infty}a_nx^n$.

It is easy to conclude by substituting appropriate expressions of $y(x)$ in the Legendre differential equation that

$$0=(1-x^2)y''-2xy'+m(m+1)y$$

$$= (1-x^2)\left(\sum_{n=2}^{\infty} n(n-1)a_n x^{n-2}\right) - 2x\left(\sum_{n=1}^{\infty} na_n x^{n-1}\right) + m(m+1)\left(\sum_{n=0}^{\infty} a_n x^n\right)$$

$$= (1-x^2)\left(\sum_{n=2}^{\infty} n(n-1)a_n x^{n-2}\right) - 2\left(\sum_{n=1}^{\infty} na_n x^{n}\right) + m(m+1)\left(\sum_{n=0}^{\infty} a_n x^n\right)$$

$$= (1-x^2)\left(\sum_{n=2}^{\infty} n(n-1)a_n x^{n-2}\right) + \sum_{n=1}^{\infty}(m(m+1)-2n)a_n x^n + m(m+1)a_0$$

$$= \sum_{n=2}^{\infty} n(n-1)a_n x^{n-2} - \sum_{n=2}^{\infty} n(n-1)a_n x^n + \sum_{n=1}^{\infty}(m(m+1)-2n)a_n x^n$$
$$+ m(m+1)a_0$$

$$= \sum_{n=0}^{\infty}(n+2)(n+1)a_{n+2}x^n - \sum_{n=2}^{\infty} n(n-1)a_n x^n + \sum_{n=1}^{\infty}(m(m+1)-2n)a_n x^n$$
$$+ m(m+1)a_0$$

$$= \sum_{n=2}^{\infty}((n+1)(n+2)a_{n+2} + (m(m+1) - n(n+1))a_n)x^n + m(m+1)a_0 + 2a_2$$
$$+ (6a_3 + (m(m+1)-2)a_1)x$$

$$= \sum_{n=2}^{\infty}((n+1)(n+2)a_{n+2} + (m+n+1)(m-n)a_n)x^n + m(m+1)a_0 + 2a_2$$
$$+ (6a_3 + (m+2)(m-1)a_1)x \tag{8.2}$$

Equating the coefficients of each power of x to zero, we have

$$a_2 = -\left(\frac{m(m+1)}{2}\right)a_0$$

$$a_3 = -\left(\frac{(m+2)(m-1)}{6}\right)a_1$$

$$a_{n+2} = -\left(\frac{(m+n+1)(m-n)}{(n+1)(n+2)}\right)a_n \qquad n \geq 2$$

In general, we have

$$a_{n+2} = -\left(\frac{(m+n+1)(m-n)}{(n+1)(n+2)}\right)a_n \qquad n \geq 0 \tag{8.3}$$

Hence the corresponding series solution is

$$y(x) = a_0\left(1+\sum_{n=1}^{\infty}(-1)^n\left(\prod_{k=0}^{2(n-1)}\left(\frac{(m+k+1)(m-k)}{(k+1)(k+2)}\right)\right)x^{2n}\right)$$
$$+\, a_1\left(x+\sum_{n=1}^{\infty}(-1)^n\left(\prod_{k=1}^{2n-1}\left(\frac{(m+k+1)(m-k)}{(k+1)(k+2)}\right)\right)x^{2n+1}\right)$$
$$= a_0\left(1+\sum_{n=1}^{\infty}(-1)^n\left(\frac{\prod_{k=0}^{2(n-1)}(m+k+1)(m-k)}{(2n)!}\right)x^{2n}\right)$$
$$+\, a_1\left(x+\sum_{n=1}^{\infty}(-1)^n\left(\frac{\prod_{k=1}^{2n-1}(m+k+1)(m-k)}{(2n+1)!}\right)x^{2n+1}\right)$$

where a_0 and a_1 are arbitrary constants. It is clear from Eq.(8.3) that $a_{m+2} = 0$. It is easy to conclude that when $m = 2r$ and $a_1 = 0$, one can have a polynomial solution to the Legendre differential equation and the polynomial solution is given by

$$y(x) = a_0\left(1+\sum_{n=1}^{r}(-1)^n\left(\frac{\prod_{k=0}^{2(n-1)}(m+k+1)(m-k)}{(2n)!}\right)x^{2n}\right) \tag{8.4}$$

If $m = 2r+1$ and $a_0 = 0$, then one can have a polynomial solution to the Legendre differential equation and the polynomial solution is given by

$$y(x) = a_1\left(x+\sum_{n=1}^{r}(-1)^n\left(\frac{\prod_{k=1}^{2n-1}(m+k+1)(m-k)}{(2n+1)!}\right)x^{2n+1}\right) \tag{8.5}$$

It is clear from Eqs.(8.4) and (8.5) that the Legendre differential equation has a polynomial solution. Since the Legendre polynomials are of great practical importance, let us consider them in different forms. Let $a_n = \frac{\prod_{i=0}^{n-1}(2n-1-2i)}{n!} = \frac{(2n)!}{2^n(n!)^2}$. For this purpose, we solve Eq.(8.3) for a_{n-2}. Hence we have

$$a_{n-2} = -\left(\frac{n(n-1)}{(m-n+2)(m+n-1)}\right)a_n \qquad n \leq m-2$$

In general, we have

$$a_{m-2n} = (-1)^n \left(\frac{(2m-2n)!}{2^m n!(m-n)!(m-2n)!} \right)$$

The resulting solution of Legendre differential Eq.(8.1) is called the *Legendre polynomial of degree m*, denoted by $P_m(x)$ and is given by

$$P_m(x) = \sum_{n=0}^{\lfloor \frac{m}{2} \rfloor} (-1)^n \left(\frac{(2m-2n)!}{2^m n!(m-n)!(m-2n)!} \right) x^{m-2n}$$

Theorem 8.1.1 If $V = (x^2-1)^m$, then $(1-x^2)U''+2x(m-r-1)U'+(2m(r+1)-r(r+1))U = 0$, where $U = D^r V$ and $r \in \mathbb{N}$.

Proof: It is easy to verify that $V' = 2mx(x^2-1)^{m-1}$. Multiplying the equation $V' = 2mx(x^2-1)^{m-1}$ by x^2-1, we have

$$(1-x^2)V' + 2mxV = 0 \tag{8.6}$$

Differentiating Eq.(8.6) twice, we have

$$(1-x^2)V''' + 2(m-2)xV'' + (4m-2)V' = 0 \tag{8.7}$$

Clearly, the given statement is true for $r = 1$ by Eq.(8.7) and replacing $U = DV = V'$. Assume that the given statement is true for $r = n$. In other words, we have

$$(1-x^2)U'' + 2(m-n-1)xU' + (2m(n+1) - n(n+1))U = 0 \tag{8.8}$$

where $U = D^n V$. Differentiating Eq.(8.8) once again, we have

$$\begin{aligned} 0 &= (1-x^2)U''' + 2((m-n-1)-1)xU'' + ((n+1)(2m-n) + 2(m-n-1))U' \\ &= (1-x^2)U''' + 2(m-n-2)xU'' + (2m(n+2) - (n+1)(n+2))U' \\ &= (1-x^2)W'' + 2(m-n-2)xW' + (2m(n+2) - (n+1)(n+2))W \end{aligned}$$

where $W = DU = D^{n+1}V$. Hence the statement is true for all $r \in \mathbb{N}$ by the method of induction.

Theorem 8.1.2 (Rodrigues' Formula) If $P_m(x)$ is the polynomial solution to the Legendre differential equation with $P_m(1) = 1$, then $P_m(x) = \left(\frac{1}{m!2^m}\right) \frac{d^m}{dx^m}(x^2-1)^m$.

Proof: According to Theorem 8.1.1 for $r = m$, we have

$$(1-x^2)U'' - 2xU' + m(m+1)U = 0 \tag{8.9}$$

where $U = D^m V$ and $V = (x^2-1)^m$. In other words, $U = D^m V$ is a polynomial solution of the Legendre differential equation.

Let $P_m(x) = cU$ is the Legendre polynomial of degree m. Hence $P_m(1) = 1$. In other words, we have

$$
\begin{aligned}
1 &= c\left(\frac{\mathrm{d}^m}{\mathrm{d}x^m}(x^2-1)^m\right)\Big|_{x=1} \\
&= c\left(\frac{\mathrm{d}^m}{\mathrm{d}x^m}((x+1)(x-1))^m\right)\Big|_{x=1} \\
&= c\left(\frac{\mathrm{d}^m}{\mathrm{d}x^m}((x+1)^m(x-1)^m)\right)\Big|_{x=1} \\
&= c\left((x-1)^m\frac{\mathrm{d}^m}{\mathrm{d}x^m}((x+1)^m)\right)\Big|_{x=1} + c\left((x+1)^m\frac{\mathrm{d}^m}{\mathrm{d}x^m}((x-1)^m)\right)\Big|_{x=1} \\
&= c2^m m!
\end{aligned}
$$

Hence $c = \frac{1}{2^m m!}$. In other words, we have

$$
P_m(x) = \left(\frac{1}{m!2^m}\right)\frac{\mathrm{d}^m}{\mathrm{d}x^m}(x^2-1)^m
$$

8.1.1 Generating Function

Theorem 8.1.3 If $P_m(x)$ is the Legendre polynomial of degree m, then $(1-2xt+t^2)^{-\frac{1}{2}} = \sum\limits_{m=0}^{\infty} P_m(x)t^m$.

Proof: Expanding $(1-2xt+t^2)^{-\frac{1}{2}}$ using binomial theorem, we have

$$
\begin{aligned}
(1-2xt+t^2)^{-\frac{1}{2}} &= (1-t(2x-t))^{-\frac{1}{2}} \\
&= 1 + \left(\frac{1}{2}\right)t(2x-t) + \left(\frac{1}{2}\cdot\frac{2}{3}\right)t^2(2x-t)^2 + \cdots \\
&\quad + \left(\frac{1.3\cdots(2m-1)}{2.4\cdots 2m}\right)t^m(2x-t)^m + \cdots \\
&= 1 + \sum_{m=1}^{\infty}\left(\prod_{i=1}^{m}\left(\frac{2i-1}{2i}\right)\right)t^m(2x-t)^m \\
&= 1 + \sum_{m=1}^{\infty}(-1)^m\left(\prod_{i=1}^{m}\left(\frac{2i-1}{2i}\right)\right)t^m(t-2x)^m \\
&= 1 + \sum_{m=1}^{\infty}(-1)^m\left(\prod_{i=1}^{m}\left(\frac{2i-1}{2i}\right)\right)t^m\left(\sum_{j=0}^{m}\binom{m}{j}t^j(-2x)^{m-j}\right)
\end{aligned}
$$

$$= 1 + \sum_{m=1}^{\infty} (-1)^m \left(\prod_{i=1}^{m} \left(\frac{2i-1}{2i} \right) \right) t^m \left(\sum_{j=0}^{m} (-1)^{m-j} \binom{m}{j} t^j (2x)^{m-j} \right)$$

$$= 1 + \sum_{m=1}^{\infty} \left(\prod_{i=1}^{m} \left(\frac{2i-1}{2i} \right) \right) t^m \left(\sum_{j=0}^{m} (-1)^{j} \binom{m}{j} t^j (2x)^{m-j} \right)$$

$$= 1 + \sum_{m=1}^{\infty} \left(\sum_{j=0}^{\lfloor \frac{m}{2} \rfloor} (-1)^j \left(\prod_{i=1}^{m-j} \left(\frac{2i-1}{2i} \right) \right) \binom{m-j}{j} (2x)^{m-2j} \right) t^m$$

$$= 1 + \sum_{m=1}^{\infty} \left(\sum_{j=0}^{\lfloor \frac{m}{2} \rfloor} (-1)^j \left(\prod_{i=1}^{m-j} \left(\frac{2i-1}{2i} \right) \right) \binom{m-j}{j} \left(\frac{2^{m-j}}{2^j} \right) x^{m-2j} \right) t^m$$

$$= 1 + \sum_{m=1}^{\infty} \left(\sum_{j=0}^{\lfloor \frac{m}{2} \rfloor} (-1)^j \left(\prod_{i=1}^{m-j} \left(\frac{2i-1}{i} \right) \right) \binom{m-j}{j} \left(\frac{x^{m-2j}}{2^j} \right) \right) t^m$$

$$= 1 + \sum_{m=1}^{\infty} \left(\sum_{j=0}^{\lfloor \frac{m}{2} \rfloor} \left(\frac{(-1)^j}{2^m} \right) \left(\prod_{i=1}^{m-j} \left(\frac{2i-1}{i} \right) \right) \binom{m-j}{j} 2^{m-j} x^{m-2j} \right) t^m$$

$$= 1 + \sum_{m=1}^{\infty} \left(\sum_{j=0}^{\lfloor \frac{m}{2} \rfloor} \left(\frac{(-1)^j}{2^m} \right) \left(\frac{\prod_{i=1}^{m-j} (2i-1)}{(m-j)!} \right) \binom{m-j}{j} 2^{m-j} x^{m-2j} \right) t^m$$

$$= 1 + \sum_{m=1}^{\infty} \left(\sum_{j=0}^{\lfloor \frac{m}{2} \rfloor} \left(\frac{(-1)^j}{2^m} \right) \left(\frac{\left(\prod_{i=1}^{m-j} (2i-1) \right) (m-j)! 2^{m-j}}{(m-j)! j! (m-2j)!} \right) x^{m-2j} \right) t^m$$

$$= 1 + \sum_{m=1}^{\infty} \left(\sum_{j=0}^{\lfloor \frac{m}{2} \rfloor} (-1)^j \left(\frac{(2m-2j)!}{2^m (m-j)! j! (m-2j)!} \right) x^{m-2j} \right) t^m$$

$$= 1 + \sum_{m=1}^{\infty} \left(\sum_{j=0}^{\lfloor \frac{m}{2} \rfloor} a_{m-2j} x^{m-2j} \right) t^m$$

$$= 1 + \sum_{m=1}^{\infty} P_m(x) t^m$$

$$= \sum_{m=0}^{\infty} P_m(x) t^m$$

Corollary 8.1.1 If $P_m(x)$ is the Legendre polynomial of degree m, then $P_m(1) = 1$.

Proof: According to Theorem 8.1.3, we have

$$\sum_{m=0}^{\infty} P_m(x)t^m = (1 - 2xt + t^2)^{-\frac{1}{2}} \tag{8.10}$$

Evaluating Eq.(8.10) at $x = 1$, we have

$$\begin{aligned}\sum_{m=0}^{\infty} P_m(1)t^m &= (1 - 2t + t^2)^{-\frac{1}{2}} \\ &= (1 - t)^{-1} \\ &= \sum_{m=0}^{\infty} t^m \end{aligned} \tag{8.11}$$

According to the series given in Eq.(8.11), we have $P_m(1) = 1$.

Corollary 8.1.2 If $P_m(x)$ is the Legendre polynomial of degree m, then $P_m(-1) = (-1)^m$.

Proof: According to Theorem 8.1.3, we have

$$\sum_{m=0}^{\infty} P_m(x)t^m = (1 - 2xt + t^2)^{-\frac{1}{2}} \tag{8.12}$$

Evaluating Eq.(8.12) at $x = -1$, we have

$$\begin{aligned}\sum_{m=0}^{\infty} P_m(-1)t^m &= (1 + 2t + t^2)^{-\frac{1}{2}} \\ &= (1 + t)^{-1} \\ &= \sum_{m=0}^{\infty} (-1)^m t^m \end{aligned} \tag{8.13}$$

According to the series given in Eq.(8.13), we have $P_m(-1) = (-1)^m$.

Corollary 8.1.3 If $P_m(x)$ is the Legendre polynomial of degree m, then

$$P_m(-x) = (-1)^m P_m(x)$$

Proof: According to Theorem 8.1.3, we have

$$(1 - 2xt + t^2)^{-\frac{1}{2}} = \sum_{m=0}^{\infty} P_m(x)t^m \tag{8.14}$$

Changing x to $-x$ in Eq.(8.14), we have

$$(1 + 2xt + t^2)^{-\frac{1}{2}} = \sum_{m=0}^{\infty} P_m(-x)t^m \tag{8.15}$$

Changing t to $-t$ in Eq.(8.14), we have

$$(1 + 2xt + t^2)^{-\frac{1}{2}} = \sum_{m=0}^{\infty} P_m(x)(-t)^m \tag{8.16}$$

Comparing Eqs.(8.15) and (8.16), we have

$$\sum_{m=0}^{\infty} P_m(-x)t^m = \sum_{m=0}^{\infty} P_m(x)(-t)^m$$

$$= \sum_{m=0}^{\infty} (-1)^m P_m(x)t^m$$

Hence $P_m(-x) = (-1)^m P_m(x)$.

Corollary 8.1.4 If $P_m(x)$ is the Legendre polynomial of degree m, then

$$P'_m(-x) = (-1)^{m+1} P'_m(x)$$

Proof: According to Theorem 8.1.3, we have

$$(1 - 2xt + t^2)^{-\frac{1}{2}} = \sum_{m=0}^{\infty} P_m(x)t^m \tag{8.17}$$

Differentiation of Eq.(8.17) with respect to x gives

$$t(1 - 2xt + t^2)^{-\frac{3}{2}} = \sum_{m=0}^{\infty} P'_m(x)t^m \tag{8.18}$$

Changing x to $-x$ in Eq.(8.18), we have

$$t(1 + 2xt + t^2)^{-\frac{3}{2}} = \sum_{m=0}^{\infty} P'_m(-x)t^m \tag{8.19}$$

Changing t to $-t$ in Eq.(8.18), we have

$$t(1 + 2xt + t^2)^{-\frac{3}{2}} = \sum_{m=0}^{\infty} (-1)P'_m(x)(-t)^m \tag{8.20}$$

Comparing Eqs.(8.19) and (8.20), we have

$$\sum_{m=0}^{\infty} P'_m(-x)t^m = \sum_{m=0}^{\infty} (-1)P'_m(x)(-t)^m$$

$$= \sum_{m=0}^{\infty} (-1)^{m+1} P'_m(x)t^m$$

Hence $P'_m(-x) = (-1)^{m+1} P'_m(x)$.

Theorem 8.1.4 If $P_{2m}(x)$ is the Legendre polynomial of degree $2m$, then

$$P_{2m}(0) = \left(-\frac{1}{2}\right)^m \left(\prod_{i=1}^{m}\left(\frac{2i-1}{i}\right)\right)$$

Proof: According to Theorem 8.1.3, we have

$$\begin{aligned}\sum_{m=0}^{\infty} P_m(0)t^m &= (1+t^2)^{-\frac{1}{2}} \\ &= \sum_{m=0}^{\infty}\binom{-\frac{1}{2}}{m}t^{2m} \\ &= \sum_{m=0}^{\infty}(-1)^m\left(\prod_{i=1}^{m}\left(\frac{2i-1}{2i}\right)\right)t^{2m} \\ &= \sum_{m=0}^{\infty}\left(-\frac{1}{2}\right)^m\left(\prod_{i=1}^{m}\left(\frac{2i-1}{i}\right)\right)t^{2m}\end{aligned}$$

In other words, we have $P_{2m}(0) = (-\frac{1}{2})^m\left(\prod_{i=1}^{m}(\frac{2i-1}{i})\right)$ by collecting the coefficients of t^{2m}.

Theorem 8.1.5 If $P_{2m+1}(x)$ is the Legendre polynomial of degree $2m+1$, then $P_{2m+1}(0) = 0$.

Proof: According to Theorem 8.1.3, we have

$$\begin{aligned}\sum_{m=0}^{\infty} P_m(0)t^m &= (1+t^2)^{-\frac{1}{2}} \\ &= \sum_{m=0}^{\infty}\binom{-\frac{1}{2}}{m}t^{2m} \\ &= \sum_{m=0}^{\infty}(-1)^m\left(\prod_{i=1}^{m}\left(\frac{2i-1}{2i}\right)\right)t^{2m} \\ &= \sum_{m=0}^{\infty}\left(-\frac{1}{2}\right)^m\left(\prod_{i=1}^{m}\left(\frac{2i-1}{i}\right)\right)t^{2m} \\ &= \sum_{m=0}^{\infty}\left(\frac{(-1)^m(2m)!}{(2^m\,m!)^2}\right)t^{2m}\end{aligned}$$

In other words, we have $P_{2m+1}(0) = 0$ by collecting the coefficients of t^{2m+1}.

Theorem 8.1.6 If $P_m(x)$ is the Legendre polynomial of degree m, then $P'_m(1) = \frac{m(m+1)}{2}$.

Proof: According to Theorem 8.1.3, we have

$$(1-2xt+t^2)^{-\frac{1}{2}} = \sum_{m=0}^{\infty} P_m(x)t^m \tag{8.21}$$

Differentiating Eq.(8.21) with respect to x gives

$$t(1-2xt+t^2)^{-\frac{3}{2}} = \sum_{m=0}^{\infty} P'_m(x)t^m \tag{8.22}$$

Evaluating Eq.(8.22) at $x = 1$, we have

$$\begin{aligned}
\sum_{m=0}^{\infty} P'_m(1)t^m &= t(1-2t+t^2)^{-\frac{3}{2}} \\
&= t(1-t)^{-3} \\
&= t\left(\sum_{m=0}^{\infty} \binom{-3}{m}(-t)^m\right) \\
&= \sum_{m=0}^{\infty} (-1)^m \binom{-3}{m} t^{m+1} \\
&= \sum_{m=0}^{\infty} (-1)^m \left(\frac{\prod_{j=0}^{m-1}(-3-j)}{m!}\right) t^{m+1} \\
&= \sum_{m=0}^{\infty} \left(\frac{\prod_{j=0}^{m-1}(3+j)}{m!}\right) t^{m+1} \\
&= \sum_{m=0}^{\infty} \left(\frac{(m+2)(m+1)}{2}\right) t^{m+1}
\end{aligned}$$

Collecting the coefficients of t^m, we have

$$P'_m(1) = \frac{m(m+1)}{2}$$

Theorem 8.1.7 If $P_m(x)$ is the Legendre polynomial of degree m, then

$$P'_m(-1) = (-1)^{m-1}\left(\frac{m(m+1)}{2}\right)$$

Proof: According to Theorem 8.1.3, we have

$$(1-2xt+t^2)^{-\frac{1}{2}} = \sum_{m=0}^{\infty} P_m(x)t^m \tag{8.23}$$

Differentiating Eq.(8.23) with respect to x gives

$$t(1-2xt+t^2)^{-\frac{3}{2}} = \sum_{m=0}^{\infty} P'_m(x)t^m \tag{8.24}$$

Evaluating Eq.(8.24) at $x=-1$, we have

$$\begin{aligned}
\sum_{m=0}^{\infty} P'_m(-1)t^m &= t(1+2t+t^2)^{-\frac{3}{2}} \\
&= t(1+t)^{-3} \\
&= t\left(\sum_{m=0}^{\infty}\binom{-3}{m}t^m\right) \\
&= \sum_{m=0}^{\infty}\binom{-3}{m}t^{m+1} \\
&= \sum_{m=0}^{\infty}\left(\frac{\prod_{j=0}^{m-1}(-3-j)}{m!}\right)t^{m+1} \\
&= \sum_{m=0}^{\infty}\left(\frac{\prod_{j=0}^{m-1}(-1)^m(3+j)}{m!}\right)t^{m+1} \\
&= \sum_{m=0}^{\infty}(-1)^m\left(\frac{(m+2)(m+1)}{2}\right)t^{m+1}
\end{aligned}$$

Collecting the coefficients of t^m, we have

$$P'_m(-1) = (-1)^{m-1}\left(\frac{m(m+1)}{2}\right)$$

8.1.2 Recurrence Relation

Theorem 8.1.8 If $P_m(x)$ is the Legendre polynomial of degree m, then

$$mP_{m-1}(x) + (m+1)P_{m+1}(x) - (2m+1)xP_m(x) = 0$$

Proof: According to Theorem 8.1.3, we have

$$(1-2xt+t^2)^{-\frac{1}{2}} = \sum_{m=0}^{\infty} P_m(x)t^m \tag{8.25}$$

Differentiation of Eq.(8.25) with respect to t, we have

$$\sum_{m=1}^{\infty} mP_m(x)t^{m-1} = (1-2xt+t^2)^{-\frac{1}{2}}$$
$$= (x-t)(1-2xt+t^2)^{-\frac{3}{2}}$$
$$= (x-t)(1-2xt+t^2)^{-1}(1-2xt+t^2)^{-\frac{1}{2}} \quad (8.26)$$

Again, using Eq.(8.26) and Theorem 8.1.3, we have

$$(x-t)\left(\sum_{m=0} P_m(x)t^m\right) = (x-t)(1-2xt+t^2)^{-\frac{1}{2}}$$
$$= (1-2xt+t^2)\left(\sum_{m=1}^{\infty} mP_m(x)t^{m-1}\right) \quad (8.27)$$

Collecting coefficients of t^m from Eq.(8.27), we have

$$xP_m(x) - P_{m-1}(x) = (m+1)P_{m+1} - 2mxP_m(x) + (m-1)P_{m-1}(x)$$

In other words, we have

$$(m+1)P_{m+1}(x) - (2m+1)xP_m(x) + mP_{m-1}(x) = 0$$

Theorem 8.1.9 If $P_m(x)$ is the Legendre polynomial of degree m, then $P'_m(x) - 2xP'_{m-1}(x) + P'_{m-2}(x) - P_{m-1}(x) = 0$.

Proof: According to Theorem 8.1.3, we have

$$(1-2xt+t^2)^{-\frac{1}{2}} = \sum_{m=0}^{\infty} P_m(x)t^m \quad (8.28)$$

Differentiation of Eq.(8.28) with respect to x gives

$$\sum_{m=0}^{\infty} P'_m(x)t^m = t(1-2xt+t^2)^{-\frac{3}{2}}$$
$$= t(1-2xt+t^2)^{-1}(1-2xt+t^2)^{-\frac{1}{2}} \quad (8.29)$$

Again, using Eq.(8.29) and Theorem 8.1.3, we have

$$(1-2xt+t^2)\left(\sum_{m=0}^{\infty} P'_m(x)t^m\right) = t(1-2xt+t^2)^{-\frac{1}{2}}$$
$$= t\left(\sum_{m=0}^{\infty} P_m(x)t^m\right)$$
$$= \sum_{m=0}^{\infty} P_m(x)t^{m+1} \quad (8.30)$$

Collecting coefficients of t^m from Eq.(8.30), we have

$$P'_m(x) - 2xP'_{m-1}(x) + P'_{m-2}(x) = P_{m-1}(x)$$

In other words, we have

$$P'_m(x) - 2xP'_{m-1}(x) + P'_{m-2}(x) - P_{m-1}(x) = 0$$

Theorem 8.1.10 If $P_m(x)$ is the Legendre polynomial of degree m, then $mP_m(x) - xP'_m(x) + P'_{m-1}(x) = 0$.

Proof: According to Theorem 8.1.3, we have

$$(1 - 2xt + t^2)^{-\frac{1}{2}} = \sum_{m=0}^{\infty} P_m(x)t^m \tag{8.31}$$

Differentiation of Eq.(8.31) with respect to x gives

$$\begin{aligned}\sum_{m=0}^{\infty} P'_m(x)t^m &= t(1 - 2xt + t^2)^{-\frac{3}{2}} \\ &= t(1 - 2xt + t^2)^{-1}(1 - 2xt + t^2)^{-\frac{1}{2}} \end{aligned} \tag{8.32}$$

Differentiation of Eq.(8.31) with respect to t gives

$$\begin{aligned}\sum_{m=1}^{\infty} mP_m(x)t^{m-1} &= (x - t)(1 - 2xt + t^2)^{-\frac{3}{2}} \\ &= (x - t)(1 - 2xt + t^2)^{-1}(1 - 2xt + t^2)^{-\frac{1}{2}} \end{aligned} \tag{8.33}$$

Comparing Eqs.(8.32) and (8.33), we have

$$(x - t)\left(\sum_{m=0}^{\infty} P'_m(x)t^m\right) = t\left(\sum_{m=1}^{\infty} mP_m(x)t^{m-1}\right)$$

Collecting the coefficients of t^m, we have

$$xP'_m(x) - P'_{m-1}(x) = mP_m(x)$$

In other words, we have

$$mP_m(x) - xP'_m(x) + P'_{m-1}(x) = 0$$

Theorem 8.1.11 If $P_m(x)$ is the Legendre polynomial of degree m, then $P'_{m-1}(x) + (2m + 1)P_m(x) - P'_{m+1}(x) = 0$.

Proof: According to Exercise 4 of this section

$$P_m(x) = P'_{m+1}(x) - 2xP'_m(x) + P'_{m-1}(x) \tag{8.34}$$

According to Theorem 8.1.10, we have

$$xP'_m(x) = mP_m(x) + P'_{m-1}(x) \tag{8.35}$$

Using Eq.(8.35) in Eq.(8.34), we have

$$P_m(x) = P'_{m+1}(x) - 2mP_m(x) - 2P'_{m-1}(x) + P'_{m-1}(x) = P'_{m+1}(x) - 2mP_m(x) - P'_{m-1}(x)$$

In other words, we have

$$P'_{m-1}(x) + (2m+1)P_m(x) - P'_{m+1}(x) = 0$$

Theorem 8.1.12 If $P_m(x)$ is the Legendre polynomial of degree m, then $(m+1)P_m(x) + xP'_m(x) - P'_{m+1}(x) = 0$.

Proof: According to Theorem 8.1.10, we have

$$mP_m(x) - xP'_m(x) + P'_{m-1}(x) = 0 \tag{8.36}$$

According to Theorem 8.1.11, we have

$$P'_{m-1}(x) + (2m+1)P_m(x) - P'_{m+1}(x) = 0 \tag{8.37}$$

Subtracting Eq.(8.36) from Eq.(8.37), we have

$$(m+1)P_m(x) - P'_{m+1}(x) + xP'_m(x) = 0$$

Theorem 8.1.13 If $P_m(x)$ is the Legendre polynomial of degree m, then $(1-x^2)P'_m(x) - mP_{m-1}(x) + mxP_m(x) = 0$.

Proof: According to Theorem 8.1.10, we have

$$mP_m(x) - xP'_m(x) + P'_{m-1}(x) = 0 \tag{8.38}$$

According to Theorem 8.1.12 and replacing $m+1$ by m, we have

$$mP_{m-1}(x) - P'_m(x) + xP'_{m-1}(x) = 0 \tag{8.39}$$

Multiplying Eq.(8.38) by x and subtracting from Eq.(8.39), we have

$$(x^2-1)P'_m(x) - xmP_m(x) + mP_{m-1}(x) = 0$$

In other words, we have

$$(1-x^2)P'_m(x) - mP_{m-1}(x) + xmP_m(x) = 0$$

8.1.3 Orthogonal Property

Theorem 8.1.14 If $P_m(x)$ and $P_n(x)$ are the Legendre polynomials of degree m and n respectively, then $\int_{-1}^{1} P_m(x)P_n(x)\,\mathrm{d}x = 0$ for $m \neq n$.

Proof: According to Eq.(8.9) and $P_m(x) = \left(\frac{1}{2^m m!}\right) \frac{\mathrm{d}^m}{\mathrm{d}x^m}(x^2-1)^m$, we have

$$(1-x^2)P''_m(x) - 2x'(x) + m(m+1)P_m(x) = 0 \tag{8.40}$$

Again, Eq.(8.40) can be rewritten as

$$\frac{\mathrm{d}}{\mathrm{d}x}((1-x^2)P'_m(x)) + m(m+1)P_m(x) = 0 \tag{8.41}$$

Similarly, for $P_n(x)$ we have

$$\frac{\mathrm{d}}{\mathrm{d}x}((1-x^2)P'_n(x)) + n(n+1)P_n(x) = 0 \tag{8.42}$$

If Eq.(8.41) is multiplied by $P_n(x)$, Eq.(8.42) is multiplied by $P_m(x)$ and the first resulting process is subtracted from the second one, then we have

$$\begin{aligned}(n-m)(m+n+1)P_m(x)P_n(x) &= (n(n+1) - m(m+1))P_m(x)P_n(x)\\ &= n(n+1)P_n(x)P_m(x) - m(m+1)P_m(x)P_n(x)\\ &= ((1-x^2)P'_m(x))'P_n(x) - ((1-x^2)P'_n(x))'P_m(x)\\ &= ((1-x^2)(P'_n(x)P_m(x) - P'_m(x)P_m(x)))' \end{aligned} \tag{8.43}$$

Again, by integration of Eq.(8.43), we have

$$\begin{aligned}(n-m)(m+n+1)\int_{-1}^{1} P_m(x)P_n(x)\,\mathrm{d}x &= \int_{-1}^{1} ((1-x^2)(P'_m(x)P_n(x) - P'_n(x)P_m(x)))'\,\mathrm{d}x\\ &= (1-x^2)(P'_m(x)P_n(x) - P'_n(x)P_m(x))\big|_{-1}^{1}\\ &= 0\end{aligned}$$

One can easily conclude that $\int_{-1}^{1} P_m(x)P_n(x)\,\mathrm{d}x = 0$.

Theorem 8.1.15 If $P_m(x)$ is the Legendre polynomial of degree m, then $\int_{-1}^{1} P_m^2(x)\,\mathrm{d}x = \frac{2}{2m+1}$.

Proof: According to Theorem 8.1.2, we have

$$P_m(x) = \left(\frac{1}{m!2^m}\right)\frac{\mathrm{d}^m}{\mathrm{d}x^m}(x^2-1)^m \tag{8.44}$$

According to Eq.(8.44) and integration by parts $0 \leq n \leq m$ times, we have

$$\begin{aligned}
\int_{-1}^{1} P_m^2(x)\,\mathrm{d}x &= \left(\frac{1}{m!2^m}\right)^2 \int_{-1}^{1} \frac{\mathrm{d}^m}{\mathrm{d}x^m}(x^2-1)^m \frac{\mathrm{d}^m}{\mathrm{d}x^m}(x^2-1)^m\,\mathrm{d}x \\
&= (-1)^n \left(\frac{1}{m!2^m}\right)^2 \int_{-1}^{1} \frac{\mathrm{d}^{m+n}}{\mathrm{d}x^{m+n}}(x^2-1)^m \frac{\mathrm{d}^{m-n}}{\mathrm{d}x^{m-n}}(x^2-1)^m\,\mathrm{d}x \\
&= (-1)^m \left(\frac{1}{m!2^m}\right)^2 \int_{-1}^{1} \frac{\mathrm{d}^{2m}}{\mathrm{d}x^{2m}}(x^2-1)^m (x^2-1)^m\,\mathrm{d}x \\
&= (-1)^m (2m)! \left(\frac{1}{m!2^m}\right)^2 \int_{-1}^{1} (x^2-1)^m\,\mathrm{d}x \\
&= (2m)! \left(\frac{1}{m!2^m}\right)^2 \int_{-1}^{1} (1-x^2)^m\,\mathrm{d}x \\
&= 2(2m)! \left(\frac{1}{m!2^m}\right)^2 \int_{0}^{1} (1-x^2)^m\,\mathrm{d}x \qquad (8.45)
\end{aligned}$$

Let $x = \sin\theta$. Clearly, Eq.(8.45) becomes

$$\begin{aligned}
\int_{-1}^{1} P_m^2(x)\,\mathrm{d}x &= 2(2m)! \left(\frac{1}{m!2^m}\right)^2 \int_{0}^{\frac{\pi}{2}} \cos^{2m+1}\theta\,\mathrm{d}\theta \\
&= 2(2m)! \left(\frac{1}{m!2^m}\right)^2 \left(\frac{\prod_{i=0}^{m-1}(2m-2i)}{\prod_{i=0}^{m}(2m+1-2i)}\right) \\
&= (2m)! \left(\frac{1}{m!2^m}\right)^2 \left(\frac{\prod_{i=0}^{m-1}(2m-2i)}{\prod_{i=1}^{m}(2m+1-2i)}\right)\left(\frac{2}{2m+1}\right) \\
&= (2m)! \left(\frac{1}{m!2^m}\right)^2 \left(\frac{m!2^m}{\prod_{i=1}^{m}(2m+1-2i)}\right)\left(\frac{2}{2m+1}\right)
\end{aligned}$$

$$= (2m)!\left(\frac{1}{m!2^m}\right)\left(\frac{1}{\prod_{i=1}^{m}(2m+1-2i)}\right)\left(\frac{2}{2m+1}\right)$$

$$= (2m)!\left(\frac{1}{\left(\prod_{j=0}^{m-1}(2m-2j)\right)\left(\prod_{i=1}^{m}(2m+1-2i)\right)}\right)\left(\frac{2}{2m+1}\right)$$

$$= (2m)!\left(\frac{1}{(2m)!}\right)\left(\frac{2}{2m+1}\right)$$

$$= \frac{2}{2m+1}$$

Theorem 8.1.16 If $P_m(x)$ is the Legendre polynomial of degree m, then

$$\int_{-1}^{1} x^m P_m(x)\,\mathrm{d}x = \frac{2^{m+1}(m!)^2}{(2m+1)!}$$

Proof: According to Rodrigues' formula, we have

$$\int_{-1}^{1} x^m P_m(x)\,\mathrm{d}x = \frac{1}{m!2^m}\int_{-1}^{1} x^m \frac{\mathrm{d}^m}{\mathrm{d}x^m}(x^2-1)^m\,\mathrm{d}x \tag{8.46}$$

If the integration in Eq.(8.46) is done using integration by parts m times, then we have

$$\int_{-1}^{1} x^m P_m(x)\,\mathrm{d}x = \frac{1}{m!2^m}\int_{-1}^{1} x^m \frac{\mathrm{d}^m}{\mathrm{d}x^m}(x^2-1)^m\,\mathrm{d}x$$

$$= \frac{1}{2^m}(-1)^m\int_{-1}^{1}(x^2-1)^m\,\mathrm{d}x$$

$$= \frac{1}{2^m}\int_{-1}^{1}(1-x^2)^m\,\mathrm{d}x$$

$$= \frac{2}{2^m}\int_{0}^{1}(1-x^2)^m\,\mathrm{d}x \tag{8.47}$$

Let $x = \sin\theta$. Clearly, Eq.(8.47) becomes

$$\int_{-1}^{1} x^m P_m(x)\,\mathrm{d}x = \frac{2}{2^m}\int_{0}^{\frac{\pi}{2}} \cos^{2m+1}\theta\,\mathrm{d}\theta$$

$$= \frac{2}{2^m}\left(\frac{\prod_{i=0}^{m-1}(2m-2i)}{\prod_{i=0}^{m}(2m+1-2i)}\right)$$

$$= \frac{1}{2^m}\left(\frac{\prod_{i=0}^{m-1}(2m-2i)}{\prod_{i=1}^{m}(2m+1-2i)}\right)\left(\frac{2}{2m+1}\right)$$

$$= \frac{1}{2^m}\left(\frac{m!2^m}{\prod_{i=1}^{m}(2m+1-2i)}\right)\left(\frac{2}{2m+1}\right)$$

$$= m!\left(\frac{1}{\prod_{i=1}^{m}(2m+1-2i)}\right)\left(\frac{2}{2m+1}\right)$$

$$= m!\left(\frac{m!2^m}{\left(\prod_{j=0}^{m-1}(2m-2j)\right)\left(\prod_{i=1}^{m}(2m+1-2i)\right)}\right)\left(\frac{2}{2m+1}\right)$$

$$= (m!)^2 2^m\left(\frac{1}{(2m)!}\right)\left(\frac{2}{2m+1}\right)$$

$$= \frac{(m!)^2 2^{m+1}}{(2m+1)!}$$

Theorem 8.1.17 If $P_m(x)$ is the Legendre polynomial of degree m, then $\int_{-1}^{1} x^k P_m(x)\,\mathrm{d}x = 0$ for $0 \le k \le m-1$.

Proof: According to Rodrigues' formula, we have

$$\int_{-1}^{1} x^k P_m(x)\,\mathrm{d}x = \frac{1}{m!2^m}\int_{-1}^{1} x^k \frac{\mathrm{d}^m}{\mathrm{d}x^m}(x^2-1)^m\,\mathrm{d}x \tag{8.48}$$

If the integration in Eq.(8.48) is done using integration by parts $k+1$ times, then we have

$$\int_{-1}^{1} x^k P_m(x)\,\mathrm{d}x = (-1)^{k+1}\int_{-1}^{1} \frac{\mathrm{d}^{k+1}}{\mathrm{d}x^{k+1}}(x^k)\frac{\mathrm{d}^{m-k-1}}{\mathrm{d}x^{m-k-1}}(x^2-1)^m\,\mathrm{d}x = 0$$

Theorem 8.1.18 If $P_m(x)$ is the Legendre polynomial of degree m, then show that

$$\int_{-1}^{1} P_{m-1}(x)P_m(x)\,\mathrm{d}x = \frac{2m}{4m^2-1}$$

Proof: According to Theorem 8.1.8, we have

$$(m+1)P_{m+1}(x) - (2m+1)xP_m(x) + mP_{m-1}(x) = 0 \tag{8.49}$$

Multiplying Eq.(8.49) by $P_{m-1}(x)$, and applying Theorems 8.1.14 and 8.1.15, we have

$$(2m+1)\int_{-1}^{1} xP_{m-1}(x)P_m(x)\,\mathrm{d}x = (m+1)\int_{-1}^{1} P_{m+1}(x)P_{m-1}(x)\,\mathrm{d}x + m\int_{-1}^{1} P_{m-1}^2(x)\,\mathrm{d}x$$

$$= \frac{2m}{2m-1}$$

In other words, we have

$$\int_{-1}^{1} P_{m-1}(x)P_m(x)\,\mathrm{d}x = \frac{2m}{4m^2-1}$$

EXAMPLE 8.1 Express $f(x) = \sum_{i=0}^{n} a_i x^i$ in terms of Legendre polynomials.

Solution: Let $f(x) = \sum_{i=0}^{n} \alpha_i P_i(x)$. Let $0 \le m \le n$ be any fixed integer. Clearly, we have

$$\int_{-1}^{1} f(x)P_m(x)\,\mathrm{d}x = \sum_{i=0}^{n} a_i \int_{-1}^{1} P_i(x)P_m(x)\,\mathrm{d}x$$

$$= \alpha_m \int_{-1}^{1} P_m^2(x)\,\mathrm{d}x$$

$$= \alpha_m\left(\frac{2}{2m+1}\right)$$

In other words, we have for $0 \le m \le n$

$$\alpha_m = \frac{2m+1}{2}\int_{-1}^{1} f(x)P_m(x)\,\mathrm{d}x$$

Hence the required expansion is given by

$$f(x) = \sum_{i=0}^{n} \left(\frac{2i+1}{2} \int_{-1}^{1} f(x) P_i(x)\, dx \right) P_i(x)$$

Exercises 8.1

1. Show that $\int_0^{\frac{\pi}{2}} \cos^{2m+1}\theta \, d\theta = \left(\frac{\prod_{i=0}^{m-1} (2m-2i)}{\prod_{i=0}^{m} (2m+1-2i)} \right)$

2. If $P_m(x)$ is the Legendre polynomial of degree m, then show that
$$(1-x^2)P'_m(x) = (m+1)\left(xP_m(x) - P_{m+1}(x)\right)$$

3. If $P_m(x)$ is the Legendre polynomial of degree m, then show that
$$(2m+1)(1-x^2)P'_m(x) = m(m+1)\left(P_{m-1}(x) - P_{m+1}(x)\right)$$

4. If $P_m(x)$ is the Legendre polynomial of degree m, then show that
$$P_m(x) = P'_{m+1}(x) - 2xP'_m(x) + P'_{m-1}(x)$$
Hint: One can replace 'm' by '$m+1$' in Theorem 8.1.9 for the required result.

5. If $P_m(x)$ is the Legendre polynomial of degree m, then show that $P'_{m+1}(x) + P'_m(x) = \sum_{i=0}^{m} (2i+1) P_i(x)$.
Hint: Apply Theorem 8.1.11 for $1 \le i \le m$ and use $P_0(x) = P'_1(x)$. Finally, add all the relations.

6. If $P_m(x)$ is the Legendre polynomial of degree m, then show that
$$\int_{-1}^{1} P_m(x)\, dx = \begin{cases} 0 & m \neq 0 \\ 2 & m = 0 \end{cases}$$
Hint: We know $P_0(x) = 1$. Apply Theorems 8.1.14 and 8.1.15.

7. If $P_m(x)$ is the Legendre polynomial of degree m, then show that
$$\int_{-1}^{1} x^2 P_{m+1}(x) P_{m-1}(x)\, dx = \frac{2m(m+1)}{(2m-1)(2m+1)(2m+3)}$$

8. If $P_m(x)$ is the Legendre polynomial of degree m, then show that
$$\int_{-1}^{1} (1-x^2) P'_m(x)\, P'_n(x) dx = \begin{cases} 0 & m \neq n \\ \frac{2m(m+1)}{2m+1} & m = n \end{cases}$$

9. If $P_m(x)$ is the Legendre polynomial of degree m, then show that

$$\int_{-1}^{1} (x^2-1)P_{m+1}(x)P'_m(x)\,\mathrm{d}x = \frac{2m(m+1)}{(2m+1)(2m+3)}$$

10. If $P_m(x)$ is the Legendre polynomial of degree m, then show that

$$\int_{-1}^{1} P_m(x)(1-2xt+t^2)^{-\frac{1}{2}}\,\mathrm{d}x = \frac{2t^m}{2m+1}$$

11. If $P_m(x)$ is the Legendre polynomial of degree m, then show that

$$\frac{1-t^2}{(1-2xt+t^2)^{\frac{3}{2}}} = \sum_{m=0}^{\infty}(2m+1)t^m P_m(x)$$

12. Solve the initial value problem $(1-x^2)y''-2xy'+12y=0$ with $y(0)=1$ and $y'(0)=0$.
13. Solve the initial value problem $(1-x^2)y''-2xy'+20y=0$ with $y(0)=1$ and $y'(0)=0$.
14. If $P_m(x)$ is the Legendre polynomial of degree m, then show that

$$\int_{-1}^{1} \frac{\mathrm{d}x}{1-2xt+t^2} = \sum_{m=0}^{\infty}\left(\frac{2}{2m+1}\right)t^{2m}$$

8.2 Hermite Differential Equation

The differential equation $y''-2xy'+2my=0$ is called the *Hermite differential equation.* The Hermite differential equation can be rewritten in the form

$$x^2y''+x(-2x^2)y'+(2mx^2)y=0 \tag{8.50}$$

Clearly, $\frac{p(x)}{x}=-2x$ and $\frac{q(x)}{x^2}=2n$ are analytic at $x=0$. Hence $x=0$ is an ordinary point. In other words, one can expect the solution in the series form $y(x)=\sum_{n=0}^{\infty}a_nx^n$. Clearly, we have

$$y'(x)=\sum_{n=1}^{\infty}na_nx^{n-1}$$

$$y''(x)=\sum_{n=2}^{\infty}n(n-1)a_nx^{n-2} \tag{8.51}$$

Substituting the expressions for $y''(x)$ and $y'(x)$ in Eq.(8.51) and $y(x)$ in the Hermite Eq.(8.50), we have

$$\begin{aligned}
0 &= x^2\left(\sum_{n=2}^{\infty} n(n-1)a_n x^{n-2}\right) - 2x^3\left(\sum_{n=1}^{\infty} na_n x^{n-1}\right) + 2mx^2\left(\sum_{n=0}^{\infty} a_n x^n\right)\\
&= \sum_{n=2}^{\infty} n(n-1)a_n x^n - 2\left(\sum_{n=1}^{\infty} na_n x^{n+2}\right) + 2m\left(\sum_{n=0}^{\infty} a_n x^{n+2}\right)\\
&= \sum_{n=2}^{\infty} n(n-1)a_n x^n - 2\left(\sum_{n=3}^{\infty}(n-2)a_{n-2} x^n\right) + 2m\left(\sum_{n=2}^{\infty} a_{n-2} x^n\right)\\
&= \sum_{n=3}^{\infty} n(n-1)a_n x^n - 2\left(\sum_{n=3}^{\infty}(n-2)a_{n-2} x^n\right) + 2m\left(\sum_{n=3}^{\infty} a_{n-2} x^n\right) + 2(a_2 + ma_0)x^2\\
&= \sum_{n=3}^{\infty} n(n-1)a_n x^n - 2\left(\sum_{n=3}^{\infty}\left((n-m-2)a_{n-2}\right) x^n\right) + 2(a_2 + ma_0)x^2\\
&= \sum_{n=3}^{\infty}\left(n(n-1)a_n - 2(n-m-2)a_{n-2}\right) x^n + 2(a_2 + ma_0)x^2 \qquad (8.52)
\end{aligned}$$

Equating the coefficient of each power of x to zero in Eq.(8.52), we have

$$a_2 = -ma_0$$

and

$$\begin{aligned}
a_n &= \left(\frac{n+m-2}{n(n-1)}\right)a_{n-2}\\
&= -\left(\frac{m-n+2}{n(n-1)}\right)a_{n-2}\\
&= -\left(\frac{m-(n-2)}{n(n-1)}\right)a_{n-2} \qquad n \geq 3
\end{aligned}$$

In general, we have

$$a_{2n} = (-1)^n\left(\frac{\prod_{i=0}^{n-1}(m-2i)}{(2n)!}\right)a_0 \qquad n \geq 1$$

and

$$a_{2n+1} = (-1)^n\left(\frac{\prod_{i=0}^{n-1}(m-2i-1)}{(2n+1)!}\right)a_1 \qquad n \geq 1$$

If $a_0 = 1$ and $a_1 = 0$, then we have

$$y(x) = 1 + \sum_{n=1}^{\infty}(-1)^n \left(\frac{\prod_{i=0}^{n-1}(m-2i)}{(2n)!} \right) x^{2n} \tag{8.53}$$

If $a_0 = 0$ and $a_1 = 1$, we have

$$y(x) = x + \sum_{n=1}^{\infty}(-1)^n \left(\frac{\prod_{i=0}^{n-1}(m-2i-1)}{(2n+1)!} \right) x^{2n+1} \tag{8.54}$$

It is clear from Eqs.(8.53) and (8.54) that for $m \in \mathbb{I}$, one of the equations becomes a polynomial. Clearly, every constant multiple of the polynomial obtained corresponding to m either from Eq.(8.53) or Eq.(8.54) is also a solution. In other words, Hermite differential equation has a polynomial solution $H_m(x)$, called *Hermite Polynomial* of degree m, whose coefficient of x^m is 2^m.

8.2.1 Generating Polynomial

Theorem 8.2.1 If $H_m(x)$ is the Hermite polynomial of degree m, then $e^{2xt-t^2} = \sum_{m=0}^{\infty} \left(\frac{H_m(x)}{m!} \right) t^m$.

Proof: It is easy to verify that

$$\begin{aligned}
e^{2xt-t^2} &= e^{t(2x-t)} \\
&= \sum_{m=0}^{\infty} \left(\frac{(t(2x-t))^m}{m!} \right) \\
&= \sum_{m=0}^{\infty} t^m \left(\frac{(2x-t)^m}{m!} \right) \\
&= \sum_{m=0}^{\infty} (-1)^m t^m \left(\frac{(t-2x)^m}{m!} \right) \\
&= \sum_{m=0}^{\infty} (-1)^m t^m \left(\frac{\sum_{i=0}^{m} \binom{m}{i} t^i (-2x)^{m-i}}{m!} \right) \\
&= \sum_{m=0}^{\infty} (-1)^m t^m \left(\frac{\sum_{i=0}^{m} \binom{m}{i} t^i (-1)^{m-i} (2x)^{m-i}}{m!} \right)
\end{aligned}$$

$$= \sum_{m=0}^{\infty} t^m \left(\frac{\sum_{i=0}^{m} (-1)^i \binom{m}{i} t^i (2x)^{m-i}}{m!} \right)$$

$$= \sum_{m=0}^{\infty} \left(\sum_{i=0}^{\lfloor \frac{m}{2} \rfloor} (-1)^i \binom{m-i}{i} \left(\frac{(2x)^{m-2i}}{(m-i)!} \right) \right) t^m$$

$$= \sum_{m=0}^{\infty} \left(\sum_{i=0}^{\lfloor \frac{m}{2} \rfloor} (-1)^i \left(\frac{(2x)^{m-2i}}{i!(m-2i)!} \right) \right) t^m$$

$$= \sum_{m=0}^{\infty} m! \left(\sum_{i=0}^{\lfloor \frac{m}{2} \rfloor} (-1)^i \left(\frac{(2x)^{m-2i}}{i!(m-2i)!} \right) \right) \left(\frac{t^m}{m!} \right)$$

$$= \sum_{m=0}^{\infty} H_m(x) \left(\frac{t^m}{m!} \right)$$

$$= \sum_{m=0}^{\infty} \left(\frac{H_m(x)}{m!} \right) t^m \tag{8.55}$$

Corollary 8.2.1 If $H_m(x)$ is the Hermite polynomial of degree m, then $H_m(-x) = (-1)^m H_m(x)$.

Proof: According to Eq.(8.55), we have

$$e^{2xt-t^2} = \sum_{m=0}^{\infty} \left(\frac{H_m(x)}{m!} \right) t^m \tag{8.56}$$

Replacing x by $-x$ in Eq.(8.56), we have

$$e^{-2xt-t^2} = \sum_{m=0}^{\infty} \left(\frac{H_m(-x)}{m!} \right) t^m \tag{8.57}$$

Replacing t by $-t$ in Eq.(8.57), we have

$$e^{2xt-t^2} = \sum_{m=0}^{\infty} (-1)^m \left(\frac{H_m(-x)}{m!} \right) t^m \tag{8.58}$$

Comparing Eqs.(8.56) and (8.58), we have

$$H_m(-x) = (-1)^m H_m(x)$$

Corollary 8.2.2 If $H_m(x)$ is the Hermite polynomial of degree m, then $H_{2m+1}(0) = 0$.

Proof: According to Eq.(8.55), we have

$$e^{2xt-t^2} = \sum_{m=0}^{\infty} \left(\frac{H_m(x)}{m!} \right) t^m \tag{8.59}$$

Evaluating Eq.(8.59) at $x = 0$ and using the expansion of e^{-t^2}, we have

$$\begin{aligned}\sum_{m=0}^{\infty}\left(\frac{H_m(0)}{m!}\right)t^m &= e^{-t^2}\\ &= \sum_{m=0}^{\infty}\left(\frac{(-t^2)^m}{m!}\right)\\ &= \sum_{m=0}^{\infty}(-1)^m\left(\frac{t^{2m}}{m!}\right)\end{aligned} \tag{8.60}$$

Comparing the coefficients of t^{2m+1} in Eq.(8.60), we have

$$H_{2m+1}(0) = 0$$

Corollary 8.2.3 If $H_m(x)$ is the Hermite polynomial of degree m, then

$$H_{2m}(0) = (-1)^m 2^m\left(\prod_{i=0}^{m-1}(2m-2i-1)\right)$$

Proof: According to Eq.(8.55), we have

$$e^{2xt-t^2} = \sum_{m=0}^{\infty}\left(\frac{H_m(x)}{m!}\right)t^m \tag{8.61}$$

Evaluating Eq.(8.61) at $x = 0$ and using the expansion of e^{-t^2}, we have

$$\begin{aligned}\sum_{m=0}^{\infty}\left(\frac{H_m(0)}{m!}\right)t^m &= e^{-t^2}\\ &= \sum_{m=0}^{\infty}\left(\frac{(-t^2)^m}{m!}\right)\\ &= \sum_{m=0}^{\infty}(-1)^m\left(\frac{t^{2m}}{m!}\right)\end{aligned} \tag{8.62}$$

Comparing the coefficients of t^{2m} in Eq.(8.62), we have

$$\begin{aligned}H_{2m}(0) &= (-1)^m\left(\frac{(2m)!}{m!}\right)\\ &= (-1)^m 2^m\left(\prod_{i=0}^{m-1}(2m-2i-1)\right)\end{aligned}$$

Theorem 8.2.2 (Rodrigues' Formula) If $H_m(x)$ is the polynomial solution of the Hermite differential equation of order m, such that the coefficient of x^m in $H_m(x)$ is 2^m, then $H_m(x) = (-1)^m e^{x^2}\left(\frac{d^m}{dx^m}\left(e^{-x^2}\right)\right)$.

Proof: According to Theorem 8.2.1, we have

$$\sum_{m=0}^{\infty}\left(\frac{H_m(x)}{m!}\right)t^m = e^{2xt-t^2}$$

$$= e^{x^2}e^{-(t-x)^2} \tag{8.63}$$

Again, differentiation of Eq.(8.63) m times with respect to t and evaluation at $t=0$ gives

$$\begin{aligned}
H_m(x) &= \frac{\partial^m}{\partial t^m}\left(e^{x^2}e^{-(t-x)^2}\right)_{t=0} \\
&= e^{x^2}\left(\frac{\partial^m}{\partial t^m}\left(e^{-(t-x)^2}\right)_{t=0}\right) \\
&= e^{x^2}\left(\frac{\partial^m}{\partial(-x)^m}\left(e^{-(t-x)^2}\right)_{t=0}\right) \\
&= e^{x^2}\left(\frac{\mathrm{d}^m}{\mathrm{d}(-x)^m}\left(e^{-x^2}\right)\right) \\
&= (-1)^m e^{x^2}\left(\frac{\mathrm{d}^m}{\mathrm{d}x^m}\left(e^{-x^2}\right)\right)
\end{aligned}$$

8.2.2 Orthogonal Properties

Theorem 8.2.3 If $H_m(x)$ is the Hermite polynomial of degree m and $m \neq n$, then

$$\int_{-\infty}^{\infty} e^{-x^2}H_m(x)H_n(x)\,\mathrm{d}x = 0$$

Proof: According to Theorem 8.2.2, we have

$$H_m(x) = (-1)^m e^{x^2}\left(\frac{\mathrm{d}^m}{\mathrm{d}x^m}\left(e^{-x^2}\right)\right)$$

Clearly, we have

$$H_m''(x) - 2xH_m(x) + 2mH_m(x) = 0 \tag{8.64}$$

Multiplying Eq.(8.64) by e^{-x^2}, one can verify that

$$\left(e^{-x^2}H_m'(x)\right)' + 2me^{-x^2}H_m(x) = 0 \tag{8.65}$$

Similarly, we have

$$\left(e^{-x^2}H_n'(x)\right)' + 2ne^{-x^2}H_n(x) = 0 \tag{8.66}$$

Multiplying Eq.(8.65) by $H_n(x)$ and Eq.(8.66) by $H_m(x)$, and then subtracting the first from the second, we have

$$(n-m)e^{-x^2}H_m(x)H_n(x) = H_n(x)\left(e^{-x^2}H'_m(x)\right)' - H_m(x)\left(e^{-x^2}H'_m(x)\right)' \tag{8.67}$$

Since $n - m \neq 0$, therefore, the integration of Eq.(8.67) from $-\infty$ to ∞ gives

$$\int_{-\infty}^{\infty} e^{-x^2}H_m(x)H_n(x)\,\mathrm{d}x = 0$$

Theorem 8.2.4 (Orthogonal Integral) If $H_r(x)$ is the Hermite polynomial of degree r, then $\int_{-\infty}^{\infty} e^{-x^2}H_r^2(x)\,\mathrm{d}x = 2^r r!\sqrt{\pi}$.

Proof: According to Theorem 8.2.1, we have

$$e^{ext-t^2} = \sum_{m=0}^{\infty}\left(\frac{H_m(x)}{m!}\right)t^m \tag{8.68}$$

Multiplying Eq.(8.68) by $e^{-x^2}H_r(x)$ and integrating from $-\infty$ to ∞, we have by Theorem 8.2.3,

$$\begin{aligned}
\frac{t^r}{r!}\int_{-\infty}^{\infty} e^{-x^2}H_r^2(x)\,\mathrm{d}x &= \sum_{m=0}^{\infty}\left(\frac{t^m}{m!}\right)\int_{-\infty}^{\infty} e^{-x^2}H_m(x)H_r(x)\,\mathrm{d}x \\
&= \int_{-\infty}^{\infty}\left(\sum_{m=0}^{\infty}\left(\frac{H_m(x)}{m!}\right)t^m H_r(x)e^{-x^2}\right)\mathrm{d}x \\
&= \int_{-\infty}^{\infty}\left(\left(\sum_{m=0}^{\infty}\left(\frac{H_m(x)}{m!}\right)t^m\right)H_r(x)e^{-x^2}\right)\mathrm{d}x \\
&= \int_{-\infty}^{\infty}\left(e^{2xt-t^2}H_r(x)e^{-x^2}\right)\mathrm{d}x \\
&= \int_{-\infty}^{\infty} e^{2xt-t^2}\left(\frac{\mathrm{d}^r}{\mathrm{d}x^r}\left(e^{-x^2}\right)\right)\mathrm{d}x
\end{aligned}$$

In other words, we have

$$\int_{-\infty}^{\infty} e^{-x^2}H_r^2(x)\,\mathrm{d}x = r!\left(\int_{-\infty}^{\infty} e^{2xt-t^2}\left(\frac{\mathrm{d}^r}{\mathrm{d}x^r}\left(e^{-x^2}\right)\right)\mathrm{d}x\right)_{t=1} \tag{8.69}$$

Evaluating the integration given in Eq.(8.69) by parts r times at $t = 1$, we have

$$\int_{-\infty}^{\infty} e^{-x^2} H_r^2(x)\,\mathrm{d}x = 2^r r! \left(\int_{-\infty}^{\infty} e^{2xt - x^2 - t^2}\,\mathrm{d}x \right)_{t=1}$$

$$= 2^r r! \left(\int_{-\infty}^{\infty} e^{-(x-t)^2}\,\mathrm{d}x \right)_{t=1}$$

$$= 2^r r! \sqrt{\pi}$$

Theorem 8.2.5 If $f(x) = \sum_{i=0}^{n} a_i x^i$ is any polynomial of degree n, then $f(x) = \sum_{m=0}^{n} \alpha_m H_m(x)$ where $\alpha_m = \left(\frac{1}{2^m m! \sqrt{\pi}}\right) \int_{-\infty}^{\infty} e^{-x^2} H_m(x) f(x)\,\mathrm{d}x$.

Proof: According to Theorems 8.2.3 and 8.2.4, we have

$$2^m m! \sqrt{\pi} \alpha_m = \alpha_m \int_{-\infty}^{\infty} e^{-x^2} H_m^2(x)\,\mathrm{d}x$$

$$= \sum_{i=0}^{n} \alpha_i \left(\int_{-\infty}^{\infty} e^{-x^2} H_m(x) H_i(x)\,\mathrm{d}x \right)$$

$$= \int_{-\infty}^{\infty} e^{-x^2} H_m(x) \left(\sum_{i=0}^{n} \alpha_i H_i(x) \right) \mathrm{d}x$$

$$= \int_{-\infty}^{\infty} e^{-x^2} H_m(x) f(x) \mathrm{d}x$$

In other words, we have

$$\alpha_m = \frac{1}{2^m m! \sqrt{\pi}} \int_{-\infty}^{\infty} e^{-x^2} H_m(x) f(x)\,\mathrm{d}x$$

8.2.3 Recurrence Property

Theorem 8.2.6 If $H_m(x)$ is the Hermite polynomial of degree m, then $H_m'(x) = 2mH_{m-1}(x)$.

Proof: According to Theorem 8.2.1, we have

$$e^{2xt - t^2} = \sum_{m=0}^{\infty} \left(\frac{H_m(x)}{m!} \right) t^m \tag{8.70}$$

Differentiating Eq.(8.70) with respect to x and using Eq.(8.70), we have

$$\sum_{m=0}^{\infty}\left(\frac{H'_m(x)}{m!}\right)t^m = 2te^{2xt-t^2}$$

$$= 2t\left(\sum_{m=0}^{\infty}\left(\frac{H_m(x)}{m!}\right)t^m\right)$$

$$= \sum_{m=0}^{\infty}\left(\frac{2H_m(x)}{m!}\right)t^{m+1} \quad (8.71)$$

Equating the coefficients of t^m in Eq.(8.71), we have

$$\frac{H'_m(x)}{m!} = \frac{2H_{m-1}(x)}{(m-1)!}$$

In other words, we have

$$H'_m(x) = 2mH_{m-1}(x)$$

Corollary 8.2.4 If $H_m(x)$ is the Hermite polynomial of degree m, then $\frac{\mathrm{d}^n}{\mathrm{d}x^n}(H_m(x)) = \left(\frac{2^n m!}{(m-n)!}\right)H_{m-n}(x)$.

Proof: Apply Theorem 8.2.6 repeatedly n times.

Theorem 8.2.7 If $H_m(x)$ is the Hermite polynomial of degree m, then $2xH_m(x) = 2mH_{m-1}(x) + H_{m+1}(x)$.

Proof: According to Theorem 8.2.1, we have

$$e^{2xt-t^2} = \sum_{m=0}^{\infty}\left(\frac{H_m(x)}{m!}\right)t^m \quad (8.72)$$

Differentiating Eq.(8.72) with respect to t and using Eq.(8.72), we have

$$\sum_{m=1}^{\infty}\left(\frac{H_m(x)}{(m-1)!}\right)t^{m-1} = 2(x-t)e^{2xt-t^2}$$

$$= 2(x-t)\left(\sum_{m=0}^{\infty}\left(\frac{H_m(x)}{m!}\right)t^m\right)$$

$$= \sum_{m=0}^{\infty}\left(\frac{2xH_m(x)}{m!} - \frac{2H_{m-1}(x)}{(m-1)!}\right)t^m \quad (8.73)$$

Equating the coefficients of t^m in Eq.(8.73), we have

$$\frac{H_{m+1}}{m!} = \frac{2xH_m(x)}{m!} - \frac{2H_{m-1}(x)}{(m-1)!}$$

In other words, we have

$$H_{m+1}(x) = 2xH_m(x) - 2mH_{m-1}(x)$$

Exercises 8.2

1. If $H_m(x)$ is the Hermite polynomial of degree m, then show that

$$H_m''(x) = 4m(m-1)H_{m-2}(x)$$

2. Express $f(x) = 1 + x + x^2$ in terms of Hermite polynomials.

3. Find the power series solution of $y'' - 2xy' + 2y = 0$, such that $y(0) = 1$ and $y'(0) = 0$.

4. If $H_m(x)$ is the Hermite polynomial of degree m, then show that

$$H_{2m+1}'(0) = 2(-1)^m \left(\frac{(2m+1)!}{m!}\right)$$

5. If $H_m(x)$ is the Hermite polynomial of degree m, then show that $H_{2m}'(0) = 0$.

6. If $H_m(x)$ is the Hermite polynomial of degree m, then show that

$$H_m(x) = \left(2x - \frac{\mathrm{d}}{\mathrm{d}x}\right)^m 1$$

7. If $H_m(x)$ is the Hermite polynomial of degree m, then show that $H_m(x) = \left(2x - \frac{\mathrm{d}}{\mathrm{d}x}\right)^m 1$.

   HINT: We know $-2x = \left(\frac{\mathrm{d}}{\mathrm{d}x} - 2x\right) 1$. One can easily verify that

$$\begin{aligned}
H_m(x) &= (-1)^m e^{x^2} \left(\frac{\mathrm{d}^m}{\mathrm{d}x^m}\left(e^{-x^2}\right)\right) \\
&= (-1)^m e^{x^2} \left(\frac{\mathrm{d}^{m-1}}{\mathrm{d}x^{m-1}}\left(\frac{\mathrm{d}}{\mathrm{d}x}\left(e^{-x^2}\right)\right)\right) \\
&= (-1)^m e^{x^2} \left(\frac{\mathrm{d}^{m-1}}{\mathrm{d}x^{m-1}}\left(\left(-2x\, e^{-x^2}\right)\right)\right) \\
&= (-1)^m e^{x^2} \left(\frac{\mathrm{d}^{m-1}}{\mathrm{d}x^{m-1}}\left(\left(e^{-x^2}\left(-2x + \frac{\mathrm{d}}{\mathrm{d}x}\right)1\right)\right)\right) \\
&\ \vdots \\
&= (-1)^m e^{x^2} \left(e^{-x^2}\left(-2x + \frac{\mathrm{d}}{\mathrm{d}x}\right)^m 1\right) \\
&= (-1)^m \left(-2x + \frac{\mathrm{d}}{\mathrm{d}x}\right)^m 1 \\
&= \left(2x - \frac{\mathrm{d}}{\mathrm{d}x}\right)^m 1
\end{aligned}$$

8.3 Chebyshev Differential Equation

In this section, we introduce the Chebyshev polynomials and obtain some of their properties. The treatment is by no means exhaustive, but is intended to display the central techniques which one can employ to derive the various relations which may be of interest. At first sight, the selection of formulae which has been made may seem unnatural, but by and large they are the ones we shall employed later in this book. The Chebyshev polynomials $T_m(x)$ are discussed in some detail. The choice of these polynomials is dictated by the fact that they have been the most used in approximation work. The differential equation $(1-x^2)y''-xy'+m^2y=0$ for $n \in \mathbb{I}$ is called the *Chebyshev differential equation.* The Chebyshev differential equation can be rewritten in the form

$$x^2y''+x\left(-\frac{x^2}{1-x^2}\right)y'+\left(\frac{x^2m^2}{1-x^2}\right)y=0 \tag{8.74}$$

Clearly, $\frac{p(x)}{x}=-\frac{x}{1-x^2}$ and $\frac{q(x)}{x^2}=\frac{m^2}{1-x^2}$ are analytic at $x=0$. Hence $x=0$ is an ordinary point. In other words, one can expect the series solution in the form $y(x)=\sum\limits_{n=0}^{\infty}a_nx^n$. Clearly, we have

$$\begin{aligned} y'(x) &= \sum_{n=1}^{\infty} na_nx^{n-1} \\ y''(x) &= \sum_{n=2}^{\infty} n(n-1)a_nx^{n-2} \end{aligned} \tag{8.75}$$

Substituting the expression $y'(x)$ and $y''(x)$ given in Eq.(8.75), and $y(x)$ in the Chebyshev differential equation $(1-x^2)y''-xy'+m^2y=0$, we have

$$\begin{aligned} 0 &= (1-x^2)\left(\sum_{n=2}^{\infty} n(n-1)a_nx^{n-2}\right)-x\left(\sum_{n=1}^{\infty} na_nx^{n-1}\right)+m^2\left(\sum_{n=0}^{\infty} a_nx^n\right) \\ &= (1-x^2)\left(\sum_{n=2}^{\infty} n(n-1)a_nx^{n-2}\right)-\sum_{n=1}^{\infty} na_nx^n+\sum_{n=0}^{\infty} m^2a_nx^n \\ &= \sum_{n=2}^{\infty} n(n-1)a_nx^{n-2}-\sum_{n=2}^{\infty} n(n-1)a_nx^n-\sum_{n=1}^{\infty} na_nx^n+\sum_{n=0}^{\infty} m^2a_nx^n \\ &= \sum_{n=0}^{\infty}(n+2)(n+1)a_{n+2}x^n-\sum_{n=2}^{\infty} n(n-1)a_nx^n-\sum_{n=1}^{\infty} na_nx^n+\sum_{n=0}^{\infty} m^2a_nx^n \\ &= \sum_{n=2}^{\infty}\left((n+2)(n+1)a_{n+2}-(n(n-1)+n-m^2)a_n\right)x^n+(m^2a_0+2a_2) \\ &\quad +(6a_3-a_1+m^2a_1)x \end{aligned}$$

$$= \sum_{n=2}^{\infty} \left((n+2)(n+1)a_{n+2} - (n^2 - m^2)a_n \right) x^n + (m^2 a_0 + 2a_2)$$

$$+ (6a_3 + (m^2 - 1)a_1)x \tag{8.76}$$

Equating the coefficients of each power of x in Eq.(8.76) to zero, we have

$$a_2 = -\left(\frac{m^2}{2}\right) a_0$$

$$a_3 = -\left(\frac{m^2 - 1}{6}\right) a_1$$

$$a_{n+2} = -\left(\frac{m^2 - n^2}{(n+2)(n+1)}\right) a_n \qquad n \geq 2$$

In general, we have

$$a_{2n} = (-1)^n \left(\frac{\prod_{i=0}^{n-1} (m^2 - (2n - 2 - 2i)^2}{(2n)!} \right) a_0 \qquad n \geq 1 \tag{8.77}$$

and

$$a_{2n+1} = (-1)^n \left(\frac{\prod_{i=0}^{n-1} (m^2 - (2n - 1 - 2i)^2}{(2n+1)!} \right) a_1 \qquad n \geq 1 \tag{8.78}$$

If $a_1 = 0$, then we have

$$y(x) = a_0 \left(1 + \sum_{n=1}^{\infty} (-1)^n \left(\frac{\prod_{i=0}^{n-1} (m^2 - (2n - 2 - 2i)^2}{(2n)!} \right) x^{2n} \right)$$

by Eq.(8.77). If $a_0 = 0$, then we have

$$y(x) = a_1 \left(x + \sum_{n=1}^{\infty} (-1)^n \left(\frac{\prod_{i=0}^{n-1} (m^2 - (2n - 1 - 2i)^2}{(2n+1)!} \right) x^{2n+1} \right)$$

by Eq.(8.78). It is clear from Eqs.(8.77) and (8.78) that $y(x)$ is a polynomial when m is a positive integer. In other words, the Chebyshev differential equation accepts the polynomial solution $T_m(x)$ of degree m, called the *Chebyshev polynomial*, satisfying $T_m(1) = 1$ by assigning suitable $a_0 = (-1)^{\frac{m}{2}}$ and $a_1 = (-1)^{\lfloor \frac{m}{2} \rfloor} m$, whichever is necessary.

To establish a connection between the Chebyshev differential equation and the Chebyshev polynomial, we use the fact that the polynomial $T_m(x)$ becomes the function $y(\theta) = \cos m\theta$ when the variable is changed from x to θ by means of $x = \cos\theta$. Let $y(\theta) = \cos m\theta$. Clearly, we have

$$\frac{d^2y}{d\theta^2} + m^2y = 0 \tag{8.79}$$

An easy calculation shows that changing the variable from θ back to x transforms Eq. (8.79) into the Chebyshev differential equation $(1-x^2)y'' - xy' + m^2y = 0$. In other words, $T_m(x)$ is that polynomial for which

$$\cos m\theta = T_m \cos\theta \tag{8.80}$$

Since $T_m(x)$ is a polynomial, it is defined for all values of x. However, if x is restricted to lie in the interval $-1 \le x \le 1$ and we write $x = \cos\theta$ where $0 \le \theta \le \pi$, then Eq.(8.80) reduces to

$$\begin{aligned} T_m(x) &= \cos m\theta \\ &= \cos\left(m\cos^{-1}x\right) \end{aligned} \tag{8.81}$$

Again, we have the following expression for $T_m(x)$ using Eq.(8.80), de Moivre's formula for theory of complex numbers and $x = \cos\theta$

$$\begin{aligned} T_m(x) &= \cos m\theta \\ &= \frac{1}{2}\left((\cos m\theta + i\cos m\theta) + (\cos m\theta - i\cos m\theta)\right) \\ &= \frac{1}{2}\left((\cos\theta + i\sin\theta)^m + (\cos\theta - i\sin\theta)^m\right) \\ &= \frac{1}{2}\left((\cos\theta + i\,(1-\cos\theta)^{\frac{1}{2}})^m + (\cos\theta - i\,(1-\cos\theta)^{\frac{1}{2}})^m\right) \\ &= \frac{1}{2}\left((\cos\theta + (\cos\theta - 1)^{\frac{1}{2}})^m + (\cos\theta - (\cos\theta - 1)^{\frac{1}{2}})^m\right) \\ &= \frac{1}{2}\left(\left(x + (x^2-1)^{\frac{1}{2}}\right)^m + \left(x - (x^2-1)^{\frac{1}{2}}\right)^m\right) \end{aligned} \tag{8.82}$$

It is easy to verify that the expression for $T_m(x)$ given in Eq.(8.82), satisfies the condition $T_m(1) = 1$. Again, we have another expression for $T_m(x)$ using Eq.(8.80), de Moivre's formula for theory of complex numbers, binomial expansion and $x = \cos\theta$.

$$\begin{aligned} T_m(x) &= \cos m\theta \\ &= \frac{1}{2}\left((\cos m\theta + i\cos m\theta) + (\cos m\theta - i\cos m\theta)\right) \\ &= \frac{1}{2}\left((\cos\theta + i\sin\theta)^m + (\cos\theta - i\sin\theta)^m\right) \end{aligned}$$

$$= \frac{1}{2}\left(\sum_{n=0}^{m}\binom{m}{n}\cos^{m-n}\theta(i\sin\theta)^n + \sum_{n=0}^{m}\binom{m}{n}\cos^{m-n}\theta(-i\sin\theta)^n\right)$$

$$= \frac{1}{2}\left(\sum_{n=0}^{m}\binom{m}{n}\cos^{m-n}\theta(i^n + (-i)^n)\sin^n\theta\right)$$

$$= \left(\sum_{n=0}^{\lfloor\frac{m}{2}\rfloor}(-1)^n\binom{m}{2n}\cos^{m-2n}\theta\sin^{2n}\theta\right)$$

$$= \left(\sum_{n=0}^{\lfloor\frac{m}{2}\rfloor}(-1)^n\binom{m}{2n}\cos^{m-2n}\theta\,(1-\cos^2\theta)^n\right)$$

$$= \left(\sum_{n=0}^{\lfloor\frac{m}{2}\rfloor}(-1)^n\binom{m}{2n}x^{m-2n}\,(1-x^2)^n\right)$$

$$= \left(\sum_{n=0}^{\lfloor\frac{m}{2}\rfloor}\binom{m}{2n}x^{m-2n}\,(x^2-1)^n\right) \quad (8.83)$$

It is easy to verify that the expression for $T_m(x)$, given in Eq.(8.83), satisfies the condition $T_m(1) = 1$.

8.3.1 Generating Function

Theorem 8.3.1 If $T_m(x)$ is the Chebyshev polynomial of degree m, then $\frac{1-tx}{1-2tx+t^2} = \sum_{m=0}^{\infty} T_m(x)t^m$.

Proof: It is easy to verify that

$$\frac{1-tx}{1-2tx+t^2} + i\left(\frac{t\sin\theta}{1-2tx+t^2}\right) = \frac{1-tx+it\sin\theta}{1-2tx+t^2}$$

$$= \frac{1-t\cos\theta+it\sin\theta}{1-2t\cos\theta+t^2}$$

$$= \frac{1-t\cos\theta+it\sin\theta}{1+t^2\cos^2\theta-2t\cos\theta+t^2-t^2\cos^2\theta}$$

$$= \frac{1-t\cos\theta+it\sin\theta}{1+t^2\cos^2\theta-2t\cos\theta+t^2\sin^2\theta}$$

$$= \frac{1-t\cos\theta+it\sin\theta}{(1-t\cos\theta)^2+t^2\sin^2\theta}$$

$$= \frac{1-t\cos\theta+it\sin\theta}{(1-t\cos\theta-it\sin\theta)(1-t\cos\theta+it\sin\theta)}$$

$$
\begin{aligned}
&= \frac{1}{1 - t\cos\theta - it\sin\theta} \\
&= \frac{1}{1 - te^{i\theta}} \\
&= \sum_{m=0}^{\infty} t^m e^{im\theta} \\
&= \sum_{m=0}^{\infty} t^m (\cos\theta + i\sin\theta)^m \\
&= \sum_{m=0}^{\infty} t^m (\cos m\theta + i\sin m\theta) \\
&= \sum_{m=0}^{\infty} t^m \cos m\theta + i\left(\sum_{m=0}^{\infty} t^m \sin m\theta\right) \\
&= \sum_{m=0}^{\infty} t^m\, T_m(x) + i\left(\sum_{m=0}^{\infty} t^m \sin m\theta\right) \\
&= \sum_{m=0}^{\infty} T_m(x)\, t^m + i\left(\sum_{m=0}^{\infty} t^m \sin m\theta\right) \qquad (8.84)
\end{aligned}
$$

Clearly, the real part of Eq.(8.84) gives the required generating function for Chebyshev polynomials.

Corollary 8.3.1 If $T_m(x)$ is the Chebyshev polynomial of degree m, then $T_m(1) = 1$.

Proof: According to Theorem 8.3.1, we have

$$\frac{1 - tx}{1 - 2tx + t^2} = \sum_{m=0}^{\infty} T_m(x)\, t^m \qquad (8.85)$$

Replacing $x = 1$ in Eq.(8.85), we have

$$
\begin{aligned}
\sum_{m=0}^{\infty} T_m(1) t^m &= \frac{1 - t}{1 - 2t + t^2} \\
&= \frac{1}{1 - t} \\
&= \sum_{m=0}^{\infty} t^m \qquad (8.86)
\end{aligned}
$$

Comparing the coefficients of t^m in Eq.(8.86), we have

$$T_m(1) = 1$$

Corollary 8.3.2 If $T_m(x)$ is the Chebyshev polynomial of degree m, then $T_m(-1) = (-1)^m$.

Proof: According to Theorem 8.3.1, we have

$$\frac{1-tx}{1-2tx+t^2} = \sum_{m=0}^{\infty} T_m(x)t^m \tag{8.87}$$

Replacing $x = -1$ in Eq.(8.87), we have

$$\begin{aligned}\sum_{m=0}^{\infty} T_m(-1)t^m &= \frac{1+t}{1+2t+t^2} \\ &= \frac{1}{1+t} \\ &= \sum_{m=0}^{\infty}(-1)^m t^m \end{aligned} \tag{8.88}$$

Comparing the coefficients of t^m in Eq.(8.88), we have

$$T_m(-1) = (-1)^m$$

Theorem 8.3.2 If $T_n(x)$ is the Chebyshev polynomial of degree m, then

$$T_m(x) = \left(\frac{(-1)^m 2^m m!}{(2m)!}\right)(1-x^2)^{\frac{1}{2}}\left(\frac{\mathrm{d}^m}{\mathrm{d}x^m}\left((1-x^2)^{m-\frac{1}{2}}\right)\right)$$

Proof: According to Theorem 8.3.3, we have

$$\int_{-1}^{1}\left(\frac{T_m(x)\psi_m(x)}{(1-x^2)^{\frac{1}{2}}}\right)\mathrm{d}x = 0 \tag{8.89}$$

where $\psi_m(x)$ is any polynomial of degree at most $m-1$. One can assume that

$$\frac{T_m(x)}{(1-x^2)^{\frac{1}{2}}} = \frac{\mathrm{d}^m}{\mathrm{d}x^m}(\phi_m(x)) \tag{8.90}$$

Substituting the expression from Eq.(8.90) in Eq.(8.89), we have

$$\int_{-1}^{1}\frac{\mathrm{d}^m}{\mathrm{d}x^m}(\phi_m(x))\psi_m(x)\mathrm{d}x = 0 \tag{8.91}$$

Integrating Eq.(8.91) by parts m times, we obtain

$$\left.\sum_{j=0}^{m-1}(-1)^j\phi_m^{(m-1-j)}(x)\psi_m^{(j)}(x)\right|_{-1}^{1} = 0$$

In other words, we have for $0 \leq j \leq m-1$

$$\phi_m^{(j)}(\pm 1) = 0 \tag{8.92}$$

Again, $T_m(x)$ is a polynomial of degree m. Hence $\frac{\mathrm{d}^{m+1}}{\mathrm{d}x^{m+1}}(T_m(x)) = 0$. In other words, we have a differential equation

$$\frac{\mathrm{d}^{m+1}}{\mathrm{d}x^{m+1}}\left((1-x^2)^{\frac{1}{2}}\left(\frac{\mathrm{d}^m}{\mathrm{d}x^m}\phi_m(x)\right)\right) = 0 \tag{8.93}$$

One can easily predict the solution of Eq.(8.93) satisfying the initial conditions given in Eq.(8.92) by

$$\phi_m(x) = \alpha(1-x^2)^{m-\frac{1}{2}}$$

where α is the integration constant. In other words, we have

$$T_m(x) = \alpha(1-x^2)^{\frac{1}{2}}\left(\frac{\mathrm{d}^m}{\mathrm{d}x^m}(1-x^2)^{m-\frac{1}{2}}\right) \tag{8.94}$$

Using the expression for $T_m(x)$ given in Eq.(8.94) and the condition $T_m(1) = 1$, one can have

$$\alpha = (-1)^m\left(\frac{1}{\prod\limits_{i=0}^{m-1}(2m-2i-1)}\right) = (-1)^m\left(\frac{2^m m!}{(2m)!}\right)$$

In other words, we have

$$T_m(x) = (-1)^m\left(\frac{2^m m!}{(2m)!}\right)(1-x^2)^{\frac{1}{2}}\left(\frac{\mathrm{d}^m}{\mathrm{d}x^m}(1-x^2)^{m-\frac{1}{2}}\right)$$

8.3.2 Orthogonal Property

Theorem 8.3.3 If $T_m(x)$ is the Chebyshev polynomial of degree m and $m \neq n$, then

$$\int_{-1}^{1}\frac{T_m(x)T_n(x)}{(1-x^2)^{\frac{1}{2}}}\mathrm{d}x = 0$$

Proof: It is easy to verify from Eq.(8.81) and $x = \cos\theta$ that for $m \neq n$

$$0 = \int_0^{\pi}\cos m\theta\cos n\theta\,\mathrm{d}\theta$$

$$= \int_{-1}^{1}\frac{T_m(x)\,T_n(x)}{(1-x^2)^{\frac{1}{2}}}\mathrm{d}x$$

Theorem 8.3.4 If $T_m(x)$ is the Chebyshev polynomial of degree m and $m \neq 0$, then $\int_{-1}^{1} \frac{T_m^2(x)}{(1-x^2)^{\frac{1}{2}}}\, dx = \frac{\pi}{2}$.

Proof: It is easy to verify from Eq.(8.81) and $x = \cos\theta$ that

$$\int_{-1}^{1} \frac{T_m^2(x)}{(1-x^2)^{\frac{1}{2}}}\, dx = \int_{0}^{\pi} \cos^2 m\theta\, d\theta$$

$$= \int_{0}^{\pi} \frac{1+\cos 2m\theta}{2}\, d\theta$$

$$= \frac{\pi}{2}$$

Theorem 8.3.5 If $T_m(x)$ is the Chebyshev polynomial of degree m, then the integration $\int_{-1}^{1} \frac{T_0^2(x)}{(1-x^2)^{\frac{1}{2}}}\, dx = \pi$.

Proof: Here,

$$\int_{-1}^{1} \frac{T_0^2(x)}{(1-x^2)^{\frac{1}{2}}}\, dx = \int_{0}^{\pi} \cos^2 0\theta\, d\theta$$

$$= \int_{0}^{\pi} d\theta$$

$$= \pi$$

8.3.3 Recurrence Relation

Theorem 8.3.6 If $T_m(x)$ is the Chebyshev polynomial of degree m, then

$$T_{m+1}(x) + T_{m-1}(x) = 2xT_m(x)$$

Proof: We know from trigonometry and Eq.(8.80) that

$$2\,T_m(x)\,T_n(x) = 2\cos m\theta \cos n\theta$$

$$= \cos(m+n)\theta + \cos(m-n)\theta$$

$$= T_{m+n}(x) + T_{m-n}(x) \tag{8.95}$$

Substituting $n = 1$ and $T_1(x) = x$ in Eq.(8.95), we have

$$2xT_m(x) = T_{m+1}(x) + T_{m-1}(x)$$

Theorem 8.3.7 If $T_m(x)$ is the Chebyshev polynomial of degree m, then

$$(1-x^2)T'_m(x) = mT_{m-1}(x) - mxT_m(x)$$

Proof: We know by Eq.(8.81) that

$$\begin{aligned} T_m(x) &= \cos m\theta \\ &= \cos\left(m\cos^{-1}x\right) \end{aligned} \tag{8.96}$$

Differentiation of Eq.(8.96) gives

$$\begin{aligned} T'_m(x) &= m\sin\left(m\cos^{-1}x\right)\left(\frac{1}{1-x^2}\right)^{\frac{1}{2}} \\ &= m\sin m\,\theta\left(\frac{1}{1-x^2}\right)^{\frac{1}{2}} \end{aligned}$$

In other words, we have

$$\begin{aligned} (1-x^2)T'_m(x) &= (1-x^2)^{\frac{1}{2}}(1-x^2)^{\frac{1}{2}}T'_m(x) \\ &= (1-x^2)^{\frac{1}{2}}(1-\cos^2\theta)^{\frac{1}{2}}T'_m(x) \\ &= (1-x^2)^{\frac{1}{2}}\sin\theta\, T'_m(x) \\ &= m\,\sin m\theta\sin\theta \\ &= \frac{m}{2}\left(\cos(m-1)\theta - \cos(m+1)\theta\right) \\ &= \frac{m}{2}\left(T_{m-1}(x) - T_{m+1}(x)\right) \end{aligned} \tag{8.97}$$

Again, one can modify Eq.(8.97) using Theorem 8.3.6 as

$$\begin{aligned} (1-x^2)T'_m(x) &= \frac{m}{2}\left(T_{m-1}(x) - T_{m+1}(x)\right) \\ &= \frac{m}{2}\left(T_{m-1}(x) - (2xT_m(x) - T_{m-1}(x))\right) \\ &= m\left(T_{m-1}(x) - xT_m(x)\right) \\ &= mT_{m-1}(x) - mxT_m(x) \end{aligned}$$

Exercises 8.3

1. Show that $T_{2n}(0) = (-1)^n$.
2. Show that $T_{2n+1}(0) = 0$.
3. If $T_m(x)$ is the Chebyshev polynomial of degree m, then show that $T_m(T_n(x)) = T_{mn}(x)$.
   Hint: We know that $T_m(\cos\theta) = \cos m\theta$ by Eq.(8.80). It is easy to verify that

$$
\begin{aligned}
T_m(T_n(\cos\theta)) &= T_m(\cos n\,\theta) \\
&= \cos mn\,\theta \\
&= T_{mn}(\cos\theta)
\end{aligned}
$$

4. If $T_m(x)$ is the Chebyshev polynomial of degree m, then show that

$$2^r x^r T_m(x) = \sum_{i=0}^{r} \binom{r}{i} T_{m-r+2i}(x)$$

Hint: One can apply the method of induction on r. If $r = 1$, then we have

$$
\begin{aligned}
2xT_m(x) &= 2\cos(\theta) T_m(x) \\
&= 2\cos(\theta)\cos(m\theta) \\
&= \cos(m-1)\theta + \cos(m+1)\theta \\
&= T_{m-1}(x) + T_{m+1}(x) \\
&= \sum_{i=0}^{1} \binom{1}{i} T_{m-1+2i}(x)
\end{aligned}
$$

Assume that $2^r x^r T_m(x) = \sum\limits_{i=0}^{r} \binom{r}{i} T_{m-r+2i}(x)$ for $0 \le r \le n$. Again, we have

$$
\begin{aligned}
2^{n+1}x^{n+1}T_m(x) &= 2x\left(2^n x^n T_m(x)\right) \\
&= 2x\left(\sum_{i=0}^{n} \binom{n}{i} T_{m-n+2i}(x)\right) \\
&= 2\cos\theta\left(\sum_{i=0}^{n} \binom{n}{i} \cos(m-n+2i)\theta\right) \\
&= \sum_{i=0}^{n} \binom{n}{i} 2\cos\theta\cos(m-n+2i)\theta \\
&= \sum_{i=0}^{n} \binom{n}{i} \left(\cos(m-n+2i-1)\theta + \cos(m-n+2i+1)\theta\right) \\
&= \cos(m-(n+1))\theta + \cos(m+n+1)\theta \\
&\quad + \sum_{i=1}^{n} \left(\binom{n}{i-1} + \binom{n}{i}\right) \cos(m-n+2i-1)\theta
\end{aligned}
$$

$$= \cos(m-(n+1))\theta + \cos(m+n+1)\theta + \sum_{i=1}^{n} \binom{n+1}{i} \cos(m-n+2i-1)\theta$$

$$= T_{m-(n+1)}(x) + \sum_{i=1}^{n} \binom{n+1}{i} T_{m-(n+1)+2i}(x) + T_{m+n+1}(x)$$

$$= \sum_{i=0}^{n+1} \binom{n+1}{i} T_{m-(n+1)+2i}(x)$$

5. If $T_m(x)$ is the Chebyshev polynomial of degree m, then show that

$$T_{m+n}(x) + T_{m-n}(x) = 2T_m(x)T_n(x)$$

Hint: Verify the proof of Theorem 8.3.6.

6. If $T_m(x)$ is the Chebyshev polynomial of degree m, then show that

$$T_m^2(x) - T_{m-1}(x)T_{m+1}(x) = 1 - x^2$$

Hint: It is easy to verify that

$$\begin{aligned} T_m^2(x) - T_{m-1}(x)T_{m+1}(x) &= \cos^2 m\theta - \cos(m-1)\theta \cos(m+1)\theta \\ &= \cos^2 m\theta - \frac{1}{2}(\cos 2m\,\theta + \cos 2\theta) \\ &= \cos^2 m\theta - \left(\cos^2 m\,\theta + \cos^2\theta - 1\right) \\ &= 1 - \cos^2\theta \\ &= 1 - x^2 \end{aligned}$$

7. If $T_m(x)$ is the Chebyshev polynomial of degree m, then show that $2T_m^2(x) - T_{2m}(x) = 1$.

   Hint: It is easy to verify that

$$\begin{aligned} 2T_m^2(x) - T_{2m}(x) &= 2\cos^2 m\theta - \cos 2m\theta \\ &= \cos^2 m\theta + \sin^2 m\theta \\ &= 1 \end{aligned}$$

8. If $T_m(x)$ is the Chebyshev polynomial of degree m, then show that the leading coefficient of $T_m(x)$ is 2^{m-1}.

   HINT: It is easy to verify that

$$\begin{aligned} \cos m\theta + i \sin m\theta &= (\cos\theta + i\sin\theta)^n \\ &= \sum_{n=0}^{m} (-i)^n \binom{m}{n} \cos^{m-n}\theta \sin^n\theta \end{aligned}$$

In other words, we have

$$\begin{aligned}
T_m(x) &= \cos m\theta \\
&= \sum_{n=0}^{\lfloor \frac{m}{2} \rfloor} (-1)^n \binom{m}{2n} \cos^{m-2n}\theta \sin^{2n}\theta \\
&= \sum_{n=0}^{\lfloor \frac{m}{2} \rfloor} (-1)^n \binom{m}{2n} x^{m-2n} (1-x^2)^n
\end{aligned}$$

Clearly, the coefficient of x^m is $\sum\limits_{n=0}^{\lfloor \frac{m}{2} \rfloor} \binom{m}{2n} = 2^{m-1}$.

9. If $T_m(x)$ is the Chebyshev polynomial of degree m, then show that

$$\frac{1-t^2}{1-2xt+t^2} = T_0(x) + 2\left(\sum_{m=1}^{\infty} T_m(x) t^m\right)$$

HINT: It is easy to verify that $x = \cos\theta = \frac{e^{i\theta}+e^{-i\theta}}{2}$. One can easily verify that

$$\begin{aligned}
\frac{1-t^2}{1-2xt+t^2} &= \frac{1-t^2}{1-(e^{i\theta}+e^{-i\theta})\,t+t^2} \\
&= \frac{1-t^2}{1-(e^{i\theta}+e^{-i\theta})\,t+t^2 e^{i\theta}e^{-i\theta}} \\
&= \frac{1-t^2}{(1-te^{i\theta})(1-te^{-i\theta})} \\
&= (1-t^2)\left(\sum_{n=0}^{\infty} t^n e^{in\theta}\right)\left(\sum_{r=0}^{\infty} t^r e^{-ir\theta}\right) \\
&= (1-t^2)\left(\sum_{n=0}^{\infty}\sum_{r=0}^{\infty} t^{n+r} e^{i(n-r)\theta}\right)] \\
&= \sum_{n=0}^{\infty}\sum_{r=0}^{\infty} t^{n+r} e^{i(n-r)\theta} - \sum_{n=0}^{\infty}\sum_{r=0}^{\infty} t^{n+r+2} e^{i(n-r)\theta} \\
&= 1 + \sum_{m=1}^{\infty} t^m \left(\sum_{k=0}^{m} e^{i(m-2k)}\right) - \sum_{m=2}^{\infty} t^m \left(\sum_{k=0}^{m-2} e^{i(m-2-2k)}\right) \\
&= T_0(x) + \sum_{m=1}^{\infty} t^m e^{im\theta} \left(\sum_{k=0}^{m} e^{-2ik\theta}\right) - \sum_{m=2}^{\infty} t^m e^{i(m-2)\theta} \left(\sum_{k=0}^{m-2} e^{-2ik\theta}\right)
\end{aligned}$$

$$= T_0(x) + \sum_{m=1}^{\infty} t^m \left(\frac{e^{im\theta} - e^{-i(m+2)\theta}}{1 - e^{-2i\theta}}\right) - \sum_{m=2}^{\infty} t^m \left(\frac{e^{i(m-2)\theta} - e^{-im\theta}}{1 - e^{-2i\theta}}\right)$$

$$= T_0(x) + 2T_1(x)t + \sum_{m=2}^{\infty} t^m \left(\frac{e^{im\theta} - e^{-i(m+2)\theta} - e^{i(m-2)\theta} + e^{-im\theta}}{1 - e^{-2i\theta}}\right)$$

$$= T_0(x) + 2T_1(x)t + \sum_{m=2}^{\infty} t^m \left(\frac{(e^{im\theta} + e^{-im\theta})(1 - e^{-2i\theta})}{1 - e^{-2i\theta}}\right)$$

$$= T_0(x) + 2T_1(x)t + \sum_{m=2}^{\infty} t^m \left(e^{im\theta} + e^{-im\theta}\right)$$

$$= T_0(x) + 2T_1(x)t + \sum_{m=2}^{\infty} T_m(x)t^m$$

$$= T_0(x) + 2\left(\sum_{m=1}^{\infty} T_m(x)t^m\right)$$

8.4 Hypergeometric Differential Equation

The differential equation $x(1-x)y'' + (c-(a+b+1)x)y' - aby = 0$ is called the *hypergeometric differential equation* where a, b and c are constants. The hypergeometric differential equation can be rewritten in the form

$$x^2y'' + x\left(\frac{c - (a+b+1)x}{1-x}\right)y' + \left(-\frac{abx}{1-x}\right)y = 0 \tag{8.98}$$

The coefficients of Eq.(8.98) may look rather strange, but we shall find that they are perfectly adapted to the use of its solution in a wide variety of situations. Clearly, $p(x) = \left(\frac{c-(a+b+1)x}{1-x}\right)$ and $q(x) = -\frac{abx}{1-x}$. Again, both $p(x) = \frac{c}{1-x} - (a+b+1)\left(\frac{x}{1-x}\right) = \sum_{n=0}^{\infty} c_n x^n$ and $q(x) = -ab\left(\frac{x}{1-x}\right) = \sum_{n=0}^{\infty} b_n x^n$ are analytic at $x = 0$, but $\frac{q(x)}{x^2}$ is not analytic at $x = 0$. Hence one can expect a series solution in the form $y(x) = \sum_{n=0}^{\infty} a_n x^{n+r}$.

In order to make $y(x)$ as a solution with $a_0 \neq 0$, one has to have

$$\begin{aligned} F(r) &= r(r-1) + rc_0 + b_0 \\ &= r(r-1) + cr \\ &= r(c-1+r) \\ &= 0 \end{aligned}$$

by Theorem 7.2.1 and the recurrence relation

$$a_n = -\frac{\sum_{k=1}^{n} G(n,\, r,\, k)a_{n-k}}{F(n+r)} \qquad n \geq 1 \tag{8.99}$$

according to Eq.(7.11), where

$$\begin{aligned}\sum_{k=1}^{n} G(n,\, r,\, k) &= \sum_{k=1}^{n} \left((n+r-k)c_k + b_k\right) a_{n-k} \\ &= \sum_{k=1}^{n} \left((n-k)c_k + b_k\right) a_{n-k} \\ &= \sum_{k=1}^{n} ((n-k)(c-a-b-1) - ab)a_{n-k}\end{aligned} \tag{8.100}$$

and

$$\begin{aligned}F(n+r) &= (n+r)(n+r-1) + (n+r)c_0 + b_0 \\ &= n(n-1) + cn \\ &= n(n+c-1)\end{aligned} \tag{8.101}$$

with $r = 0$. According to Eqs.(8.99), (8.100) and (8.101), we have

$$\begin{aligned}a_1 &= \left(\frac{ab}{c}\right) a_0 \\ a_2 &= \left(\frac{a(a+1)b(b+1)}{2c(c+1)}\right) a_0 \\ &\vdots\end{aligned}$$

and

$$a_n = \left(\prod_{i=0}^{n-1} \left(\frac{(a+i)(b+i)}{c+i}\right)\right) \frac{a_0}{n!} \qquad n \geq 1 \tag{8.102}$$

In other words, one of the series solution to Eq.(8.98) is given by

$$y(x) = a_0 \left(1 + \sum_{n=1}^{\infty} \left(\prod_{i=0}^{n-1} \left(\frac{(a+i)(b+i)}{c+i}\right)\right) \frac{x^n}{n!}\right) \tag{8.103}$$

The series given in Eq.(8.103) is called *hypergeometric series*, and denoted by $F(a,\, b,\, c,\, x)$. It is clear from Eq.(8.102) that when either a or b is negative integer or zero, then the hypergeometric differential equation accepts polynomial solution.

Exercises 8.4

1. Show that $\sum\limits_{k=1}^{n} G(n, r, k) = -(a+n-1)(b+n-1)a_{n-1}$ for the hypergeometric equation given in Eq. (8.98).
2. If $F(a, b, c, x)$ is the hypergeometric series, then show that $F(1, b, b, x) = \frac{1}{1-x}$.
3. If $F(a, b, c, x)$ is the hypergeometric series, then show that $F(-r, b, b, x) = (1+x)^r$.
4. If $F(a, b, c, x)$ is the hypergeometric series, then show that $xF(1, 1, 2, -x) = \ln(1+x)$.
5. If $F(a, b, c, x)$ is the hypergeometric series, then show that $xF\left(\frac{1}{2}, \frac{1}{2}, \frac{3}{2}, x^2\right) = \sin^{-1} x$.
6. If $F(a, b, c, x)$ is the hypergeometric series, then show that $xF\left(\frac{1}{2}, 1, \frac{3}{2}, -x^2\right) = \tan^{-1} x$.
7. If $F(a, b, c, x)$ is the hypergeometric series, then show that $\lim\limits_{b\to\infty} F\left(a, b, a, \frac{x}{b}\right) = e^x$.
8. If $F(a, b, c, x)$ is the hypergeometric series, then show that $x\left(\lim\limits_{a\to\infty} F\left(a, a, \frac{3}{2}, -\frac{x^2}{4a^2}\right)\right)$ $= \sin x$.
9. If $F(a, b, c, x)$ is the hypergeometric series, then show that $\lim\limits_{a\to\infty} F\left(a, a, \frac{1}{2}, -\frac{x^2}{4a^2}\right)$ $= \cos x$.
10. If $F(a, b, c, x)$ is the hypergeometric series, then show that $F'(a, b, c, x) = \left(\frac{ab}{c}\right)$ $F(a+1, b+1, c+1, x)$.
11. If $F(a, b, c, x)$ is the hypergeometric series, then show that $F\left(m, -m, \frac{1}{2}, \frac{1-x}{2}\right) = T_m(x)$, the Chebyshev polynomial of degree m.
12. If $F(a, b, c, x)$ is the hypergeometric series, then show that $x\left(\lim\limits_{a\to\infty} F\left(a, a, \frac{3}{2}, \frac{x^2}{4a^2}\right)\right)$ $= \sinh(x)$.
13. If $F(a, b, c, x)$ is the hypergeometric series, then show that $\lim\limits_{a\to\infty} F\left(a, a, \frac{1}{2}, \frac{x^2}{4a^2}\right)$ $= \cosh(x)$.
14. Find the solution of the differential equation $x(1-x)y'' + \left(\frac{1}{4} - \frac{5x}{3}\right)y' - \frac{y}{9} = 0$.
15. Find the solution of the differential equation $x(1-x)y'' - \left(\frac{1}{2} + \frac{5x}{2}\right)y' - \frac{y}{2} = 0$.

8.5 Bessel Differential Equation

The differential equation $x^2y'' + xy' + (x^2 - m^2)y = 0$, where m is a non-negative constant, is called the *Bessel differential equation.* Clearly, $p(x) = 1$ and $q(x) = x^2 - m^2$. Again, both $p(x) = \sum\limits_{n=0}^{\infty} c_n x^n$ and $q(x) = \sum\limits_{n=0}^{\infty} b_n x^n$ are analytic at $x = 0$, but both $\frac{p(x)}{x}$ and $\frac{q(x)}{x^2}$ are not analytic at $x = 0$. Hence one can expect a series solution in the form $y(x) = \sum\limits_{n=0}^{\infty} a_n x^{n+r}$.

In order to make $y(x)$ as a solution with $a_0 \neq 0$, one has to have

$$\begin{aligned}F(r) &= r(r-1) + rc_0 + b_0 \\ &= r(r-1) + r - m^2 \\ &= r^2 - m^2 \\ &= (r+m)(r-m) \\ &= 0\end{aligned}$$

by Theorem 7.2.1 and the recurrence relation

$$a_n = -\frac{\sum_{k=1}^{n} G(n, r, k) a_{n-k}}{F(n+r)} \qquad n \geq 1$$

according to Eq.(7.11), where

$$\begin{aligned}\sum_{k=1}^{n} G(n, r, k) &= \sum_{k=1}^{n} \left((n+r-k)c_k + b_k\right) a_{n-k} \\ &= \sum_{k=1}^{n} b_k a_{n-k} \\ &= a_{n-2}\end{aligned}$$

and

$$\begin{aligned}F(n+r) &= (n+r)(n+r-1) + (n+r)c_0 + b_0 \\ &= (n+r)(n+r-1) + n + r - m^2 \\ &= (n+r)(n+r) - m^2 \\ &= (n+m)^2 - m^2 \\ &= n(2m+n)\end{aligned}$$

where $r = m$. In general, one can simply the recurrence relation to

$$a_n = -\left(\frac{1}{n(2m+n)}\right) a_{n-2} \qquad n \geq 2$$

In other words, we have

$$a_{2n} = (-1)^n \left(\prod_{i=0}^{n-1} \left(\frac{1}{(2n-2i)(2m+2+2i)}\right)\right) a_0$$

$$= (-1)^n \left(\frac{1}{2^{2n} n!}\right) \left(\frac{1}{\prod_{i=1}^{n}(p+i)}\right) a_0 \qquad n \geq 1$$

$$a_{2n+1} = 0 \qquad n \geq 0 \tag{8.104}$$

Clearly, one of the series solutions of the Bessel differential equation when $r = m$, according to Eq.(8.104), is given by

$$y(x) = a_0 x^m \left(1 + \sum_{n=1}^{\infty}(-1)^n \left(\frac{1}{2^{2n} n!}\right) \left(\frac{1}{\prod_{i=1}^{n}(p+i)}\right) x^{2n}\right) \tag{8.105}$$

If $a_0 = \frac{1}{2^m m!}$, the series solution given in Eq.(8.105) is called the *Bessel function of the first kind of order* m, and is denoted by $J_m(x)$. In other words, we have

$$J_m(x) = \left(\frac{x^m}{2^m m!}\right) \left(\sum_{n=0}^{\infty}(-1)^n \left(\frac{1}{2^{2n} n!}\right) \left(\frac{1}{\prod_{i=1}^{n}(p+i)}\right) x^{2n}\right)$$

$$= \sum_{n=0}^{\infty}(-1)^n \left(\frac{1}{2^{2n+m} n!(m+n)!}\right) x^{2n+m}$$

$$= \sum_{n=0}^{\infty}(-1)^n \left(\frac{1}{n!(m+n)!}\right) \left(\frac{x}{2}\right)^{2n+m} \tag{8.106}$$

The series given in Eq.(8.106), first arose in Daniel Bernoulli's investigation of the oscillations of a hanging chain, and appeared again in Euler's theory of vibrations of a circular membrane and Bessel's studies of planetary motion. More recently, the Bessel functions have turned out to have very diverse applications in physics and engineering, in connection with the propagation of waves, elasticity, and fluid motion, and especially in many problems of potential theory and diffusion involving cylindrical symmetry.

Theorem 8.5.1 (Neumann Function) If $J_m(x)$ is the Bessel function of the first kind of order m, then $Y_m(x) = J_m(x) \int \frac{dx}{x J_m^2(x)}$ is the *Bessel function of second kind of order* m.

Proof: Clearly, $J_m(x)$ is a solution of Bessel differential equation

$$y'' + \left(\frac{1}{x}\right) y' + \left(1 - \frac{m^2}{x^2}\right) y = 0 \tag{8.107}$$

Let $Y_m(x) = v(x) J_m(x)$ be the second independent solution of Eq.(8.107). According to the formula given in Eq.(5.29), we have

$$v(x) = \int \frac{e^{-\int p(x)\, dx}}{J_m^2(x)}\, dx$$

$$= \int \frac{e^{-\int \left(\frac{1}{x}\right) \mathrm{d}x}}{J_m^2(x)} \mathrm{d}x$$

$$= \int \frac{\mathrm{d}x}{x J_m^2(x)}$$

In other words, the Bessel function of second kind of order m is given by

$$Y_m(x) = J_m(x) \int \frac{\mathrm{d}x}{x J_m^2(x)}$$

8.5.1 Generating Function

Theorem 8.5.2 If $J_m(x)$ is the Bessel function of the first kind with integer order m, then $e^{\frac{x(t^2-1)}{2t}} = \sum\limits_{m=-\infty}^{\infty} J_m(x) t^m$.

Proof: It is easy to verify that

$$e^{\frac{x(t^2-1)}{2t}} = e^{\frac{xt}{2}} e^{-\frac{x}{2t}}$$

$$= \left(\sum_{i=0}^{\infty} \frac{\left(\frac{xt}{2}\right)^i}{i!}\right) \left(\sum_{j=0}^{\infty} \frac{\left(-\frac{x}{2t}\right)^j}{j!}\right)$$

$$= \left(\sum_{i=0}^{\infty} \frac{(xt)^i}{i! 2^i}\right) \left(\sum_{j=0}^{\infty} \frac{(-1)^j x^j}{j!(2t)^j}\right)$$

$$= \left(\sum_{i=0}^{\infty} \frac{x^i t^i}{i! 2^i}\right) \left(\sum_{j=0}^{\infty} \frac{(-1)^j x^j t^{-j}}{j! 2^j}\right)$$

$$= \sum_{m=0}^{\infty} \left(\sum_{n=0}^{\infty} \left(\frac{x^{m+n} t^{m+n}}{(m+n)! 2^{m+n}}\right) \left(\frac{(-1)^n x^n t^{-n}}{n! 2^n}\right)\right) +$$

$$\sum_{m=1}^{\infty} \left(\sum_{n=0}^{\infty} \left(\frac{x^n t^n}{n! 2^n}\right) \left(\frac{(-1)^{m+n} x^{m+n} t^{-m-n}}{(m+n)! 2^{m+n}}\right)\right)$$

$$= \sum_{m=0}^{\infty} \left(\sum_{n=0}^{\infty} \frac{(-1)^n x^{m+2n}}{n!(m+n)! 2^{m+2n}}\right) t^m + \sum_{m=1}^{\infty} \left(\sum_{n=0}^{\infty} \frac{(-1)^{m+n} x^{m+2n}}{n!(m+n)! 2^{m+2n}}\right) t^{-m}$$

$$= \sum_{m=0}^{\infty} J_m(x) t^m + \sum_{m=1}^{\infty} (-1)^m J_m(x) t^{-m}$$

$$= \sum_{m=0}^{\infty} J_m(x)t^m + \sum_{m=1}^{\infty} J_{-m}(x)t^{-m}$$

$$= \sum_{m=-\infty}^{\infty} J_m(x)t^m$$

Theorem 8.5.3 (Integral Formula) If $J_m(x)$ is the Bessel function of the first kind of integer order m, then $J_n(x) = \left(\frac{1}{\pi}\right)\int_0^{\pi} \cos(n\theta - x\sin\theta)\,\mathrm{d}\theta$.

Proof: According to Theorem 8.5.2, we have

$$e^{\frac{x(t^2-1)}{2t}} = \sum_{m=-\infty}^{\infty} J_m(x)t^m \tag{8.108}$$

Replacing $t = e^{i\theta}$ in Eq.(8.108) and using the relation $x\left(\frac{e^{i\theta}-e^{-i\theta}}{2}\right) = i\,x\sin\theta$, we have

$$\begin{aligned}
e^{i\,x\sin\theta} &= e^{\frac{x}{2}(t-\frac{1}{t})} \\
&= e^{\frac{x(t^2-1)}{2t}} \\
&= \sum_{m=-\infty}^{\infty} J_m(x)t^m \\
&= \sum_{m=-\infty}^{\infty} J_m(x)e^{im\theta} \\
&= \sum_{m=-\infty}^{\infty} J_m(x)\cos m\theta + i\left(\sum_{m=-\infty}^{\infty} J_m(x)\sin m\theta\right)
\end{aligned} \tag{8.109}$$

Again, applying Theorem 8.5.11 to Eq.(8.109), we have

$$\begin{aligned}
\cos(x\sin\theta) + i\sin(x\sin\theta) &= e^{i\,x\sin\theta} \\
&= \sum_{m=-\infty}^{\infty} J_m(x)\cos m\theta + i\left(\sum_{m=-\infty}^{\infty} J_m(x)\sin m\theta\right) \\
&= J_0(x) + 2\left(\sum_{m=1}^{\infty} J_{2m}(x)\cos 2m\theta\right) \\
&\quad + 2i\left(\sum_{m=0}^{\infty} J_{2m+1}(x)\sin(2m+1)\theta\right)
\end{aligned} \tag{8.110}$$

The series given in Eq.(8.110) is known as *Jacobi* series. Equating real and imaginary parts of Eq.(8.110), we have

$$\begin{aligned}\cos(x\sin\theta) &= J_0(x) + 2\left(\sum_{m=1}^{\infty} J_{2m}(x)\cos 2m\theta\right)\\ &= \frac{a_0}{2} + \sum_{k=1}^{\infty} a_k \cos k\,\theta + \sum_{k=1}^{\infty} b_k \sin k\,\theta \qquad (8.111)\end{aligned}$$

and

$$\begin{aligned}\sin(x\sin\theta) &= 2\left(\sum_{m=0}^{\infty} J_{2m+1}(x)\sin((2m+1)\theta)\right)\\ &= \frac{A_0}{2} + \sum_{k=1}^{\infty} A_k \cos k\,\theta + \sum_{k=1}^{\infty} B_k \sin k\,\theta \qquad (8.112)\end{aligned}$$

It is clear from Eqs.(8.111) and (8.112) that

$$\begin{aligned}a_k &= \frac{1}{\pi}\int_{-\pi}^{\pi} \cos(x\sin\theta)\cos k\theta\,\mathrm{d}\theta\\ &= \begin{cases} 0, & k = 2m+1\\ 2J_k(x), & k = 2m\end{cases} \qquad (8.113)\end{aligned}$$

and

$$\begin{aligned}B_k &= \frac{1}{\pi}\int_{-\pi}^{\pi} \sin(x\sin\theta)\sin k\theta\,\mathrm{d}\theta\\ &= \begin{cases} 2J_k(x), & k = 2m+1\\ 0, & k = 2m\end{cases} \qquad (8.114)\end{aligned}$$

whereas

$$\begin{aligned}b_k &= \left(\frac{1}{\pi}\right)\int_{-\pi}^{\pi} \cos(x\sin\theta)\sin k\theta\,\mathrm{d}\theta\\ &= 0 \quad \text{for } k \in \mathbb{I}\end{aligned}$$

and

$$\begin{aligned}A_k &= \left(\frac{1}{\pi}\right)\int_{-\pi}^{\pi} \sin(x\sin\theta)\cos k\theta\,\mathrm{d}\theta\\ &= 0 \quad \text{for } k \in \mathbb{N}\end{aligned}$$

Adding Eqs.(8.113) and (8.114), one can conclude that

$$\begin{aligned}2J_k(x) &= \frac{1}{\pi}\int_{-\pi}^{\pi} \cos(x\sin\theta)\cos k\theta\,\mathrm{d}\theta + \frac{1}{\pi}\int_{-\pi}^{\pi} \sin(x\sin\theta)\sin k\theta\,\mathrm{d}\theta \\ &= \frac{1}{\pi}\int_{-\pi}^{\pi} \cos(n\theta - x\sin\theta)\,\mathrm{d}\theta \\ &= \frac{2}{\pi}\int_{0}^{\pi} \cos(n\theta - x\sin\theta)\,\mathrm{d}\theta\end{aligned}$$

In other words, we have

$$J_n(x) = \frac{1}{\pi}\int_{0}^{\pi} \cos(n\theta - x\sin\theta)\,\mathrm{d}\theta \tag{8.115}$$

The integral formula given in Eq.(8.115) is called the *Bessel integral formula.*

Theorem 8.5.4 If $J_m(x)$ is the Bessel function of first kind of integer order m, then $J_m(x) = \left(\frac{(-1)^m}{\pi}\right)\int_{-1}^{1}\left(\frac{e^{iux}}{(1-u^2)^{\frac{1}{2}}}\right) T_m(u)\,du.$

Proof: Replacing $\theta = \frac{\pi}{2} - \theta$ in Eqs.(8.111) and (8.112), we have

$$\begin{aligned}\cos(x\cos\theta) &= J_0(x) + 2\left(\sum_{m=1}^{\infty}(-1)^m J_{2m}(x)\cos 2m\theta\right) \\ &= \frac{a_0}{2} + \sum_{k=1}^{\infty} a_k \cos k\,\theta\end{aligned} \tag{8.116}$$

and

$$\begin{aligned}\sin(x\cos\theta) &= 2\left(\sum_{m=0}^{\infty}(-1)^m J_{2m+1}(x)\cos((2m+1)\theta)\right) \\ &= A_0 + \sum_{k=1}^{\infty} A_k \cos k\,\theta\end{aligned} \tag{8.117}$$

It is clear from Eqs.(8.116) and (8.117) that

$$\begin{aligned}a_k &= \frac{1}{\pi}\int_{-\pi}^{\pi} \cos(x\cos\theta)\cos k\theta\,\mathrm{d}\theta \\ &= \begin{cases} 0, & k = 2m+1 \\ (-1)^{\frac{k}{2}} 2J_k(x), & k = 2m \end{cases}\end{aligned} \tag{8.118}$$

and

$$A_k = \frac{1}{\pi}\int_{-\pi}^{\pi} \sin(x\cos\theta)\cos k\theta \,\mathrm{d}\theta$$

$$= \begin{cases} (-1)^{\frac{k-1}{2}} 2J_k(x), & k = 2m+1 \\ 0, & k = 2m \end{cases} \tag{8.119}$$

Adding Eqs.(8.118) and i times (8.119) where $i^2 = -1$, one can represent in a single integration for all integer m by

$$J_m(x) = \left(\frac{(-1)^m}{\pi}\right)\int_0^{\pi} e^{i\,x\cos\theta}\cos m\theta \,\mathrm{d}\theta$$

$$= \left(\frac{(-1)^m}{\pi}\right)\int_{-1}^{1}\left(\frac{e^{ixu}}{(1-u^2)^{\frac{1}{2}}}\right) T_m(u)\,\mathrm{d}u$$

8.5.2 Orthogonal Property

Theorem 8.5.5 If λ_r is a zero of $J_m(x)$, the Bessel function of first kind of order m and $n \neq r$, then $\int_0^1 xJ_m(\lambda_r x)J_m(\lambda_n x)\,\mathrm{d}x = 0$.

Proof: Let $u(x) = J_m(\lambda_r x)$ and $v(x) = J_m(\lambda_n x)$. Clearly, $u(1) = J_m(\lambda_r) = 0$ and $v(1) = J_m(\lambda_n) = 0$. It is easy to verify by Exercise 11 given in Exercises 8.5 that

$$u'' + \frac{u'}{x} + \left(\lambda_r^2 - \frac{m^2}{x^2}\right)u = 0 \tag{8.120}$$

and

$$v'' + \frac{v'}{x} + \left(\lambda_n^2 - \frac{m^2}{x^2}\right)v = 0 \tag{8.121}$$

Multiplying Eq.(8.120) by $v(x)$ and Eq.(8.121) by $u(x)$, and then subtracting we get

$$\frac{\mathrm{d}}{\mathrm{d}x}(u'v - v'u) + \frac{1}{x}(u'v - v'u) = (\lambda_n^2 - \lambda_r^2)uv \tag{8.122}$$

Again, multiplying Eq.(8.122) by x gives

$$\frac{\mathrm{d}}{\mathrm{d}x}(x(u'v - v'u)) = (\lambda_n^2 - \lambda_r^2)xuv \tag{8.123}$$

Integrating Eq.(8.123), we have

$$\begin{aligned}(\lambda_n^2 - \lambda_r^2)\int_0^1 xuv\,\mathrm{d}x &= \int_0^1 (\lambda_n^2 - \lambda_r^2)xuv\,\mathrm{d}x\\ &= (x(u'v - v'u))\big|_0^1\\ &= 0 \end{aligned} \tag{8.124}$$

In other words, we have

$$\begin{aligned}\int_0^1 xJ_m(\lambda_r x)J_m(\lambda_n x)\,\mathrm{d}x &= \int_0^1 xuv\,\mathrm{d}x\\ &= 0\end{aligned}$$

as $\lambda_n^2 - \lambda_r^2 \neq 0$.

Theorem 8.5.6 If λ_r is a zero of $J_m(x)$, the Bessel function of first kind of order m, then $\int_0^1 xJ_m^2(\lambda_r x)\,\mathrm{d}x = \frac{J_{m+1}^2(\lambda_r)}{2}$.

Proof: Let $u(x) = J_m(\lambda_r x)$. Clearly, $u(1) = J_m(\lambda_r) = 0$. It is easy to verify by Exercise 11 given in Exercises 8.5 that

$$u'' + \frac{u'}{x} + \left(\lambda_r^2 - \frac{m^2}{x^2}\right)u = 0 \tag{8.125}$$

Multiplying Eq.(8.125) by $2x^2u'$, we have

$$\begin{aligned}0 &= 2x^2u'u'' + 2x^2u'\left(\frac{u'}{x}\right) + 2x^2u'\left(\lambda_r^2 - \frac{m^2}{x^2}\right)u\\ &= 2x^2u'u'' + 2x\left(u'\right)^2 + 2\lambda_r^2x^2u'u - 2m^2u'u\\ &= \frac{\mathrm{d}}{\mathrm{d}x}\left(x^2(u')^2\right) + \frac{\mathrm{d}}{\mathrm{d}x}\left(\lambda_r^2x^2u^2\right) - 2\lambda_r^2xu^2 - \frac{\mathrm{d}}{\mathrm{d}x}\left(m^2u^2\right)\end{aligned} \tag{8.126}$$

Integration of Eq.(8.126) gives

$$\begin{aligned}2\lambda_r^2\int_0^1 xJ_m^2(\lambda_r x)\,\mathrm{d}x &= 2\lambda_r^2\int_0^1 xu^2\,\mathrm{d}x\\ &= x^2(u')^2 + \lambda_r^2x^2u^2 - m^2u^2\big|_0^1\\ &= (u'(1))^2 + (\lambda_r x^2 - m^2)u^2(x)\big|_0^1\\ &= \lambda_r^2\left(J_m'(\lambda_r)\right)^2\\ &= \lambda_r^2J_{m+1}^2(\lambda_r)\end{aligned}$$

by Exercise 9 and Exercise 10 given in Exercises 8.5. In other words, we have

$$\int_0^1 xJ_m^2(\lambda_r x)\,\mathrm{d}x = \frac{J_{m+1}^2(\lambda_r)}{2}$$

8.5.3 Recurrence Relation

Theorem 8.5.7 (Higher Order) If $J_m(x)$ is the Bessel function of first kind of order m, then $\frac{\mathrm{d}}{\mathrm{d}x}\left(x^{-m}J_m(x)\right) = -x^{-m}J_{m+1}(x)$.

Proof: According to Eq.(8.106), we have

$$J_m(x) = \sum_{n=0}^{\infty}(-1)^n\left(\frac{1}{n!(m+n)!}\right)\left(\frac{x}{2}\right)^{2n+m} \tag{8.127}$$

It is easy to verify by Eq.(8.127) that

$$\begin{aligned}
\frac{\mathrm{d}}{\mathrm{d}x}\left(x^{-m}J_m(x)\right) &= \frac{\mathrm{d}}{\mathrm{d}x}\left(x^{-m}\left(\sum_{n=0}^{\infty}(-1)^n\left(\frac{1}{n!(m+n)!}\right)\left(\frac{x}{2}\right)^{2n+m}\right)\right)\\
&= \frac{\mathrm{d}}{\mathrm{d}x}\left(\sum_{n=0}^{\infty}2^m(-1)^n\left(\frac{1}{n!(m+n)!}\right)\left(\frac{x}{2}\right)^{2n}\right)\\
&= \sum_{n=1}^{\infty}n2^m(-1)^n\left(\frac{1}{n!(m+n)!}\right)\left(\frac{x}{2}\right)^{2n-1}\\
&= x^{-m}\left(\sum_{n=1}^{\infty}(-1)^n\left(\frac{1}{(n-1)!(m+n)!}\right)\left(\frac{x}{2}\right)^{2n-1+m}\right)\\
&= -x^{-m}\left(\sum_{n=0}^{\infty}(-1)^n\left(\frac{1}{n!(m+n+1)!}\right)\left(\frac{x}{2}\right)^{2n+1+m}\right)\\
&= -x^{-m}J_{m+1}(x)
\end{aligned}$$

Theorem 8.5.8 (Lower Order) If $J_m(x)$ is the Bessel function of first kind of order m, then $\frac{\mathrm{d}}{\mathrm{d}x}\left(x^{m}J_m(x)\right) = x^{m}J_{m-1}(x)$.

Proof: According to Eq.(8.106), we have

$$J_m(x) = \sum_{n=0}^{\infty}(-1)^n\left(\frac{1}{n!(m+n)!}\right)\left(\frac{x}{2}\right)^{2n+m} \tag{8.128}$$

It is easy to verify by Eq.(8.128) that

$$\begin{aligned}\frac{\mathrm{d}}{\mathrm{d}x}\left(x^m J_m(x)\right) &= \frac{\mathrm{d}}{\mathrm{d}x}\left(x^m\left(\sum_{n=0}^{\infty}(-1)^n\left(\frac{1}{n!(m+n)!}\right)\left(\frac{x}{2}\right)^{2n+m}\right)\right)\\ &= \frac{\mathrm{d}}{\mathrm{d}x}\left(\sum_{n=0}^{\infty}2^m(-1)^n\left(\frac{1}{n!(m+n)!}\right)\left(\frac{x}{2}\right)^{2n+2m}\right)\\ &= \sum_{n=0}^{\infty}(m+n)2^m(-1)^n\left(\frac{1}{n!(m+n)!}\right)\left(\frac{x}{2}\right)^{2n+2m-1}\\ &= x^m\left(\sum_{n=0}^{\infty}(-1)^n\left(\frac{1}{n!(m+n-1)!}\right)\left(\frac{x}{2}\right)^{2n-1+m}\right)\\ &= -x^m\left(\sum_{n=0}^{\infty}(-1)^n\left(\frac{1}{n!(m+n-1)!}\right)\left(\frac{x}{2}\right)^{2n-1+m}\right)\\ &= x^m J_{m-1}(x)\end{aligned}$$

Theorem 8.5.9 If $J_m(x)$ is the Bessel function of first kind of order m, then $2\,J'_m(x) = J_{m-1}(x) - J_{m+1}(x)$.

Proof: According to Theorems 8.5.7 and 8.5.8, we have

$$\begin{aligned}-x^{-m}J_{m+1}(x) &= \frac{\mathrm{d}}{\mathrm{d}x}\left(x^{-m}J_m(x)\right)\\ &= -mx^{-(m+1)}J_m(x) + x^{-m}J'_m(x)\end{aligned} \tag{8.129}$$

and

$$\begin{aligned}x^m J_{m-1}(x) &= \frac{\mathrm{d}}{\mathrm{d}x}\left(x^m J_m(x)\right)\\ &= mx^{m-1}J_m(x) + x^m J'_m(x)\end{aligned} \tag{8.130}$$

In other words, Eqs.(8.129) and (8.130) get reduced to

$$-J_{m+1}(x) = -mx^{-1}J_m(x) + J'_m(x) \tag{8.131}$$

and

$$J_{m-1}(x) = mx^{-1}J_m(x) + J'_m(x) \tag{8.132}$$

Adding Eqs.(8.131) and (8.132), we have

$$J_{m-1}(x) - J_{m+1}(x) = 2J'_m(x)$$

Theorem 8.5.10 If $J_m(x)$ is the Bessel function of first kind of order m, then

$$2m\,J_m(x) = xJ_{m-1}(x) + xJ_{m+1}(x)$$

Proof: According to Theorems 8.5.7 and 8.5.8, we have

$$\begin{aligned}-x^{-m}J_{m+1}(x) &= \frac{\mathrm{d}}{\mathrm{d}x}\left(x^{-m}J_m(x)\right)\\ &= -mx^{-(m+1)}J_m(x) + x^{-m}J'_m(x) \end{aligned} \tag{8.133}$$

and

$$\begin{aligned}x^{m}J_{m-1}(x) &= \frac{\mathrm{d}}{\mathrm{d}x}\left(x^{m}J_m(x)\right)\\ &= mx^{m-1}J_m(x) + x^{m}J'_m(x) \end{aligned} \tag{8.134}$$

In other words, Eqs.(8.133) and (8.134) get reduced to

$$-J_{m+1}(x) = -mx^{-1}J_m(x) + J'_m(x) \tag{8.135}$$

and

$$J_{m-1}(x) = mx^{-1}J_m(x) + J'_m(x) \tag{8.136}$$

Subtracting Eq.(8.135) from Eq.(8.136), we have

$$J_{m-1}(x) + J_{m+1}(x) = 2mx^{-1}J'_m(x)$$

8.5.4 Properties of $J_m(x)$

Theorem 8.5.11 If $J_m(x)$ is the Bessel function of first kind of integer order m, then

$$J_{-m}(x) = (-1)^m J_m(x)$$

Proof: According to Eq.(8.106), we have

$$\begin{aligned}J_{-m}(x) &= \sum_{n=0}^{\infty}(-1)^n\left(\frac{1}{n!(-m+n)!}\right)\left(\frac{x}{2}\right)^{2n-m}\\ &= \sum_{n=-m}^{\infty}(-1)^{m+n}\left(\frac{1}{(m+n)!n!}\right)\left(\frac{x}{2}\right)^{m+2n}\\ &= \sum_{n=0}^{\infty}(-1)^{m+n}\left(\frac{1}{(m+n)!n!}\right)\left(\frac{x}{2}\right)^{m+2n}\\ &= (-1)^m\left(\sum_{n=0}^{\infty}(-1)^{n}\left(\frac{1}{(m+n)!n!}\right)\left(\frac{x}{2}\right)^{m+2n}\right)\\ &= (-1)^m J_m(x)\end{aligned}$$

Theorem 8.5.12 If $J_m(x)$ is the Bessel function of first kind of integer order m, then

$$J_m(-x) = (-1)^m J_m(x)$$

Proof: According to Theorem 8.5.2, we have

$$\sum_{m=-\infty}^{\infty} J_m(-x)t^m = e^{\frac{(-x)(t^2-1)}{2t}}$$

$$= e^{\frac{x(t^2-1)}{-2t}}$$

$$= e^{\frac{x((-t)^2-1)}{2(-t)}}$$

$$= \sum_{m=-\infty}^{\infty} J_m(x)(-t)^m$$

$$= \sum_{m=-\infty}^{\infty} (-1)^m J_m(x)t^m \tag{8.137}$$

Comparing the coefficients of t^m in Eq.(8.137), we have

$$J_m(-x) = (-1)^m J_m(x)$$

Theorem 8.5.13 If $J_m(x)$ is the Bessel function of first kind of order m, then

$$J_{\frac{1}{2}}(x) = \left(\frac{2}{\pi x}\right)^{\frac{1}{2}} \sin x$$

Proof: According to Eq.(8.106), we have

$$J_{\frac{1}{2}}(x) = \sum_{n=0}^{\infty}(-1)^n \left(\frac{1}{(\frac{1}{2}+n)!n!}\right)\left(\frac{x}{2}\right)^{\frac{1}{2}+2n}$$

$$= \sum_{n=0}^{\infty}(-1)^n \left(\frac{1}{(\frac{1+2n}{2})!n!}\right)\left(\frac{x}{2}\right)^{\frac{1}{2}+2n}$$

$$= \sum_{n=0}^{\infty}(-1)^n \left(\frac{1}{(-\frac{1}{2})!\left(\prod_{i=0}^{n}\left(\frac{1+2n-2i}{2}\right)\right)n!}\right)\left(\frac{x}{2}\right)^{\frac{1}{2}+2n}$$

$$= \sum_{n=0}^{\infty}(-1)^n \left(\frac{2^{n+1}}{(-\frac{1}{2})!\left(\prod_{i=0}^{n}(1+2n-2i)\right)n!}\right)\left(\frac{x}{2}\right)^{\frac{1}{2}+2n}$$

$$= \left(\frac{2}{x}\right)^{\frac{1}{2}}\left(\sum_{n=0}^{\infty}(-1)^n \left(\frac{1}{(-\frac{1}{2})!2^n\left(\prod_{i=0}^{n-1}(1+2n-2i)\right)n!}\right)x^{2n+1}\right)$$

$$= \left(\frac{2}{\pi x}\right)^{\frac{1}{2}} \left(\sum_{n=0}^{\infty}(-1)^n \left(\frac{1}{2^n \left(\prod_{i=0}^{n-1}(1+2n-2i)\right) n!}\right) x^{2n+1}\right)$$

$$= \left(\frac{2}{\pi x}\right)^{\frac{1}{2}} \left(\sum_{n=0}^{\infty}(-1)^n \left(\frac{1}{\left(\prod_{i=0}^{n}(1+2n-2i)\right)\left(\prod_{i=1}^{n}(2i)\right)}\right) x^{2n+1}\right)$$

$$= \left(\frac{2}{\pi x}\right)^{\frac{1}{2}} \left(\sum_{n=0}^{\infty}(-1)^n \left(\frac{1}{\left(\prod_{i=1}^{2n+1} i\right)}\right) x^{2n+1}\right)$$

$$= \left(\frac{2}{\pi x}\right)^{\frac{1}{2}} \left(\sum_{n=0}^{\infty}(-1)^n \frac{x^{2n+1}}{(2n+1)!}\right)$$

$$= \left(\frac{2}{\pi x}\right)^{\frac{1}{2}} \sin x$$

Theorem 8.5.14 If $J_m(x)$ is the Bessel function of first kind of order m, then

$$J_{-\frac{1}{2}}(x) = \left(\frac{2}{\pi x}\right)^{\frac{1}{2}} \cos x$$

Proof: According to Eq.(8.106), we have

$$J_{-\frac{1}{2}}(x) = \sum_{n=0}^{\infty}(-1)^n \left(\frac{1}{(-\frac{1}{2}+n)!n!}\right)\left(\frac{x}{2}\right)^{-\frac{1}{2}+2n}$$

$$= \sum_{n=0}^{\infty}(-1)^n \left(\frac{1}{(\frac{2n-1}{2})!n!}\right)\left(\frac{x}{2}\right)^{-\frac{1}{2}+2n}$$

$$= \sum_{n=0}^{\infty}(-1)^n \left(\frac{1}{(-\frac{1}{2})!\left(\prod_{i=0}^{n-1}\left(\frac{2n-1-2i}{2}\right)\right) n!}\right)\left(\frac{x}{2}\right)^{-\frac{1}{2}+2n}$$

$$= \sum_{n=0}^{\infty}(-1)^n \left(\frac{2^n}{(-\frac{1}{2})!\left(\prod_{i=0}^{n-1}(2n-1-2i)\right) n!}\right)\left(\frac{x}{2}\right)^{-\frac{1}{2}+2n}$$

$$= \left(\frac{2}{\pi x}\right)^{\frac{1}{2}} \left(\sum_{n=0}^{\infty}(-1)^n \left(\frac{1}{2^n\left(\prod_{i=0}^{n-1}(2n-1-2i)\right)n!}\right)x^{2n}\right)$$

$$= \left(\frac{2}{\pi x}\right)^{\frac{1}{2}} \left(\sum_{n=0}^{\infty}(-1)^n \left(\frac{1}{2^n\left(\prod_{i=0}^{n-1}(2n-1-2i)\right)n!}\right)x^{2n}\right)$$

$$= \left(\frac{2}{\pi x}\right)^{\frac{1}{2}} \left(\sum_{n=0}^{\infty}(-1)^n \left(\frac{1}{\left(\prod_{i=0}^{n-1}(2n-1-2i)\right)\left(\prod_{i=1}^{n}(2i)\right)}\right)x^{2n}\right)$$

$$= \left(\frac{2}{\pi x}\right)^{\frac{1}{2}} \left(\sum_{n=0}^{\infty}(-1)^n \left(\frac{1}{\left(\prod_{i=1}^{2n} i\right)}\right)x^{2n}\right)$$

$$= \left(\frac{2}{\pi x}\right)^{\frac{1}{2}} \left(\sum_{n=0}^{\infty}(-1)^n \frac{x^{2n}}{(2n)!}\right)$$

$$= \left(\frac{2}{\pi x}\right)^{\frac{1}{2}} \cos x$$

Theorem 8.5.15 The differential equation $x^2y''+x(a+2bx^p)y'+(c+dx^{2q}+b(a+p-1)x^p+b^2x^{2p})y=0$ in which $d \neq 0$ can be transformed to Bessel equation $u^2v''+uv'+\left(u^2-\beta^2\right)v=0$ where $\beta = \frac{\sqrt{(1-a)^2-4c}}{2q}$ by substituting $y = \left(x^{\frac{1-a}{2}}e^{-\left(\frac{b}{p}\right)x^p}\right)v$ and $x = \left(\frac{qu}{\sqrt{|d|}}\right)^{\frac{1}{q}}$.

Proof: Let $y = \left(x^{\frac{1-a}{2}}e^{-\left(\frac{b}{p}\right)x^p}\right)v$. It is easy to verify that

$$y' = \left(\frac{1-a}{2}\right)x^{-\left(\frac{1+a}{2}\right)}e^{-\left(\frac{b}{p}\right)x^p}v - bx^{p-1}x^{\left(\frac{1-a}{2}\right)}e^{-\left(\frac{b}{p}\right)x^p}v + x^{\left(\frac{1-a}{2}\right)}e^{-\left(\frac{b}{p}\right)x^p}v'$$

and

$$\begin{aligned}y'' = &-\left(\frac{1-a^2}{4}\right)x^{-\left(\frac{3+a}{2}\right)}e^{-\left(\frac{b}{p}\right)x^p}v - b\left(\frac{1-a}{2}\right)x^{-\left(\frac{1+a}{2}\right)}e^{-\left(\frac{b}{p}\right)x^p}v \\ &- b(p-1)x^{p-2}x^{\frac{1-a}{2}}e^{-\left(\frac{b}{p}\right)x^p}v - b\left(\frac{1-a}{2}\right)x^{-\left(\frac{1+a}{2}\right)}x^{p-1}e^{-\left(\frac{b}{p}\right)x^p}v \\ &+ b^2x^{2(p-1)}x^{\left(\frac{1-a}{2}\right)}e^{-\left(\frac{b}{p}\right)x^p}v + (1-a)x^{-\left(\frac{1+a}{2}\right)}e^{-\left(\frac{b}{p}\right)x^p}v' \\ &- 2bx^{\left(\frac{1-a}{2}\right)}x^{p-1}e^{-\left(\frac{b}{p}\right)x^p}v' + x^{\left(\frac{1+a}{2}\right)}e^{-\left(\frac{b}{p}\right)x^p}v''\end{aligned}$$

Hence the given differential equation reduces to

$$\begin{aligned}0 &= x^{\frac{1-a}{2}}e^{-\left(\frac{b}{p}\right)x^p}x^2v'' + x^{\frac{1-a}{2}}e^{-\left(\frac{b}{p}\right)x^p}xv' + x^{\frac{1-a}{2}}e^{-\left(\frac{b}{p}\right)x^p}\left(dx^{2q} - \left(\frac{(1-a)^2-4c}{4}\right)\right)v \\ &= x^{\frac{1-a}{2}}e^{-\left(\frac{b}{p}\right)x^p}\left(x^2v'' + xv' + \left(dx^{2q} - \left(\frac{(1-a)^2-4c}{4}\right)\right)v\right)\end{aligned}$$

In other words, one can have

$$x^2v'' + xv' + \left(dx^{2q} - \left(\frac{(1-a)^2-4c}{4}\right)\right)v = 0 \tag{8.138}$$

Let $x = \left(\frac{qu}{\sqrt{|d|}}\right)^{\frac{1}{q}}$. It is easy to verify that

$$\begin{aligned}\frac{dv}{dx} &= \left(\sqrt{|d|}x^{q-1}\right)\frac{dv}{du} \\ &= \sqrt{|d|}\left(\frac{qu}{\sqrt{|d|}}\right)^{\frac{q-1}{q}}\frac{dv}{du}\end{aligned}$$

and

$$\frac{d^2v}{dx^2} = d\left(\frac{q^2u^2}{|d|}\right)^{\left(1-\frac{1}{q}\right)}\frac{d^2v}{du^2} + d\left(1-\frac{1}{q}\right)\left(\frac{q^2}{|d|}\right)^{\left(1-\frac{1}{q}\right)}u^{\left(1-\frac{2}{q}\right)}\frac{dv}{du}$$

Hence Eq. (8.138) reduces to the Bessel equation in terms of the variables u and v

$$u^2v'' + uv' + \left(u - \beta^2\right)v = 0$$

EXAMPLE 8.2 Solve the differential equation $x^2y'' + xy' + \left(\lambda^2x^2 - m^2\right)y = 0$ by reducing into Bessel's differential equation.

Solution: According to Theorem 8.5.15, the corresponding reduced differential equation is $u^2y'' + uy' + \left(u^2 - m^2\right)y = 0$ where $u = \lambda x$.

EXAMPLE 8.3 Solve the differential equation $y'' + \left(\lambda^2 - \frac{4m^2-1}{4x^2}\right)y = 0$ by reducing into Bessel's differential equation.

Solution: According to Theorem 8.5.15, the corresponding reduced differential equation is $u^2v'' + uv' + \left(u^2 - m^2\right)v = 0$.

Theorem 8.5.16 Let $v \in \mathbb{R}$. If a and b are two positive roots of $J_v(x)$, then both $J_{v-1}(x)$ and $J_{v+1}(x)$ must have at least one root between a and b.

Proof: Let $f(x) = x^vJ_v(x)$. Clearly, $f(a) = f(b) = 0$. According to Rolle's theorem, $f'(\alpha) = 0$ for some $a < \alpha < b$. Again, $f'(x) = x^vJ_{v-1}(x)$ by Theorem 8.5.8. Hence $J_{v-1}(\alpha) = 0$ for $a < \alpha < b$. Let $g(x) = x^{-v}J_v(x)$. Clearly, $g(a) = g(b) = 0$. According to Rolle's Theorem, $g'(\beta) = 0$ for some $a < \beta < b$. Again, $g'(x) = -x^{-v}J_{v+1}(x)$ by Theorem 8.5.7. Hence $J_{v+1}(\beta) = 0$ for $a < \beta < b$.

Exercises 8.5

1. If $J_m(x)$ is the Bessel function of first kind of integer order m, then show that

$$\cos x = J_0(x) + 2\left(\sum_{m=1}^{\infty}(-1)^m J_{2m}(x)\right)$$

2. If $J_m(x)$ is the Bessel function of first kind of integer order m, then show that

$$\sin x = 2\left(\sum_{m=0}^{\infty}(-1)^m J_{2m+1}(x)\right)$$

3. If $J_m(x)$ is the Bessel function of first kind of integer order m, then show that

$$1 = J_0(x) + 2\left(\sum_{m=1}^{\infty} J_{2m}(x)\right)$$

4. If $J_m(x)$ is the Bessel function of first kind with integer order, then show that

$$J_m(x+y) = \sum_{n=-\infty}^{\infty} J_{m-n}(x)J_n(y)$$

5. If $J_m(x)$ is the Bessel function of first kind with order m, then show that

$$J_{\frac{3}{2}}(x) = \left(\frac{2}{\pi x}\right)^{\frac{1}{2}}\left(\frac{\sin x}{x} - \cos x\right)$$

Hint: Use Theorems 8.5.10, 8.5.13 and 8.5.14 with $m = \frac{1}{2}$.

6. If $J_m(x)$ is the Bessel function of first kind with order m, then show that

$$x^2 J_m''(x) = (m^2 - m - x)J_m(x) + xJ_{m+1}(x)$$

7. If $J_m(x)$ is the Bessel function of first kind with order m, then show that

$$J_{\frac{1}{2}}(x) = J_{-\frac{1}{2}}(x)\tan x$$

8. If $J_m(x)$ is the Bessel function of first kind with order m, then show that

$$J_{-\frac{1}{2}}(x) = J_{\frac{1}{2}}(x)\cot x$$

9. If $J_m(x)$ is the Bessel function of first kind with order m, then show that

$$(\lambda_r x^2 - m^2)u(x)\Big|_0^1 = 0$$

where λ_r is a zero of $J_m(x)$ and $u(x) = J_m(\lambda_r x)$.

Hint: According the Exercise 11, we have

$$\begin{aligned}(\lambda_r^2 x^2 - m^2)u^2(x)\Big|_0^1 &= -\left(x^2 u(x)u''(x) + xu(x)u'(x)\right)\Big|_0^1\\ &= 0\end{aligned}$$

10. If $J_m(x)$ is the Bessel function of first kind with order m and $u(x) = J_m(ax)$, then show that

$$u'(1) = aJ'_m(a)$$

11. If $J_m(x)$ is the Bessel polynomial of first kind of order m and $u(x) = J_m(ax)$ for some positive a, then show that

$$u'' + \frac{u'}{x} + \left(a^2 - \frac{m^2}{x^2}\right) u = 0$$

12. If $J_m(x)$ is the Bessel polynomial of first kind of order m, then show that

$$J'_m(x) = -\left(\frac{n}{x}\right) J_m(x) + J_{n-1}(x)$$

Hint: Apply Theorem 8.5.7.

13. If $J_m(x)$ is the Bessel polynomial of first kind of order m, then show that

$$J'_m(x) = \left(\frac{n}{x}\right) J_m(x) J_{n+1}(x)$$

Hint: Apply Theorem 8.5.8.

14. If $J_m(x)$ is the Bessel function of first kind of integer order m, then show that

$$J_0^2(x) + 2\left(\sum_{m=1}^{\infty} J_m^2(x)\right) = 1$$

15. If $m \neq n$ are integers, then show that

$$\int_0^{\pi} \cos m\theta \cos n\theta \, \mathrm{d}\theta = 0$$

16. If $m \neq n$ are integers, then show that

$$\int_0^{\pi} \sin m\theta \sin n\theta \, \mathrm{d}\theta = 0$$

17. If m is an integer, then show that

$$\int_0^{\pi} \cos^2 m\theta \, \mathrm{d}\theta = \frac{\pi}{2}$$

18. If m is an integer, then show that

$$\int_0^{\pi} \sin^2 m\theta \, \mathrm{d}\theta = \frac{\pi}{2}$$

19. If m is an integer and $J_m(x)$ is the Bessel function of first kind of order m, then show that

$$\int_0^{\pi} \sin(x \sin\theta) \sin(2m+1)\theta \, d\theta = \pi J_{2m+1}(x)$$

20. If m is an integer and $J_m(x)$ is the Bessel function of first kind of order m, then show that

$$\int_0^{\pi} \cos(x \sin\theta) \cos 2m\theta \, d\theta = \pi J_{2m}(x)$$

21. Show that $J_0(x) = \frac{1}{2\pi} \int_0^{2\pi} e^{ix\cos\theta} \, d\theta$.

HINTS: According to Eq. (8.109), we have

$$J_0(x) = \frac{1}{2\pi} \int_0^{2\pi} \left(\sum_{m=-\infty}^{\infty} J_m(x) e^{ixm} \right) d\theta$$

$$= \frac{1}{2\pi} \int_0^{2\pi} e^{ix\sin\theta} \, d\theta$$

$$= \frac{1}{2\pi} \int_{-\frac{\pi}{2}}^{\frac{3\pi}{2}} e^{ix\sin(\theta+\frac{\pi}{2})} \, d\theta$$

$$= \frac{1}{2\pi} \int_{-\frac{\pi}{2}}^{\frac{3\pi}{2}} e^{ix\cos\theta} \, d\theta$$

$$= \frac{1}{2\pi} \int_0^{\frac{3\pi}{2}} e^{ix\cos\theta} \, d\theta + \frac{1}{2\pi} \int_{-\frac{\pi}{2}}^{0} e^{ix\cos\theta} \, d\theta$$

$$= \frac{1}{2\pi} \int_0^{\frac{3\pi}{2}} e^{ix\cos\theta} \, d\theta + \frac{1}{2\pi} \int_{\frac{3\pi}{2}}^{2\pi} e^{ix\cos(\theta+2\pi)} \, d\theta$$

$$= \frac{1}{2\pi}\int_{0}^{\frac{3\pi}{2}} e^{ix\cos\theta}\,d\theta + \frac{1}{2\pi}\int_{\frac{3\pi}{2}}^{2\pi} e^{ix\cos\theta}\,d\theta$$

$$= \frac{1}{2\pi}\int_{0}^{2\pi} e^{ix\cos\theta}\,d\theta$$

22. If $I_m(x)$ is a solution of $x^2y'' + xy' - (x^2 - m^2)y = 0$ called *modified Bessel differential equation*, then show that $i^m I_m(x) = J_m(ix)$.
23. Find the general solution of the following differential equations in terms of Bessel's function using Theorem 8.5.15.
    (a) $x^2y'' + xy' + (\lambda^2x^2 - r^2)y = 0$.
    (b) $x^2y'' + xy' + 4(x^4 - r^2)y = 0$.
    (c) $4x^2y'' + 4xy' + (x - r^2)y = 0$.
    (d) $4xy'' + 4y' + y = 0$.
    (e) $xy'' - y' + xy = 0$.
    (f) $4x^2y'' + (4x^2 + 1)y = 0$.
    (g) $4x^2y'' + (4x + 3)y = 0$.
    (h) $x^2y'' - 3xy' + 4(x^4 - 3)y = 0$.
    (i) $y'' + xy = 0$.
    (j) $xy'' + (1 + 2n)y' + xy = 0$.

8.6 Laguerre Differential Equation

The differential equation $xy'' + (1 - x)y' + my = 0$ is called the *Laguerre differential equation* of order $m \in \mathbb{I}$. The Laguerre differential equation can be rewritten in the form

$$x^2y'' + x(1 - x)y' + mxy = 0 \tag{8.139}$$

Clearly, $p(x) = 1 - x$ and $q(x) = mx$. Again, both $p(x) = \sum_{n=0}^{\infty} c_nx^n$ and $q(x) = \sum_{n=0}^{\infty} b_nx^n$ are analytic at $x = 0$, but both $\frac{p(x)}{x}$ and $\frac{q(x)}{x^2}$ are not analytic at $x = 0$. Hence one can expect a solution in the form $y(x) = \sum_{n=0}^{\infty} a_nx^{n+r}$.

In order to make $y(x)$ as a solution with $a_0 \neq 0$, one has to have

$$\begin{aligned} F(r) &= r(r-1) + rc_0 + b_0 \\ &= r(r-1) + r \\ &= r^2 \\ &= 0 \end{aligned}$$

by Theorem 7.2.1 and the recurrence relation

$$a_n = -\frac{\sum_{k=1}^{n} G(n, r, k)a_{n-k}}{F(n+r)} \qquad n \geq 1$$

according to Eq.(7.11), where

$$\begin{aligned}\sum_{k=1}^{n} G(n, r, k) &= \sum_{k=1}^{n} \left((n+r-k)c_k + b_k\right) a_{n-k} \\ &= \sum_{k=1}^{n} \left((n-k)c_k + b_k\right) a_{n-k} \\ &= ((n-1)c_1 + b_1)a_{n-1} \\ &= (m-n+1)a_{n-1}\end{aligned}$$

and

$$\begin{aligned}F(n+r) &= (n+r)(n+r-1) + (n+r)c_0 + b_0 \\ &= n(n-1) + n \\ &= n^2\end{aligned}$$

In general, one can simplify the recurrence relation to

$$\begin{aligned}a_n &= -\left(\frac{m-n+1}{n^2}\right) a_{n-1} \\ &= -\left(\frac{m-(n-1)}{n^2}\right) a_{n-1} \qquad n \geq 1\end{aligned}$$

In other words, we have

$$\begin{aligned}a_n &= (-1)^n \left(\prod_{i=0}^{n-1} \left(\frac{m-i}{(i+1)^2}\right)\right) a_0 \\ &= (-1)^n \binom{m}{n} \left(\frac{1}{n!}\right) a_0 \qquad n \geq 1\end{aligned} \tag{8.140}$$

Clearly, one of the solutions to the Laguerre differential equation is

$$\begin{aligned}y(x) &= a_0 \left(1 + \sum_{n=1}^{\infty} (-1)^n \left(\prod_{i=0}^{n-1} \left(\frac{m-i}{(i+1)^2}\right)\right) x^n\right) \\ &= a_0 \left(1 + \sum_{n=1}^{\infty} (-1)^n \binom{m}{n} \left(\frac{1}{n!}\right) x^n\right)\end{aligned} \tag{8.141}$$

It is clear from Eqs.(8.140) and (8.141) that the Laguerre differential equation accepts polynomial solution. The polynomial $L_m(x)$ of degree m, which is a solution of Laguerre differential equation of order m is called the *Laguerre polynomial* when $a_0 = m!$. In other words, the leading coefficient of $L_m(x)$ is $(-1)^m$.

8.6.1 Generating Function

Theorem 8.6.1 If $L_m(x)$ is the Laguerre polynomial of degree m, then

$$e^{-\left(\frac{xt}{1-t}\right)} = (1-t)\sum_{m=0}^{\infty}\left(\frac{L_m(x)}{m!}\right)t^m$$

Proof: It is easy to verify that

$$\begin{aligned}
(1-t)^{-1}e^{-\left(\frac{xt}{1-t}\right)} &= (1-t)^{-1}\left(\sum_{m=0}^{\infty}\left(-\frac{xt}{1-t}\right)^m\left(\frac{1}{m!}\right)\right)\\
&= \sum_{m=0}^{\infty}(-xt)^m\left(\frac{1}{1-t}\right)^{m+1}\left(\frac{1}{m!}\right)\\
&= \sum_{m=0}^{\infty}(-xt)^m\left(\frac{1}{m!}\right)\left(\sum_{i=0}^{\infty}\binom{-m-1}{i}(-t)^i\right)\\
&= \sum_{m=0}^{\infty}(-xt)^m\left(\frac{1}{m!}\right)\left(\sum_{i=0}^{\infty}\binom{m+i}{i}t^i\right)\\
&= \sum_{m=0}^{\infty}\left(\sum_{n=0}^{m}(-1)^n x^n\left(\frac{1}{n!}\right)\binom{n+m-n}{m-n}\right)t^m\\
&= \sum_{m=0}^{\infty}\left(\sum_{n=0}^{m}(-1)^n x^n\left(\frac{m!}{n!}\right)\binom{m}{n}\right)\left(\frac{t^m}{m!}\right)\\
&= \sum_{m=0}^{\infty}m!\left(\sum_{n=0}^{m}(-1)^n x^n\left(\frac{1}{n!}\right)\binom{m}{n}\right)\left(\frac{t^m}{m!}\right)\\
&= \sum_{m=0}^{\infty}\left(\frac{L_m(x)}{m!}\right)t^m
\end{aligned}$$

Theorem 8.6.2 (Rodrigue Formula) If $L_m(x)$ is the Laguerre polynomial of degree m, then $L_m(x) = e^x\left(\frac{\mathrm{d}^m}{\mathrm{d}x^m}(x^m e^{-x})\right)$.

Proof: According to the Leibnitz rule of differentiation, we have

$$e^x\left(\frac{\mathrm{d}^m}{\mathrm{d}x^m}(x^m e^{-x})\right) = e^x\left(\sum_{n=0}^{m}\binom{m}{n}\left(\frac{\mathrm{d}^n}{\mathrm{d}x^n}(e^{-x})\right)\left(\frac{\mathrm{d}^{m-n}}{\mathrm{d}x^{m-n}}(x^m)\right)\right)$$

$$= e^x \left(\sum_{n=0}^{m} \binom{m}{n} (-1)^n e^{-x} \left(\prod_{i=0}^{m-n-1} (m-i) \right) x^n \right)$$

$$= \sum_{n=0}^{m} (-1)^n \binom{m}{n} \left(\prod_{i=0}^{m-n-1} (m-i) \right) x^n$$

$$= \sum_{n=0}^{m} (-1)^n \binom{m}{n} \left(\frac{m!}{n!} \right) x^n$$

$$= m! \left(\sum_{n=0}^{m} (-1)^n \binom{m}{n} \left(\frac{1}{n!} \right) x^n \right)$$

$$= L_m(x)$$

8.6.2 Orthogonal Property

Theorem 8.6.3 If $L_m(x)$ is the Laguerre polynomial of degree m and $m \neq n$, then

$$\int_0^\infty e^{-x} L_m(x) L_n(x)\, \mathrm{d}x = 0$$

Proof: Clearly, we have

$$L_m''(x) + (1-x)L_m'(x) + mL_m(x) = 0 \tag{8.142}$$

and

$$xL_n''(x) + (1-x)L_n'(x) + nL_n(x) = 0 \tag{8.143}$$

Let $U(x) = L_m'(x)L_n(x) - L_n'(x)L_m(x)$. Clearly, we have $U'(x) = L_m''(x)L_n(x) - L_n''(x)L_m(x)$. Multiplying Eq.(8.142) by $L_n(x)$ and Eq.(8.143) by $L_m(x)$, and subtracting the second from the first, we have

$$xU' + (1-x)U = (n-m)L_m(x)L_n(x) \tag{8.144}$$

Clearly, Eq.(8.144) is a first order nonhomogeneous differential equation with $p(x) = \frac{1-x}{x}$ and $f(x) = \left(\frac{n-m}{x}\right) L_m(x)L_n(x)$. According to the formula given in Eq.(2.8), the solution of Eq.(8.144) is

$$(n-m)\int_0^\infty e^{-x} L_m(x)L_n(x)\mathrm{d}x = \int_0^\infty (xe^{-x}) \left(\frac{n-m}{x}\right) L_m(x)L_n(x)\mathrm{d}x$$

$$= \int_0^\infty e^{\int p(x)\, \mathrm{d}x} f(x)\, \mathrm{d}x$$

$$= \int_0^\infty \left((L'_m(x)L_n(x) - L'_n(x)L_m(x))xe^{-x}\right)' \, \mathrm{d}x$$

$$= (L'_m(x)L_n(x) - L'_n(x)L_m(x))xe^{-x}\Big|_0^\infty$$

$$= 0$$

In other words, we have

$$\int_0^\infty e^{-x} L_m(x) L_n(x) \mathrm{d}x = 0$$

as $n - m \neq 0$.

Theorem 8.6.4 If $L_m(x)$ is the Laguerre polynomial of degree m, then

$$(m!)^2 = \int_0^\infty e^{-x} L_m^2(x) \, \mathrm{d}x$$

Proof: According to Theorem 8.6.1, we have

$$\left(\frac{1}{(1-v)(1-u)}\right) e^{-x\left(\frac{v}{1-v}+\frac{u}{1-u}\right)} = \left(\frac{1}{(1-v)(1-u)}\right) e^{-\left(\frac{xv}{1-v}+\frac{xu}{1-u}\right)}$$

$$= \left(\sum_{m=0}^{\infty}\left(\frac{L_m(x)}{m!}\right) v^m\right)\left(\sum_{n=0}^{\infty}\left(\frac{L_n(x)}{n!}\right) u^m\right)$$

$$= \sum_{m=0}^{\infty}\sum_{n=0}^{\infty}\left(\frac{v^m u^n}{m!n!}\right) L_m(x) L_n(x) \tag{8.145}$$

Integration of Eq.(8.145) by multiplying e^{-x}, we have

$$\sum_{m=0}^{\infty}\sum_{n=0}^{\infty}\left(\frac{v^m u^n}{m!n!}\right) \int_0^\infty e^{-x} L_m(x) L_n(x) \, \mathrm{d}x = \int_0^\infty \left(\sum_{m=0}^{\infty}\sum_{n=0}^{\infty}\left(\frac{v^m u^n}{m!n!}\right) e^{-x} L_m(x) L_n(x)\right) \mathrm{d}x$$

$$= \int_0^\infty e^{-x} \left(\sum_{m=0}^{\infty}\sum_{n=0}^{\infty}\left(\frac{v^m u^n}{m!n!}\right) L_m(x) L_n(x)\right) \mathrm{d}x$$

$$= \int_0^\infty e^{-x} \left(\frac{1}{(1-v)(1-u)}\right) e^{-x\left(\frac{v}{1-v}+\frac{u}{1-u}\right)} \, \mathrm{d}x$$

$$= \frac{1}{(1-v)(1-u)} \int_0^\infty \left(e^{-x\left(1+\frac{v}{1-v}+\frac{u}{1-u}\right)} \right) dx$$

$$= \frac{1}{(1-u)(1-v)+v(1-u)+u(1-v)}$$

$$= \frac{1}{1-uv}$$

$$= \sum_{i=0}^{\infty} (uv)^i \tag{8.146}$$

Comparing the coefficients of uv in Eq. (8.146), we have

$$\int_0^\infty e^{-x} L_m^2(x)\, dx = (m!)^2$$

8.6.3 Recurrence Relation

Theorem 8.6.5 If $L_m(x)$ is the Laguerre polynomial of degree m, then

$$(2m+1-x)L_m(x) - L_{m+1}(x) = m^2 L_{m+1}(x)$$

Proof: According to Theorem 8.6.1, we have

$$\left(\frac{1}{1-t}\right) e^{-\frac{xt}{1-t}} = \sum_{m=0}^{\infty} \left(\frac{L_m(x)}{m!}\right) t^m \tag{8.147}$$

Differentiation of Eq.(8.147) with respect to t gives

$$\sum_{m=1}^{\infty} \left(\frac{L_m(x)}{(m-1)!}\right) t^{m-1} = \left(\frac{1}{1-t}\right)^2 e^{-\frac{xt}{1-t}} - \left(\frac{x}{(1-t)^3}\right) e^{-\frac{xt}{1-t}}$$

$$= \left(\frac{1}{1-t}\right) \left(\sum_{m=0}^{\infty} \left(\frac{L_m(x)}{m!}\right) t^m \right)$$

$$- \left(\frac{x}{(1-t)^2}\right) \left(\sum_{m=0}^{\infty} \left(\frac{L_m(x)}{m!}\right) t^m \right) \tag{8.148}$$

Multiplying Eq.(8.148) by $(1-t)^2$, we have

$$(1-t)^2 \left(\sum_{m=1}^{\infty} \left(\frac{L_m(x)}{(m-1)!}\right) t^{m-1} \right) = (1-t) \left(\sum_{m=0}^{\infty} \left(\frac{L_m(x)}{m!}\right) t^m \right)$$

$$- x \left(\sum_{m=0}^{\infty} \left(\frac{L_m(x)}{m!}\right) t^m \right)$$

Comparing the coefficients of t^m, we have

$$(2m+1-x)L_m(x) = L_{m+1}(x) + m^2 L_{m-1}(x)$$

Theorem 8.6.6 If $L_m(x)$ is the Laguerre polynomial of degree m, then

$$L'_m(x) = mL'_{m-1}(x) - mL_{m-1}(x)$$

Proof: According to Theorem 8.6.1, we have

$$\left(\frac{1}{1-t}\right)e^{-\frac{xt}{1-t}} = \sum_{m=0}^{\infty}\left(\frac{L_m(x)}{m!}\right)t^m \tag{8.149}$$

Differentiation of Eq.(8.149) with respect to x gives

$$\begin{aligned}\sum_{m=1}^{\infty}\left(\frac{L'_m(x)}{m!}\right)t^m &= \left(-\frac{t}{1-t}\right)\left(e^{-\frac{xt}{1-t}}\right)\\ &= \left(-\frac{t}{1-t}\right)\left(\sum_{m=1}^{\infty}\left(\frac{L_m(x)}{m!}\right)t^m\right)\end{aligned}$$

In other words, we have

$$(1-t)\left(\sum_{m=1}^{\infty}\left(\frac{L'_m(x)}{m!}\right)t^m\right) = -t\left(\sum_{m=1}^{\infty}\left(\frac{L_m(x)}{m!}\right)t^m\right) \tag{8.150}$$

Comparing the coefficients of t^m in Eq.(8.150), we have

$$L'_m(x) = mL'_{m-1}(x) - mL_{m-1}(x)$$

Exercises 8.6

1. If $u(x)$ and $v(x)$ are any two differentiable functions, then show that the *Leibnitz rule of differentiation* is
$$\frac{d^n}{dx^n}(uv) = \sum_{r=0}^{n}\binom{n}{r}\left(\frac{d^r}{dx^r}(u)\right)\left(\frac{d^{n-r}}{dx^{n-r}}(v)\right)$$
2. If $L_m(x)$ is the Laguerre polynomial of degree m, then show that $L_0(x) = 1$.
3. If $L_m(x)$ is the Laguerre polynomial of degree m, then show that $L_m(0) = m!$.
4. If $L_m(x)$ is the Laguerre polynomial of degree m, then show that $L'_m(0) = -m!m$.
5. If $L_m(x)$ is the Laguerre polynomial of degree m, then show that $L''_m(0) = \frac{m!m(m-1)}{2}$.

Chapter 9

Laplace Transform

There are numerous integral transforms that have been developed over the years, many of which are highly specialized. The most versatile of integral transforms is the Laplace transform. Let S be the set of all exponentially bounded real valued functions. Define a transform $\mathcal{L} : S \to \mathbb{R}$ by $\mathcal{L}(f(t)) = \int_0^\infty e^{-st} f(t)\,dt$. Clearly, we have $\mathcal{L}(f(t)) \leq \frac{m}{s-a}$ whenever $f(t) \leq me^{at}$ and $s > a \geq 0$. The above defined transform, called the *Laplace transform*, has the following properties:

1. $\mathcal{L}(f(t) + g(t)) = \mathcal{L}(f(t)) + \mathcal{L}(g(t))$
2. $\mathcal{L}(\alpha f(t)) = \alpha \mathcal{L}(f(t))$

9.1 Shifting Formula

One can find the Laplace transform of some exponentially bounded function $f(t)$ directly by definition. Sometimes it becomes complicated and time consuming while evaluating the integration. There are certain rules by which one can evaluate the complicated integration in terms of some simple known integration called *shifting formula*. The first idea behind the shifting formula is the *unit step function* $u_a(t)$. Let $a \geq 0$. Define a function $u_a : [0, \infty) \to \mathbb{R}$ by

$$u_a(t) = \begin{cases} 0 & t < a \\ 1 & t > a \end{cases}$$

According to the definition of the Laplace transform, we have

$$\begin{aligned} \mathcal{L}(u_a(t)) &= \int_0^\infty e^{-st} u_a(t)\,dt \\ &= \int_0^a e^{-st} u_a(t)\,dt + \int_a^\infty e^{-st} u_a(t)\,dt \end{aligned}$$

$$= \int_a^\infty e^{-st} u_a(t)\, dt$$

$$= \int_a^\infty e^{-st}\, dt$$

$$= \frac{e^{-sa}}{s}$$

Once the Laplace transform $\mathcal{L}(f(t))$ of some exponentially bounded function $f(t)$ is known, the behavioural relation between $f(t)$ and $\mathcal{L}(f(t))$ can be established through initial value and final value theorems.

Theorem 9.1.1 (First Shifting Formula) Let $s \geq a$, and $F(s) = \mathcal{L}(f(t))$. If $g(t) = e^{at} f(t)$, then $\mathcal{L}(g(t)) = F(s-a)$.

Proof: According to the definition of the Laplace transform, we have

$$\mathcal{L}(g(t)) = \int_0^\infty g(t) e^{-st}\, dt$$

$$= \int_0^\infty e^{at} e^{-st} f(t)\, dt$$

$$= \int_0^\infty e^{(a-s)t} f(t)\, dt$$

$$= \int_0^\infty e^{-(s-a)t} f(t)\, dt$$

$$= F(s-a)$$

EXAMPLE 9.1 Find the Laplace transform of $g(t) = e^{at}$.

Solution: Let $f(t) = 1$. According to the definition of Laplace transform, we have

$$F(s) = \mathcal{L}(f(t))$$

$$= \int_0^\infty e^{-st} f(t)\, dt$$

$$= \int_0^\infty e^{-st}\, dt$$

$$= \frac{1}{s}$$

In other words, $\mathcal{L}(f(t)) = \frac{1}{s} = F(s)$. Again, $g(t) = e^{at} = e^{at}f(t)$. According to the first shifting Theorem 9.1.1, we have

$$\begin{aligned}\mathcal{L}(g(t)) &= F(s-a)\\ &= \frac{1}{s-a}\end{aligned}$$

EXAMPLE 9.2 Find the Laplace transform of $g(t) = e^{-at}$.

Solution: Let $f(t) = 1$. According to the definition of Laplace transform, we have

$$\begin{aligned}F(s) &= \mathcal{L}(f(t))\\ &= \int_0^\infty e^{-st} f(t)\,\mathrm{d}t\\ &= \int_0^\infty e^{-st}\,\mathrm{d}t\\ &= \frac{1}{s}\end{aligned}$$

In other words, $\mathcal{L}(f(t)) = \frac{1}{s} = F(s)$. Again, $g(t) = e^{-at} = e^{-at}f(t)$. According to the first shifting Theorem 9.1.1, we have

$$\begin{aligned}\mathcal{L}(g(t)) &= F(s+a)\\ &= \frac{1}{s+a}\end{aligned}$$

EXAMPLE 9.3 Find the Laplace transform of $g(t) = 1$.

Solution: Let $f(t) = e^{at}$. According to Example 9.1, we have

$$\begin{aligned}F(s) &= \mathcal{L}(f(t))\\ &= \frac{1}{s-a}\end{aligned}$$

Again, $g(t) = 1 = e^{-at}e^{at} = e^{-at}f(t)$. According to the first shifting Theorem 9.1.1, we have

$$\begin{aligned}\mathcal{L}(g(t)) &= F(s+a)\\ &= \frac{1}{s-a+a}\\ &= \frac{1}{s}\end{aligned}$$

Theorem 9.1.2 (Second Shifting Formula) Let $a \geq 0$ and $F(s) = \mathcal{L}(f(t))$. If $g(t) = f(t-a)u_a(t)$, then $\mathcal{L}(g(t)) = e^{-sa}F(s)$.

Proof: According to the definition of the Laplace transform, we have

$$\begin{aligned}
\mathcal{L}(g(t)) &= \int_0^\infty g(t)e^{-st}\,\mathrm{dt} \\
&= \int_0^a g(t)e^{-st}\,\mathrm{dt} + \int_a^\infty g(t)e^{-st}\,\mathrm{dt} \\
&= \int_a^\infty e^{-st}f(t-a)\,\mathrm{dt} \\
&= \int_0^\infty e^{-s(u+a)}f(u)\,\mathrm{du} \\
&= e^{-sa}\left(\int_0^\infty e^{-su}f(u)\,\mathrm{du}\right) \\
&= e^{-sa}F(s)
\end{aligned}$$

EXAMPLE 9.4 Find the Laplace transform of the function $f(t)$ defined by

$$f(t) = \begin{cases} 2, & t < a \\ t, & t > a \end{cases}$$

using the second shifting formula.

Solution: One can express the function $f(t)$ in terms of the function $u_a(t)$. Hence we have $f(t) = 2u_0(t) - 2u_a(t) + tu_a(t)$. According to the definition of the Laplace transform, we have

$$\begin{aligned}
\mathcal{L}(f(t)) &= \mathcal{L}\left(2u_0(t) - 2u_a(t) + tu_a(t)\right) \\
&= 2\mathcal{L}\left(u_0(t)\right) - 2\mathcal{L}\left(u_a(t)\right) + \mathcal{L}\left(tu_a(t)\right) \\
&= 2\mathcal{L}\left(u_0(t)\right) - 2\mathcal{L}\left(u_a(t)\right) + \mathcal{L}\left((t-a)u_a(t)\right) + \mathcal{L}(au_a(t)) \\
&= \frac{2}{s} - \frac{2}{s}e^{-sa} + \frac{1}{s^2}e^{-sa} + \frac{a}{s}e^{-sa}
\end{aligned}$$

EXAMPLE 9.5 Find the Laplace transform of the function $f(t)$ defined by

$$f(t) = \begin{cases} 0, & 0 < t < a \\ \alpha, & a < t < b \\ 0, & t < b \end{cases}$$

using the second shifting formula.

Solution: One can express the function $f(t)$ by $f(t) = \alpha u_a(t) - \alpha u_b(t)$. According to the definition of the Laplace transform, we have

$$\begin{aligned}
\mathcal{L}(f(t)) &= \mathcal{L}\left(\alpha u_a(t) - \alpha u_b(t)\right) \\
&= \alpha \mathcal{L}\left(u_a(t)\right) - \alpha \mathcal{L}\left(u_b(t)\right) \\
&= \frac{\alpha}{s} e^{-sa} - \frac{\alpha}{s} e^{-sb} \\
&= \frac{\alpha}{s}\left(e^{-as} - e^{-bs}\right)
\end{aligned}$$

EXAMPLE 9.6 Find the Laplace transform of the function

$$f(t) = \begin{cases} \frac{\alpha}{a} t & 0 < t < a \\ \frac{\alpha}{a}(2a - t) & a < t < 2a \\ 0 & \text{otherwise} \end{cases}$$

where α and a are constants, using the definition of the Laplace transform.

Solution: According to the definition of the Laplace transform, we have

$$\begin{aligned}
\mathcal{L}(f(t)) &= \int_0^\infty f(t)\, e^{-st}\, dt \\
&= \int_0^a f(t)\, e^{-st}\, dt + \int_a^{2a} f(t)\, e^{-st}\, dt + \int_{2a}^\infty f(t)\, e^{-st}\, dt \\
&= \int_0^a \left(\frac{\alpha}{a}\right) t\, e^{-st}\, dt + \int_a^{2a} \left(\frac{\alpha}{a}\right)(2a - t)\, e^{-st}\, dt \\
&= -\left(\frac{\alpha}{s}\right) e^{-sa} + \frac{\alpha}{as} \int_0^a e^{-st}\, dt + \left(\frac{\alpha}{s}\right) e^{-sa} - \frac{\alpha}{sa} \int_a^{2a} e^{-st}\, dt \\
&= \frac{\alpha}{as^2} - \left(\frac{2\alpha}{as^2}\right) e^{-sa} + \left(\frac{\alpha}{as^2}\right) e^{-2sa} \\
&= \frac{\alpha}{as^2}\left(1 - 2e^{-sa} + e^{-2sa}\right) \\
&= \frac{\alpha}{as^2}\left(1 - e^{-sa}\right)^2
\end{aligned}$$

EXAMPLE 9.7 Find the Laplace transform of the function $f(t)$ defined by

$$f(t) = \begin{cases} \frac{\alpha}{a}\, t & 0 < t < a \\ \frac{\alpha}{a}\,(2a - t) & a < t < 2a \\ 0 & \text{otherwise} \end{cases}$$

where α and a are constants, using the second shifting formula.

Solution: It is easy to verify that $f(t) = \left(\frac{\alpha}{a}\right) tu_0(t) - 2\left(\frac{\alpha}{a}\right)(t-a)u_a(t) + \left(\frac{\alpha}{a}\right)(t-2a)u_{2a}(t)$. According to the second shifting Theorem 9.1.2, we have

$$\begin{aligned} \mathcal{L}(f(t)) &= \left(\frac{\alpha}{a}\right)\frac{1}{s^2} - 2\left(\frac{\alpha}{a}\right)\left(\frac{1}{s^2}\right)e^{-as} + e^{-2as}\left(\frac{\alpha}{a}\right)\left(\frac{1}{s^2}\right) \\ &= \frac{\alpha}{as^2}\left(1 - 2e^{-as} + e^{-2as}\right) \\ &= \frac{\alpha}{as^2}\left(1 - e^{-as}\right)^2 \end{aligned}$$

EXAMPLE 9.8 Find the Laplace transform of the function $f(t)$ defined by

$$f(t) = \begin{cases} 0, & 0 < t < 1 \\ t^2, & 1 < t < 2 \\ 0, & 2 < t \end{cases}$$

Solution: It is easy to verify that

$$\begin{aligned} f(t) &= t^2u_1(t) - t^2u_2(t) \\ &= u_1(t)\left((t-1)^2 + 2(t-1) + 1\right) - u_2(t)\left((t-2)^2 + 4(t-2) - 4\right) \\ &= (t-1)^2u_1(t) + 2(t-1)u_1(t) + u_1(t) - (t-2)^2u_2(t) \\ &\quad - 4(t-2)u_2(t) - 4u_2(t) \end{aligned}$$

According to the second shifting Theorem 9.1.2, we have

$$\begin{aligned} \mathcal{L}(f(t)) &= \mathcal{L}\left(u_1(t)\left((t-1)^2 + 2(t-1) + 1\right)\right) - \mathcal{L}\left(u_2(t)\left((t-2)^2 + 4(t-2) + 4\right)\right) \\ &= e^{-s}\left(\frac{2}{s^3} + \frac{2}{s^2} - \frac{1}{s}\right) - e^{-2s}\left(\frac{2}{s^3} + \frac{4}{s^2} + \frac{4}{s}\right) \end{aligned}$$

EXAMPLE 9.9 Find the Laplace transform of the function $f(t)$ defined by

$$f(t) = \begin{cases} 0, & 0 < t < a \\ \frac{1}{\epsilon}, & a < t < a + \epsilon \\ 0, & t < a + \epsilon \end{cases}$$

using the second shifting formula.

Solution: According to Example 9.5, we have

$$\mathcal{L}(f(t)) = \frac{1}{s\epsilon}\left(e^{-as} - e^{-(a+\epsilon)s}\right)$$
$$= \frac{e^{-as}}{s\epsilon}\left(1 - e^{-s\epsilon}\right)$$

Theorem 9.1.3 If $F(s) = \mathcal{L}(f(t))$ and $n \in \mathbb{N}$, then $\mathcal{L}(t^n f(t)) = (-1)^n \left(\frac{\mathrm{d}^n}{\mathrm{d}s^n}\right) F(s)$.

Proof: One can prove this theorem by the method of induction on n. According to the definition of the Laplace transform, we have

$$F(s) = \int_0^\infty e^{-st} f(t)\,\mathrm{d}t \tag{9.1}$$

Again, differentiation of Eq.(9.1) with respect to s gives

$$F'(s) = \int_0^\infty -e^{-st} t f(t)\,\mathrm{d}t$$
$$= -\int_0^\infty e^{-st} t f(t)\,\mathrm{d}t$$
$$= -\mathcal{L}(t f(t))$$

Hence the given statement is true for $n = 1$. Assume that the given statement is true for $n = r$. In other words, we have

$$(-1)^r \left(\frac{\mathrm{d}^r}{\mathrm{d}s^r}\right) F(s) = \mathcal{L}(t^r f(t))$$
$$= \int_0^\infty e^{-st} t^r f(t)\,\mathrm{d}t \tag{9.2}$$

Again, differentiation of Eq.(9.2) with respect to s gives

$$(-1)^{r+1} \left(\frac{\mathrm{d}^{r+1}}{\mathrm{d}s^{r+1}}\right) F(s) = -\frac{\mathrm{d}}{\mathrm{d}s}\int_0^\infty e^{-st} t^r f(t)\,\mathrm{d}t$$
$$= -\int_0^\infty -e^{-st} t^{r+1} f(t)\,\mathrm{d}t$$
$$= \int_0^\infty e^{-st} t^{r+1} f(t)\,\mathrm{d}t$$
$$= \mathcal{L}(t^{r+1} f(t))$$

Corollary 9.1.1 If $F(s) = \mathcal{L}(f(t))$, then $\mathcal{L}(tf(t)) = -F'(s)$.

Proof: If one takes $n = 1$ in Theorem 9.1.3, then we have $\mathcal{L}(tf(t)) = -F'(s)$.

Theorem 9.1.4 If $F(s) = \mathcal{L}(f(t))$, then $\mathcal{L}\left(\frac{f(t)}{t}\right) = \int\limits_s^\infty F(x)\,\mathrm{d}x$.

Proof: According to the definition of the Laplace transform, we have

$$F(x) = \int\limits_0^\infty e^{-xt} f(t)\,\mathrm{d}t$$

Again, we have

$$\begin{aligned}
\int\limits_s^\infty F(x)\,\mathrm{d}x &= \int\limits_s^\infty \left(\int\limits_0^\infty e^{-xt} f(t)\,\mathrm{d}t\right) \mathrm{d}x \\
&= \int\limits_0^\infty f(t) \left(\int\limits_s^\infty e^{-xt}\,\mathrm{d}x\right) \mathrm{d}t \\
&= \int\limits_0^\infty f(t) \left(-\left(\frac{e^{-xt}}{t}\right)\Big|_s^\infty\right) \mathrm{d}t \\
&= \int\limits_0^\infty f(t) \left(\frac{e^{-st}}{t}\right) \mathrm{d}t \\
&= \int\limits_0^\infty e^{-st} \left(\frac{f(t)}{t}\right) \mathrm{d}t \\
&= \mathcal{L}\left(\frac{f(t)}{t}\right)
\end{aligned}$$

Theorem 9.1.5 If $\mathcal{L}(f(t)) = F(s)$, then $\mathcal{L}\left(\int\limits_0^t f(u)\,\mathrm{d}u\right) = \frac{F(s)}{s}$.

Proof: Clearly, we have according to the definition of $F(s)$,

$$F(s) = \int\limits_0^\infty e^{-su} f(u)\,\mathrm{d}u$$

According to the definition of the Laplace transform, we have

$$\mathcal{L}\left(\int_0^t f(u)\,\mathrm{d}u\right) = \int_0^\infty e^{-st}\left(\int_0^t f(u)\,\mathrm{d}u\right)\mathrm{d}t$$

$$= \int_0^\infty \int_u^\infty e^{-st} f(u)\,\mathrm{d}t\,\mathrm{d}u$$

$$= \int_0^\infty \left(\int_u^\infty e^{-st}\,\mathrm{d}t\right) f(u)\,\mathrm{d}u$$

$$= \int_0^\infty \left(\frac{e^{-su}}{s}\right) f(u)\,\mathrm{d}u$$

$$= \left(\frac{1}{s}\right)\int_0^\infty e^{-su} f(u)\,\mathrm{d}u$$

$$= \left(\frac{1}{s}\right) F(s)$$

$$= \frac{F(s)}{s}$$

EXAMPLE 9.10 Find the Laplace transform of $g(t) = \int_0^t e^{au}\,\mathrm{d}u$.

Solution: Let $f(t) = e^{at}$. Clearly, $\mathcal{L}(f(t)) = \frac{1}{s-a} = F(s)$ according to Example 9.1. According to Theorem 9.1.5, we have

$$\mathcal{L}\left(\int_0^t e^{au}\,\mathrm{d}u\right) = \frac{F(s)}{s}$$

$$= \frac{1}{s(s-a)}$$

EXAMPLE 9.11 Find the Laplace transform of $g(t) = \int_0^t e^{-au}\,\mathrm{d}u$.

Solution: Let $f(t) = e^{-at}$. Clearly, $\mathcal{L}(f(t)) = \frac{1}{s+a} = F(s)$ according to Example 9.2. According to Theorem 9.1.5, we have

$$\mathcal{L}\left(\int_0^t e^{-au}\,\mathrm{d}u\right) = \frac{F(s)}{s}$$

$$= \frac{1}{s(s+a)}$$

Theorem 9.1.6 The Laplace transform of the *Dirac delta* function $\delta_a(t)$ with impulse at time $t = a > 0$, defined by

$$\delta_a(t) = \begin{cases} 0 & t \neq a \\ \infty & t = a \end{cases}$$

is e^{-as}.

Proof: Define the function $\delta_a(\epsilon,\, t)$ by

$$\delta_a(\epsilon,\, t) = \begin{cases} 0 & 0 < t < a \\ \frac{1}{\epsilon} & a < t < a + \epsilon \\ 0 & a + \epsilon < t \end{cases}$$

According to Example 9.9, we have

$$\mathcal{L}(\delta_a(\epsilon,\, t)) = \left(\frac{1}{s\epsilon}\right) e^{-as}\left(1 - e^{-s\epsilon}\right)$$

In other words, we have

$$\begin{aligned} \mathcal{L}(\delta_a(t)) &= \lim_{\epsilon \to 0} \mathcal{L}(\delta_a(\epsilon,\, t)) \\ &= e^{-as} \end{aligned}$$

Corollary 9.1.2 The Laplace transform of the *Dirac delta* function $\delta_0(t)$, with impulse at time $t = 0$, defined by

$$\delta_0(t) = \begin{cases} 0 & t \neq 0 \\ \infty & t = 0 \end{cases}$$

is 1.

Proof: According to Theorem 9.1.6, it is easy to verify that

$$\begin{aligned} \mathcal{L}(\delta_0(t)) &= \lim_{a \to 0} \mathcal{L}(\delta_a(t)) \\ &= \lim_{a \to 0} e^{-as} \\ &= 1 \end{aligned}$$

Theorem 9.1.7 If $F(s) = \mathcal{L}(f(t))$, then $\mathcal{L}\left(f\left(at\right)\right) = \frac{1}{a}F\left(\frac{s}{a}\right)$.

Proof: According to the definition of the Laplace transform, we have

$$\begin{aligned} \mathcal{L}(f(at)) &= \int_0^{\infty} e^{-st} f(at)\, \mathrm{d}t \\ &= \frac{1}{a}\int_0^{\infty} e^{-\left(\frac{s}{a}\right)t} f(u)\, \mathrm{d}u \\ &= \frac{1}{a}F\left(\frac{s}{a}\right) \end{aligned}$$

Theorem 9.1.8 (Initial Value) If $F(s) = \mathcal{L}(f(t))$, then $\lim_{t\to 0} f(t) = \lim_{s\to\infty} sF(s)$.

Proof: According to the definition of $\mathcal{L}(f'(t))$, we have

$$\mathcal{L}\left(f'(t)\right) = \int_0^\infty e^{-st} f'(t)\,dt \tag{9.3}$$

Hence one can conclude that $\lim_{s\to\infty} \mathcal{L}\left(f'(t)\right) = 0$ by Eq.(9.3). It is easy to verify that

$$\begin{aligned}\mathcal{L}\left(f'(t)\right) &= \int_0^\infty e^{-st} f'(t)\,dt \\ &= s\int_0^\infty e^{-st} f(t)\,dt - f(0) \\ &= sF(s) - f(0) \qquad (9.4)\end{aligned}$$

Hence one can conclude that $\lim_{s\to\infty} \mathcal{L}\left(f'(t)\right) = \lim_{s\to\infty} sF(s) - f(0)$ by Eq.(9.4). Comparing both limiting values, one can conclude that $\lim_{s\to\infty} sF(s) = f(0) = \lim_{t\to 0} f(t)$.

Theorem 9.1.9 (Final Value) If $F(s) = \mathcal{L}(f(t))$, then $\lim_{t\to\infty} f(t) = \lim_{s\to 0} sF(s)$.

Proof: According to the definition of $\mathcal{L}(f'(t))$, we have

$$\mathcal{L}\left(f'(t)\right) = \int_0^\infty e^{-st} f'(t)\,dt \tag{9.5}$$

Hence one can conclude that $\lim_{s\to 0} \mathcal{L}\left(f'(t)\right) = \lim_{t\to\infty} f(t) - f(0)$ by Eq.(9.5). It is easy to verify that

$$\begin{aligned}\mathcal{L}\left(f'(t)\right) &= \int_0^\infty e^{-st} f'(t)\,dt \\ &= s\int_0^\infty e^{-st} f(t)\,dt - f(0) \\ &= sF(s) - f(0) \qquad (9.6)\end{aligned}$$

Hence one can conclude that $\lim_{s\to 0} \mathcal{L}\left(f'(t)\right) = \lim_{s\to 0} sF(s) - f(0)$ by Eq.(9.6). Comparing both limiting values, one can conclude that $\lim_{t\to\infty} f(t) = \lim_{s\to 0} sF(s)$.

Exercises 9.1

1. Find the Laplace transform of the function

$$f(t) = \begin{cases} 1, & t < \pi \\ 0, & \pi < t < 2\pi \\ \cos t, & t > 2\pi \end{cases}$$

2. Find the Laplace transform of the function

$$f(t) = \begin{cases} 1, & t < \pi \\ t, & \pi < t < 2\pi \\ \cos t, & t > 2\pi \end{cases}$$

3. Find the Laplace transform of the function

$$f(t) = \begin{cases} 1, & t < \pi \\ 0, & \pi < t < 2\pi \\ \sin t, & t > 2\pi \end{cases}$$

4. Find the Laplace transform of the function

$$f(t) = \begin{cases} 1, & t < \pi \\ t, & \pi < t < 2\pi \\ \sin t, & t > 2\pi \end{cases}$$

5. Find the Laplace transform of the function

$$f(t) = \begin{cases} 0, & t < \pi \\ t - \pi, & t > \pi \end{cases}$$

6. Find the Laplace transform of the function

$$f(t) = \begin{cases} 0, & t < \pi \\ (t - \pi)^2, & t > \pi \end{cases}$$

7. Find the Laplace transform of the function

$$f(t) = \begin{cases} 0, & t < \pi \\ t - \pi, & \pi < t < 2\pi \\ \frac{t}{2}, & t > 2\pi \end{cases}$$

8. Find the Laplace transform of the function

$$f(t) = \begin{cases} 0, & t < \pi \\ \pi - t, & \pi < t < 2\pi \\ t - \pi, & t > 2\pi \end{cases}$$

9. Find the Laplace transform of the function

$$f(t) = \begin{cases} 0, & t < 2 \\ 2t, & 2 < t < 4 \\ 0, & t > 4 \end{cases}$$

10. Find the Laplace transform of the function

$$f(t) = \begin{cases} \sin t, & t < \pi \\ \sin 2t, & t > 2\pi \end{cases}$$

11. Find the Laplace transform of the function

$$f(t) = \begin{cases} \cos t, & t < \pi \\ \cos 2t, & t > 2\pi \end{cases}$$

9.2 Laplace Transform of Some Functions

In this section, we have derived the Laplace transforms of some standard functions. If one knows the Laplace transform of some standard functions, then with the help of the shifting formulae one can calculate the Laplace transform of any arbitrary function.

EXAMPLE 9.12 If $f(t) = t^{-\frac{1}{2}}$, then $\mathcal{L}(f(t)) = \sqrt{\frac{\pi}{s}}$.

Solution: According to the definition of the Laplace transform, we have

$$\begin{aligned} \mathcal{L}(f(t)) &= \mathcal{L}(t^{-\frac{1}{2}}) \\ &= \int_0^\infty e^{-st} t^{-\frac{1}{2}}\, dt \end{aligned} \tag{9.7}$$

Replacing $st = x$ in Eq.(9.7), we have

$$\mathcal{L}(t^{-\frac{1}{2}}) = s^{-\frac{1}{2}} \int_0^\infty e^{-x} x^{-\frac{1}{2}} dx \tag{9.8}$$

Substituting $x = u^2$ in Eq.(9.8) and using Exercise 1 of Exercises 9.2, we have

$$\begin{aligned}\mathcal{L}(t^{-\frac{1}{2}}) &= 2s^{-\frac{1}{2}}\int_0^\infty e^{-u^2}\,\mathrm{d}u \\ &= 2s^{-\frac{1}{2}}\left(\frac{\sqrt{\pi}}{2}\right) \\ &= s^{-\frac{1}{2}}\sqrt{\pi} \\ &= \sqrt{\frac{\pi}{s}}\end{aligned}$$

Theorem 9.2.1 If $f(t) = t^{n+\frac{1}{2}}$ where $n \in \mathbb{N}$, then $\mathcal{L}(f(t)) = \frac{\Gamma(n+1+\frac{1}{2})}{s^{n+1+\frac{1}{2}}}$ where the *gamma function* is defined by $\Gamma(n) = \int_0^\infty u^{n-1}e^{-u}\,\mathrm{d}u$.

Proof: One can prove this theorem by the method of induction on n. If $n = 0$, then $\mathcal{L}\left(t^{\frac{1}{2}}\right) = \left(\frac{1}{2s}\right)\sqrt{\frac{\pi}{s}} = \frac{\Gamma(1+\frac{1}{2})}{s^{1+\frac{1}{2}}}$ by Exercise 2 of Exercises 9.2. Assume that the statement is true for $n = r$. In other words, we have

$$\begin{aligned}\mathcal{L}\left(t^{r+\frac{1}{2}}\right) &= \int_0^\infty e^{-st}t^{r+\frac{1}{2}}\,\mathrm{d}t \\ &= \frac{\Gamma\left(r+1+\frac{1}{2}\right)}{s^{r+1+\frac{1}{2}}} \qquad (9.9)\end{aligned}$$

It is easy to verify from Eq.(9.9) that

$$\begin{aligned}\mathcal{L}\left(t^{r+1+\frac{1}{2}}\right) &= \int_0^\infty e^{-st}t^{r+1+\frac{1}{2}}\,\mathrm{d}t \\ &= -\frac{1}{s}\int_0^\infty t^{r+1+\frac{1}{2}}\,\mathrm{d}\left(e^{-st}\right) \\ &= \frac{r+1+\frac{1}{2}}{s}\int_0^\infty e^{-st}t^{r+\frac{1}{2}}\,\mathrm{d}t \\ &= \frac{r+1+\frac{1}{2}}{s}\left(\frac{\Gamma\left(r+1+\frac{1}{2}\right)}{s^{r+1+\frac{1}{2}}}\right) \\ &= \frac{\Gamma\left(r+1+1+\frac{1}{2}\right)}{s^{r+1+1+\frac{1}{2}}}\end{aligned}$$

EXAMPLE 9.13 Show that if $f(t) = \sin\sqrt{t}$, then $\mathcal{L}(f(t)) = \frac{1}{2s}\sqrt{\frac{\pi}{s}}e^{-\frac{1}{4s}}$.

Solution: According to expansion of $\sin\sqrt{t}$, we have

$$\begin{aligned} f(t) &= \sin\sqrt{t} \\ &= \sum_{n=0}^{\infty}(-1)^n\left(\frac{t^{n+\frac{1}{2}}}{(2n+1)!}\right) \end{aligned} \tag{9.10}$$

Applying the Laplace transform to Eq.(9.10), we have

$$\begin{aligned} \mathcal{L}(f(t)) &= \mathcal{L}\left(\sum_{n=0}^{\infty}(-1)^n\left(\frac{t^{n+\frac{1}{2}}}{(2n+1)!}\right)\right) \\ &= \sum_{n=0}^{\infty}(-1)^n\mathcal{L}\left(\frac{t^{n+\frac{1}{2}}}{(2n+1)!}\right) \\ &= \sum_{n=0}^{\infty}(-1)^n\left(\frac{1}{(2n+1)!}\right)\mathcal{L}\left(t^{n+\frac{1}{2}}\right) \\ &= \sum_{n=0}^{\infty}(-1)^n\left(\frac{\Gamma\left(n+1+\frac{1}{2}\right)}{(2n+1)!}\right)\left(\frac{1}{s^{n+1+\frac{1}{2}}}\right) \\ &= \frac{1}{2s}\sqrt{\frac{\pi}{s}}\left(\sum_{n=0}^{\infty}(-1)^n\left(\frac{1}{n!}\right)\left(\frac{1}{4s}\right)^n\right) \\ &= \frac{1}{2s}\sqrt{\frac{\pi}{s}}e^{-\frac{1}{4s}} \end{aligned}$$

EXAMPLE 9.14 Show that if $g(t) = \frac{\cos\sqrt{t}}{\sqrt{t}}$, then $\mathcal{L}(g(t)) = \sqrt{\frac{\pi}{s}}e^{-\frac{1}{4s}}$.

Solution: Let $f(t) = \sin\sqrt{t}$. Clearly, $f'(t) = \frac{\cos\sqrt{t}}{2\sqrt{t}}$ and $g(t) = 2f'(t)$. Again, $\mathcal{L}(f(t)) = F(s) = \frac{1}{2s}\sqrt{\frac{\pi}{s}}e^{-\frac{1}{4s}}$ by Example 9.13. It is easy to verify that

$$\begin{aligned} \mathcal{L}(g(t)) &= \mathcal{L}\left(\frac{\cos\sqrt{t}}{\sqrt{t}}\right) \\ &= 2\mathcal{L}(f'(t)) \\ &= 2(sF(s) - f(0)) \\ &= 2sF(s) \\ &= 2s\left(\frac{1}{2s}\sqrt{\frac{\pi}{s}}e^{-\frac{1}{4s}}\right) \\ &= \sqrt{\frac{\pi}{s}}e^{-\frac{1}{4s}} \end{aligned}$$

EXAMPLE 9.15 Find the Laplace transform of $g(t) = \frac{\cos\sqrt{at}}{\sqrt{at}}$.

Solution: Let $f(t) = \frac{\cos\sqrt{t}}{\sqrt{t}}$. Clearly, $f(at) = g(t)$ and $\mathcal{L}(f(t)) = \sqrt{\frac{\pi}{s}}e^{-\frac{1}{4s}} = F(s)$ by Example 9.14. According to Theorem 9.1.7, we have

$$\begin{aligned}\mathcal{L}(g(t)) &= \mathcal{L}(f(at)) \\ &= \frac{1}{a}F\left(\frac{s}{a}\right) \\ &= \sqrt{\frac{\pi}{as}}e^{-\frac{a}{4s}}\end{aligned}$$

EXAMPLE 9.16 Find the Laplace transform of $g(t) = \frac{\cos a\sqrt{t}}{\sqrt{t}}$.

Solution: Clearly, $g(t) = a\left(\frac{\cos\sqrt{a^2t}}{\sqrt{a^2t}}\right)$. Let $f(t) = a\left(\frac{\cos\sqrt{t}}{\sqrt{t}}\right)$. Clearly, $f(a^2t) = g(t)$ and $\mathcal{L}(f(t)) = a\sqrt{\frac{\pi}{s}}e^{-\frac{1}{4s}} = F(s)$ by Example 9.14. According to Theorem 9.1.7, we have

$$\begin{aligned}\mathcal{L}(g(t)) &= \mathcal{L}(f(a^2t)) \\ &= \frac{1}{a^2}F\left(\frac{s}{a^2}\right) \\ &= \sqrt{\frac{\pi}{s}}e^{-\frac{a^2}{4s}}\end{aligned}$$

EXAMPLE 9.17 Show that if $f(t) = \frac{2}{\sqrt{\pi}}\int\limits_0^{\sqrt{t}} e^{-u^2}\,du$ is the *error function*, then $\mathcal{L}(f(t)) = \frac{1}{s\sqrt{s+1}}$.

Solution: It is easy to verify that

$$\begin{aligned}f(t) &= \frac{2}{\sqrt{\pi}}\int\limits_0^{\sqrt{t}} e^{-u^2}\,du \\ &= \frac{2}{\sqrt{\pi}}\int\limits_0^{\sqrt{t}}\left(\sum_{n=0}^{\infty}(-1)^n\left(\frac{u^{2n}}{n!}\right)\right)du \\ &= \frac{2}{\sqrt{\pi}}\left(\sum_{n=0}^{\infty}(-1)^n\left(\frac{t^{n+\frac{1}{2}}}{n!(2n+1)}\right)\right) \qquad (9.11)\end{aligned}$$

Applying the Laplace transform to Eq.(9.11), we have

$$\mathcal{L}(f(t)) = \mathcal{L}\left(\frac{2}{\sqrt{\pi}}\left(\sum_{n=0}^{\infty}(-1)^n\left(\frac{t^{n+\frac{1}{2}}}{n!(2n+1)}\right)\right)\right)$$

$$= \frac{2}{\sqrt{\pi}}\mathcal{L}\left(\sum_{n=0}^{\infty}(-1)^n\left(\frac{t^{n+\frac{1}{2}}}{n!(2n+1)}\right)\right)$$

$$= \frac{2}{\sqrt{\pi}}\left(\sum_{n=0}^{\infty}(-1)^n\mathcal{L}\left(\frac{t^{n+\frac{1}{2}}}{n!(2n+1)}\right)\right)$$

$$= \frac{2}{\sqrt{\pi}}\left(\sum_{n=0}^{\infty}\left(\frac{(-1)^n}{n!(2n+1)}\right)\mathcal{L}\left(t^{n+\frac{1}{2}}\right)\right)$$

$$= \frac{2}{\sqrt{\pi}}\left(\sum_{n=0}^{\infty}\left(\frac{(-1)^n}{n!(2n+1)}\right)\left(\frac{\Gamma\left(n+1+\frac{1}{2}\right)}{s^{n+1+\frac{1}{2}}}\right)\right)$$

$$= \frac{1}{s^{\frac{3}{2}}} + \sum_{n=1}^{\infty}(-1)^n\left(\prod_{i=1}^{n}\left(\frac{2i-1}{2i}\right)\right)\left(\frac{1}{s^{n+1+\frac{1}{2}}}\right)$$

$$= \frac{1}{s^{\frac{3}{2}}}\left(1+\sum_{n=1}^{\infty}(-1)^n\left(\prod_{i=1}^{n}\left(\frac{2i-1}{2i}\right)\right)\left(\frac{1}{s^n}\right)\right)$$

$$= \frac{1}{s^{\frac{3}{2}}}\left(1+\frac{1}{s}\right)^{-\frac{1}{2}}$$

$$= \frac{1}{s\sqrt{s+1}}$$

EXAMPLE 9.18 Show that if $f(t) = J_0(t)$, then $\mathcal{L}(f(t)) = \frac{1}{\sqrt{s^2+1}}$.

Solution: According to the definition of $J_0(t)$ given in Eq.(8.106), we have

$$\begin{aligned} f(t) &= J_0(t) \\ &= 1+\sum_{n=1}^{\infty}(-1)^n\left(\prod_{i=1}^{n}\left(\frac{1}{(2i)^2}\right)\right)t^{2n} \end{aligned} \tag{9.12}$$

Applying the Laplace transform to Eq.(9.12), we have

$$\mathcal{L}(f(t)) = \mathcal{L}\left(1+\sum_{n=1}^{\infty}(-1)^n\left(\prod_{i=1}^{n}\left(\frac{1}{(2i)^2}\right)\right)t^{2n}\right)$$

$$= \frac{1}{s}+\mathcal{L}\left(\sum_{n=1}^{\infty}(-1)^n\left(\prod_{i=1}^{n}\left(\frac{1}{(2i)^2}\right)\right)t^{2n}\right)$$

$$= \frac{1}{s}+\sum_{n=1}^{\infty}(-1)^n\left(\prod_{i=1}^{n}\left(\frac{1}{(2i)^2}\right)\right)\mathcal{L}\left(t^{2n}\right)$$

$$= \frac{1}{s} + \sum_{n=1}^{\infty}(-1)^n \left(\prod_{i=1}^{n}\left(\frac{1}{(2i)^2}\right)\right)\left(\frac{(2n)!}{s^{2n+1}}\right)$$

$$= \frac{1}{s}\left(1 + \sum_{n=1}^{\infty}(-1)^n \left(\prod_{i=1}^{n}\left(\frac{2i-1}{2i}\right)\right)\left(\frac{1}{s^2}\right)^n\right)$$

$$= \frac{1}{s}\left(1 + \frac{1}{s^2}\right)^{-\frac{1}{2}}$$

$$= \frac{1}{\sqrt{s^2+1}}$$

EXAMPLE 9.19 Find the Laplace transform of $J_0(at)$.

Solution: Clearly, $\mathcal{L}(J_0(t)) = \frac{1}{\sqrt{s^2+1}} = F(s)$ according to Example 9.18. Again, we have by Theorem 9.1.7,

$$\mathcal{L}(J_0(at)) = \frac{1}{a}F\left(\frac{s}{a}\right)$$

$$= \frac{1}{\sqrt{s^2+a^2}}$$

EXAMPLE 9.20 Show that if $f(t) = J_0\left(a\sqrt{t}\right)$, then $\mathcal{L}(f(t)) = \frac{1}{s}e^{-\frac{a^2}{4s}}$.

Solution: According to the definition of $J_0\left(a\sqrt{t}\right)$ given in Eq.(8.106), we have

$$f(t) = J_0\left(a\sqrt{t}\right)$$

$$= 1 + \sum_{n=1}^{\infty}(-1)^n \left(\prod_{i=1}^{n}\left(\frac{1}{(2i)^2}\right)\right) a^{2n}t^n \tag{9.13}$$

Applying the Laplace transform to Eq.(9.12), we have

$$\mathcal{L}(f(t)) = \mathcal{L}\left(1 + \sum_{n=1}^{\infty}(-1)^n \left(\prod_{i=1}^{n}\left(\frac{1}{(2i)^2}\right)\right) a^{2n}t^n\right)$$

$$= \frac{1}{s} + \mathcal{L}\left(\sum_{n=1}^{\infty}(-1)^n \left(\prod_{i=1}^{n}\left(\frac{1}{(2i)^2}\right)\right) a^{2n}t^n\right)$$

$$= \frac{1}{s} + \sum_{n=1}^{\infty}(-1)^n \left(\prod_{i=1}^{n}\left(\frac{1}{(2i)^2}\right)\right) a^{2n}\mathcal{L}\left(t^n\right)$$

$$= \frac{1}{s} + \sum_{n=1}^{\infty} (-a^2)^n \left(\prod_{i=1}^{n} \left(\frac{1}{(2i)^2} \right) \right) \left(\frac{n!}{s^{n+1}} \right)$$

$$= \frac{1}{s} \left(1 + \sum_{n=1}^{\infty} (-1)^n \left(\frac{1}{n!} \right) \left(\frac{a^2}{4s} \right)^n \right)$$

$$= \frac{1}{s} e^{-\frac{a^2}{4s}}$$

Theorem 9.2.2 If $f(t) = t^n$ where $n \in \mathbb{N}$, then $\mathcal{L}(f(t)) = \frac{n!}{s^{n+1}}$.

Proof: According to the definition of the Laplace transform, we have

$$\mathcal{L}(t^n) = \int_0^{\infty} e^{-st} f(t)\,\mathrm{d}t$$

$$= \int_0^{\infty} e^{-st} t^n\,\mathrm{d}t$$

$$= -\frac{1}{s} \int_0^{\infty} t^n \mathrm{d}(e^{-st})$$

$$= -\frac{1}{s} t^n e^{-st} \Big|_0^{\infty} + \frac{n}{s} \int_0^{\infty} t^{n-1} e^{-st}\,\mathrm{d}t$$

$$= \frac{n}{s} \int_0^{\infty} t^{n-1} e^{-st}\,\mathrm{d}t$$

$$\vdots$$

$$= \frac{n!}{s^n} \int_0^{\infty} e^{-st}\,\mathrm{d}t$$

$$= \frac{n!}{s^{n+1}}$$

EXAMPLE 9.21 Show that if $f(t) = \sin at$, then $\mathcal{L}(f(t)) = \frac{a}{s^2+a^2}$.

Solution: We know that $\sin at = \frac{e^{iat} - e^{-iat}}{2i}$. According to the property of the Laplace transform, we have

$$\mathcal{L}(\sin at) = \mathcal{L}\left(\frac{e^{iat} - e^{-iat}}{2i} \right)$$

$$= \mathcal{L}\left(\frac{e^{iat}}{2i}\right) - \mathcal{L}\left(\frac{e^{-iat}}{2i}\right)$$

$$= \frac{1}{2i(s-ia)} - \frac{1}{2i(s+ia)}$$

$$= \frac{s+ia-(s-ia)}{2i(s^2+a^2)}$$

$$= \frac{a}{s^2+a^2}$$

EXAMPLE 9.22 Show that if $f(t) = \cos at$, then $\mathcal{L}(f(t)) = \frac{s}{s^2+a^2}$.

Solution: We know that $\cos at = \frac{e^{iat}+e^{-iat}}{2}$. According to the property of the Laplace transform, we have

$$\mathcal{L}(\cos at) = \mathcal{L}\left(\frac{e^{iat}+e^{-iat}}{2}\right)$$

$$= \mathcal{L}\left(\frac{e^{iat}}{2}\right) + \mathcal{L}\left(\frac{e^{-iat}}{2}\right)$$

$$= \frac{1}{2(s-ia)} + \frac{1}{2(s+ia)}$$

$$= \frac{s+ia+(s-ia)}{2(s^2+a^2)}$$

$$= \frac{s}{s^2+a^2}$$

EXAMPLE 9.23 Find the Laplace transform of $g(t) = 2\left(\frac{1-\cos t}{t}\right)$.

Solution: Let $f(t) = 2(1-\cos t)$. Clearly, $\mathcal{L}(f(t)) = 2\left(\frac{1}{s(s^2+1)}\right) = F(s)$. Again, $g(t) = \frac{f(t)}{t}$. According to Theorem 9.1.4, we have

$$\mathcal{L}(g(t)) = \mathcal{L}\left(\frac{f(t)}{t}\right)$$

$$= \int_s^\infty F(x)\,\mathrm{d}x$$

$$= \int_s^\infty 2\left(\frac{1}{x(x^2+1)}\right)\mathrm{d}x$$

$$= 2\int_s^\infty \left(\frac{1}{x(x^2+1)}\right)\mathrm{d}x$$

$$= 2\int_s^\infty \left(\frac{1}{x} - \frac{x}{x^2+1}\right) dx$$

$$= 2\int_s^\infty \frac{1}{x} dx - \int_s^\infty \frac{2x}{x^2+1} dx$$

$$= 2\ln x\big|_s^\infty - \ln(x^2+1)\big|_s^\infty$$

$$= \ln x^2\big|_s^\infty - \ln(x^2+1)\big|_s^\infty$$

$$= \ln\left(\frac{x^2}{x^2+1}\right)\Bigg|_s^\infty$$

$$= -\ln\left(\frac{s^2}{s^2+1}\right)$$

$$= \ln\left(\left(\frac{s^2}{s^2+1}\right)^{-1}\right)$$

$$= \ln\left(\frac{s^2+1}{s^2}\right)$$

$$= \ln\left(1 + \frac{1}{s^2}\right)$$

EXAMPLE 9.24 Show that if $g(t) = e^{bt}\sin at$, then $\mathcal{L}(g(t)) = \frac{a}{(s-b)^2+a^2}$.

Solution: Let $f(t) = \sin at$. Clearly, $\mathcal{L}(f(t)) = \frac{a}{s^2+a^2} = F(s)$. According to the first shifting Theorem 9.1.1, we have

$$\mathcal{L}(g(t)) = \mathcal{L}(e^{bt} f(t))$$
$$= F(s-b)$$
$$= \frac{a}{(s-b)^2+a^2}$$

EXAMPLE 9.25 Show that if $g(t) = e^{bt}\cos at$, then $\mathcal{L}(g(t)) = \frac{s-b}{(s-b)^2+a^2}$.

Solution: Let $f(t) = \cos at$. Clearly, $\mathcal{L}(f(t)) = \frac{s}{s^2+a^2} = F(s)$. According to the first shifting Theorem 9.1.1, we have

$$\mathcal{L}(g(t)) = \mathcal{L}(e^{bt} f(t))$$
$$= F(s-b)$$
$$= \frac{s-b}{(s-b)^2+a^2}$$

EXAMPLE 9.26 Show that if $f(t) = \cosh at$, then $\mathcal{L}(f(t)) = \frac{s}{s^2-a^2}$.

Solution: We know that $\cosh at = \frac{e^{at}+e^{-at}}{2}$. According to the property of the Laplace transform, we have

$$\begin{aligned}\mathcal{L}(\cosh at) &= \mathcal{L}\left(\frac{e^{at}+e^{-at}}{2}\right)\\ &= \mathcal{L}\left(\frac{e^{at}}{2}\right)+\mathcal{L}\left(\frac{e^{-at}}{2}\right)\\ &= \frac{1}{2(s-a)}+\frac{1}{2(s+a)}\\ &= \frac{s+a+(s-a)}{2(s^2-a^2)}\\ &= \frac{s}{s^2-a^2}\end{aligned}$$

EXAMPLE 9.27 Show that if $f(t) = \sinh at$, then $\mathcal{L}(f(t)) = \frac{a}{s^2-a^2}$.

Solution: We know that $\sinh at = \frac{e^{at}-e^{-at}}{2}$. According to the property of the Laplace transform, we have

$$\begin{aligned}\mathcal{L}(\sinh at) &= \mathcal{L}\left(\frac{e^{at}-e^{-at}}{2}\right)\\ &= \mathcal{L}\left(\frac{e^{at}}{2}\right)-\mathcal{L}\left(\frac{e^{-at}}{2}\right)\\ &= \frac{1}{2(s-a)}-\frac{1}{2(s+a)}\\ &= \frac{s+a-(s-a)}{2(s^2-a^2)}\\ &= \frac{a}{s^2-a^2}\end{aligned}$$

EXAMPLE 9.28 Show that if $g(t) = e^{bt}\sinh at$, then show that $\mathcal{L}(g(t)) = \frac{a}{(s-b)^2-a^2}$.

Solution: Let $f(t) = \sinh at$. Clearly, $\mathcal{L}(f(t)) = \frac{a}{s^2-a^2} = F(s)$. According to the first shifting Theorem 9.1.1, we have

$$\begin{aligned}\mathcal{L}(g(t)) &= \mathcal{L}(e^{bt}f(t))\\ &= F(s-b)\\ &= \frac{a}{(s-b)^2-a^2}\end{aligned}$$

EXAMPLE 9.29 If $g(t) = e^{bt}\cosh at$, then $\mathcal{L}(g(t)) = \frac{s-b}{(s-b)^2-a^2}$.

Solution: Let $f(t) = \cosh at$. Clearly, $\mathcal{L}(f(t)) = \frac{s}{s^2-a^2} = F(s)$. According to the first shifting Theorem 9.1.1, we have

$$\begin{aligned}\mathcal{L}(g(t)) &= \mathcal{L}(e^{bt}f(t))\\ &= F(s-b)\\ &= \frac{s-b}{(s-b)^2-a^2}\end{aligned}$$

Theorem 9.2.3 If $g(t) = e^{at}t^n$, then show that $\mathcal{L}(g(t)) = \frac{n!}{(s-a)^{n+1}}$.

Proof: Let $f(t) = t^n$. Clearly, $\mathcal{L}(f(t)) = \frac{n!}{s^{n+1}} = F(s)$. According to the first shifting Theorem 9.1.1, we have

$$\begin{aligned}\mathcal{L}(g(t)) &= \mathcal{L}\left(e^{at}t^n\right)\\ &= F(s-a)\\ &= \frac{n!}{(s-a)^{n+1}}\end{aligned}$$

Theorem 9.2.4 If $g(t) = t^n\sin at$, then show that $\mathcal{L}(g(t)) = \frac{n!s^{n+1}}{(s^2+a^2)^{n+1}}\left(\sum\limits_{r=0}^{\lfloor\frac{n+1}{2}\rfloor}(-1)^r\left(\frac{a}{s}\right)^{2r+1}\right)$.

Proof: Let $f(t) = t^n$. Clearly, $\mathcal{L}(f(t)) = \frac{n!}{s^{n+1}}$. We know that, $\sin at = \frac{e^{iat}-e^{-iat}}{2i}$. Again, $g(t) = \left(\frac{e^{iat}-e^{-iat}}{2i}\right)f(t)$. According to the first shifting Theorem 9.1.1 and the property of the Laplace transform, we have

$$\begin{aligned}\mathcal{L}(g(t)) &= \mathcal{L}\left(\left(\frac{e^{iat}-e^{-iat}}{2i}\right)t^n\right)\\ &= \mathcal{L}\left(\left(\frac{e^{iat}}{2i}\right)t^n\right) - \mathcal{L}\left(\left(\frac{e^{-iat}}{2i}\right)t^n\right)\\ &= \frac{1}{2i}\mathcal{L}\left(e^{iat}t^n\right) - \left(\frac{1}{2i}\right)\mathcal{L}\left(e^{-iat}t^n\right)\\ &= \frac{1}{2i}\left(\mathcal{L}\left(e^{iat}t^n\right) - \mathcal{L}\left(e^{-iat}t^n\right)\right)\\ &= \frac{1}{2i}\left(\left(\frac{n!}{(s-ia)^{n+1}}\right) - \left(\frac{n!}{(s+ia)^{n+1}}\right)\right)\\ &= \frac{n!}{2i}\left(\frac{(s+ia)^{n+1}-(s-ia)^{n+1}}{(s^2+a^2)^{n+1}}\right)\end{aligned}$$

$$= \frac{n!}{(s^2+a^2)^{n+1}} \left(\frac{(s+ia)^{n+1} - (s-ia)^{n+1}}{2i} \right)$$

$$= \frac{n!s^{n+1}}{(s^2+a^2)^{n+1}} \left(\frac{\left(1+i\left(\frac{a}{s}\right)\right)^{n+1} - \left(1-i\left(\frac{a}{s}\right)\right)^{n+1}}{2i} \right)$$

$$= \frac{n!s^{n+1}}{(s^2+a^2)^{n+1}} \left(\sum_{r=0}^{\lfloor \frac{n+1}{2} \rfloor} (-1)^r \left(\frac{a}{s}\right)^{2r+1} \right)$$

Theorem 9.2.5 If $g(t) = t^n \cos at$, then show that $\mathcal{L}(g(t)) = \frac{n!s^{n+1}}{(s^2+a^2)^{n+1}} \left(\sum_{r=0}^{\lfloor \frac{n+1}{2} \rfloor} (-1)^r \left(\frac{a}{s}\right)^{2r} \right)$.

Proof: Let $f(t) = t^n$. Clearly, $\mathcal{L}(f(t)) = \frac{n!}{s^{n+1}}$. We know that $\cos at = \frac{e^{iat}+e^{-iat}}{2}$. Again, $g(t) = \left(\frac{e^{iat}+e^{-iat}}{2}\right) f(t)$. According to the first shifting Theorem 9.1.1, and the property of the Laplace transform, we have

$$\mathcal{L}(g(t)) = \mathcal{L}\left(\left(\frac{e^{iat}+e^{-iat}}{2}\right) t^n\right)$$

$$= \mathcal{L}\left(\left(\frac{e^{iat}}{2}\right) t^n\right) + \mathcal{L}\left(\left(\frac{e^{-iat}}{2}\right) t^n\right)$$

$$= \frac{1}{2}\mathcal{L}\left(e^{iat}t^n\right) + \frac{1}{2}\mathcal{L}\left(e^{-iat}t^n\right)$$

$$= \frac{1}{2}\left(\mathcal{L}\left(e^{iat}t^n\right) + \mathcal{L}\left(e^{-iat}t^n\right)\right)$$

$$= \frac{1}{2}\left(\left(\frac{n!}{(s-ia)^{n+1}}\right) + \left(\frac{n!}{(s+ia)^{n+1}}\right)\right)$$

$$= \frac{n!}{2}\left(\frac{(s+ia)^{n+1} + (s-ia)^{n+1}}{(s^2+a^2)^{n+1}}\right)$$

$$= \frac{n!}{(s^2+a^2)^{n+1}} \left(\frac{(s+ia)^{n+1} + (s-ia)^{n+1}}{2} \right)$$

$$= \frac{n!s^{n+1}}{(s^2+a^2)^{n+1}} \left(\frac{\left(1+i\left(\frac{a}{s}\right)\right)^{n+1} + \left(1-i\left(\frac{a}{s}\right)\right)^{n+1}}{2} \right)$$

$$= \frac{n!s^{n+1}}{(s^2+a^2)^{n+1}} \left(\sum_{r=0}^{\lfloor \frac{n+1}{2} \rfloor} (-1)^r \left(\frac{a}{s}\right)^{2r} \right)$$

EXAMPLE 9.30 Find the Laplace transform of $g(t) = \frac{1}{2a^3}(\sin at - at\cos at)$.

Solution: Let $F(s) = \mathcal{L}(\cos at)$. Clearly, $F(s) = \frac{s}{s^2+a^2}$. According to Corollary 9.1.1, and using the properties of the Laplace transform, we have

$$\begin{aligned}
\mathcal{L}(g(t)) &= \mathcal{L}\left(\frac{1}{2a^3}(\sin at - at\cos at)\right) \\
&= \frac{1}{2a^3}\mathcal{L}(\sin at) - \frac{1}{2a^2}\mathcal{L}(t\cos at) \\
&= \frac{1}{2a^2(s^2+a^2)} + \frac{1}{2a^2}F'(s) \\
&= \frac{1}{2a^2(s^2+a^2)} + \frac{1}{2a^2}\left(\frac{a^2-s^2}{(s^2+a^2)^2}\right) \\
&= \frac{1}{(s^2+a^2)^2}
\end{aligned}$$

EXAMPLE 9.31 Find the Laplace transform of $g(t) = \frac{1}{2a}(\sin at + at\cos at)$.

Solution: Let $F(s) = \mathcal{L}(\cos at)$. Clearly, $F(s) = \frac{s}{s^2+a^2}$. According to Corollary 9.1.1, and using the properties of the Laplace transform, we have

$$\begin{aligned}
\mathcal{L}(g(t)) &= \mathcal{L}\left(\frac{1}{2a}(\sin at + at\cos at)\right) \\
&= \frac{1}{2a}\mathcal{L}(\sin at) + \frac{1}{2}\mathcal{L}(t\cos at) \\
&= \frac{1}{2(s^2+a^2)} - \frac{1}{2}F'(s) \\
&= \frac{1}{2(s^2+a^2)} - \frac{1}{2}\left(\frac{a^2-s^2}{(s^2+a^2)^2}\right) \\
&= \frac{s^2}{(s^2+a^2)^2}
\end{aligned}$$

EXAMPLE 9.32 Find the Laplace transform of $g(t) = \frac{1}{2a}(t\sin at)$.

Solution: Let $F(s) = \mathcal{L}(\sin at)$. Clearly, $F(s) = \frac{a}{s^2+a^2}$. According to Corollary 9.1.1, and using the properties of the Laplace transform, we have

$$\begin{aligned}
\mathcal{L}(g(t)) &= \mathcal{L}\left(\frac{1}{2a}(t\sin at)\right) \\
&= \frac{1}{2a}\mathcal{L}(t\sin at)
\end{aligned}$$

$$= -\frac{1}{2a}F'(s)$$

$$= -\frac{1}{2a}\left(\frac{-2sa}{(s^2+a^2)^2}\right)$$

$$= \frac{s}{(s^2+a^2)^2}$$

EXAMPLE 9.33 Find the Laplace transform of $g(t) = t\cos at$.

Solution: Let $F(s) = \mathcal{L}(\cos at)$. Clearly, $F(s) = \frac{s}{s^2+a^2}$. According to Corollary 9.1.1, and using the properties of the Laplace transform, we have

$$\mathcal{L}(g(t)) = \mathcal{L}(t\cos at)$$
$$= -F'(s)$$
$$= -\frac{a^2-s^2}{(s^2+a^2)^2}$$
$$= \frac{s^2-a^2}{(s^2+a^2)^2}$$

Exercises 9.2

1. Show that $\int\limits_0^\infty e^{-u^2}\,\mathrm{d}u = \frac{\sqrt{\pi}}{2}$.

   Hint: Let $a = \int\limits_0^\infty e^{-u^2}\,\mathrm{d}u$. Clearly, $a^2 = \int\limits_0^\infty e^{-u^2}\,\mathrm{d}u \int\limits_0^\infty e^{-v^2}\,\mathrm{d}v = \int\limits_0^\infty\int\limits_0^\infty e^{-(u^2+v^2)}\,\mathrm{d}u\,\mathrm{d}v$. Replace $u = r\sin\theta$ and $v = r\cos\theta$. Hence $u^2+v^2 = r^2$ and $\mathrm{d}u\,\mathrm{d}v = r\,\mathrm{d}r\,\mathrm{d}\theta$. In other words, $a^2 = \int\limits_0^{\frac{\pi}{2}}\int\limits_0^\infty re^{-r^2}\,\mathrm{d}r\,\mathrm{d}\theta = \frac{\pi}{2}\int\limits_0^\infty re^{-r^2}\,\mathrm{d}r = -\frac{\pi}{4}\int\limits_0^\infty -2re^{-r^2}\,\mathrm{d}r = \frac{\pi}{4}$.

2. Find the Laplace transform of $g(t) = t^{\frac{1}{2}}$.

   Hint: It is easy to verify that $g(t) = tt^{-\frac{1}{2}} = tf(t)$ where $f(t) = t^{-\frac{1}{2}}$. Apply Corollary 9.1.1 and Example 9.12.

3. Find the Laplace transform of $g(t) = \sin^2 t$.
4. Find the Laplace transform of $g(t) = \cos^2 t$.
5. Find the Laplace transform of $g(t) = t\cos 2t$.
6. Find the Laplace transform of $g(t) = t\,e^{2t}$.
7. Find the Laplace transform of $g(t) = t^2\,e^t$.
8. Find the Laplace transform of $g(t) = t\cosh t$.
9. Find the Laplace transform of $g(t) = t\sin 2t$.

10. Find the Laplace transform of $g(t) = t^2 \cos at$.
11. Find the Laplace transform of $g(t) = t^2 \sin at$.
12. Find the Laplace transform of $g(t) = t\, e^{-2t} \sin at$.
13. Find the Laplace transform of $g(t) = t\, e^{-2t} \cos at$.
14. If $L_m(t) = \left(\frac{e^t}{m!}\right) \frac{\mathrm{d}^m}{\mathrm{d}t^m} \left(t^m e^{-t}\right)$, then show that $\mathcal{L}\left(L_m(t)\right) = \frac{(s-1)^m}{s^{m+1}}$.

    Hint: Apply Corollary 9.1.1 and Theorem 9.4.3 with induction on m.

9.3 Inverse Laplace Transform

Inverse Laplace transform is the process of finding a function $f(t)$ such that $\mathcal{L}(f(t)) = F(s)$ where $F(s)$ is given one. There are two different approaches for finding the inverse Laplace transform, and they are as follows:

1. Partial fraction
2. Convolution.

These two approaches are discussed in detail in the following sections. We can present the inverse Laplace transform of some standard functions according to Section 9.2 in Table 9.1 for future reference while applying the partial fraction approach to any arbitrary function $F(s)$.

9.3.1 Partial Fraction

Let $F(s) = \frac{p(s)}{q(s)}$ be any function such that $\lim_{s\to\infty} F(s) = 0$. Express $F(s)$ as the sum of the terms of the type $\frac{\alpha}{s+a}$, $\frac{\alpha s+\beta}{(s+a)^2}$ and $\frac{\alpha s+\beta}{s^2 \pm a^2}$. The inverse Laplace transform of $F(s)$ is the sum of the inverse Laplace transforms of each of the expressed terms.

EXAMPLE 9.34 Find the inverse Laplace transform of the function $H(s)$ defined by

$$H(s) = \frac{9}{s^2(s^2-9)}$$

using partial fraction.

Solution: It is easy to verify that

$$\begin{aligned} H(s) &= \frac{9}{s^2(s^2-9)} \\ &= \frac{1}{s^2-9} - \frac{1}{s^2} \\ &= \frac{1}{3}\left(\frac{3}{s^2-9}\right) - \frac{1}{s^2} \end{aligned}$$

Table 9.1: Inverse Laplace transform of standard functions

$F(s)$	$f(t) = \mathcal{L}^{-1}(F(s))$
e^{-as}	$\delta_a(t)$
$\frac{e^{-as}}{s}$	$u_a(t)$
$\frac{\Gamma(n+1+\frac{1}{2})}{s^{n+1+\frac{1}{2}}}$	$t^{n+\frac{1}{2}}$
$\frac{n!}{s^{n+1}}$	t^n
$\frac{1}{s-a}$	e^{at}
$\frac{n!}{(s-a)^{n+1}}$	$e^{at}\, t^n$
$\frac{a}{s^2+a^2}$	$\sin at$
$\frac{s}{s^2+a^2}$	$\cos at$
$\frac{a}{(s-b)^2+a^2}$	$e^{bt} \sin at$
$\frac{s-b}{(s-b)^2+a^2}$	$e^{bt} \cos at$
$\frac{a}{s^2-a^2}$	$\sinh at$
$\frac{s}{s^2-a^2}$	$\cosh at$
$\frac{a}{(s-b)^2-a^2}$	$e^{bt} \sinh at$
$\frac{s-b}{(s-b)^2-a^2}$	$e^{bt} \cosh at$
$\frac{1}{(s^2+a^2)^2}$	$\frac{1}{2a^3}(\sin at - at \cos at)$
$\frac{s}{(s^2+a^2)^2}$	$\frac{1}{2a}(t \sin at)$
$\frac{s^2-a^2}{(s^2+a^2)^2}$	$t \cos at$
$\frac{s^2}{(s^2+a^2)^2}$	$\frac{1}{2a}(\sin at + at \cos at)$
$\frac{1}{s}\ln\left(1+\frac{1}{s^2}\right)$	$2\int_0^t \frac{1-\cos u}{u} \mathrm{d}u$
$\ln\left(1+\frac{1}{s^2}\right)$	$2\left(\frac{1-\cos t}{t}\right)$
$\frac{1}{s}e^{-\frac{a^2}{4s}}$	$J_0(a\sqrt{t})$
$\frac{1}{\sqrt{s^2+a^2}}$	$J_0(at)$
$\frac{s}{(s^2+a^2)^{\frac{3}{2}}}$	$tJ_0(at)$
$\frac{1}{s^3(s^2+1)}$	$\frac{t^2}{2} + \cos t - 1$
$\frac{1}{s\sqrt{s+1}}$	$\frac{2}{\sqrt{\pi}} \int_0^{\sqrt{t}} e^{-u^2}\, \mathrm{d}u$
$\sqrt{\frac{\pi}{s}}e^{-\frac{1}{4s}}$	$\frac{\cos\sqrt{t}}{\sqrt{t}}$
$\frac{1}{2s}\sqrt{\frac{\pi}{s}}e^{-\frac{1}{4s}}$	$\sin\sqrt{t}$

In other words, we have

$$\begin{aligned}\mathcal{L}^{-1}(H(s)) &= \mathcal{L}^{-1}\left(\frac{1}{3}\left(\frac{3}{s^2-9}\right)-\frac{1}{s^2}\right)\\ &= \frac{1}{3}\mathcal{L}^{-1}\left(\frac{3}{s^2-9}\right)-\mathcal{L}^{-1}\left(\frac{1}{s^2}\right)\\ &= \frac{1}{3}\sinh 3t - t\end{aligned}$$

EXAMPLE 9.35 Find the inverse Laplace transform of the function $H(s)$ defined by

$$H(s) = \frac{9}{s^2(s^2+9)}$$

using partial fraction.

Solution: It is easy to verify that

$$\begin{aligned}H(s) &= \frac{9}{s^2(s^2+9)}\\ &= \frac{1}{s^2}-\frac{1}{s^2+9}\\ &= \frac{1}{s^2}-\frac{1}{3}\left(\frac{3}{s^2+9}\right)\end{aligned}$$

In other words, we have

$$\begin{aligned}\mathcal{L}^{-1}(H(s)) &= \mathcal{L}^{-1}\left(\frac{1}{s^2}-\frac{1}{3}\left(\frac{3}{s^2+9}\right)\right)\\ &= \mathcal{L}^{-1}\left(\frac{1}{s^2}\right)-\frac{1}{3}\mathcal{L}^{-1}\left(\frac{3}{s^2+9}\right)\\ &= t-\frac{1}{3}\sin 3t\end{aligned}$$

EXAMPLE 9.36 Find the inverse Laplace transform of $F(s) = \frac{s-3}{s^2-1}$.

Solution: Clearly, we have

$$\begin{aligned}F(s) &= \frac{s-3}{s^2-1}\\ &= \frac{2}{s+1}-\frac{1}{s-1}\end{aligned}$$

In other words, we have

$$\begin{aligned}\mathcal{L}^{-1}(F(s)) &= \mathcal{L}^{-1}\left(\frac{2}{s+1} - \frac{1}{s-1}\right)\\ &= \mathcal{L}^{-1}\left(\frac{2}{s+1}\right) - \mathcal{L}^{-1}\left(\frac{1}{s-1}\right)\\ &= 2\mathcal{L}^{-1}\left(\frac{1}{s+1}\right) - \mathcal{L}^{-1}\left(\frac{1}{s-1}\right)\\ &= 2e^{-t} - e^{t}\end{aligned}$$

EXAMPLE 9.37 Find the inverse Laplace transform of $F(s) = \frac{s+3}{s^2-1}$.
Solution: Clearly, we have

$$\begin{aligned}F(s) &= \frac{s+3}{s^2-1}\\ &= \frac{2}{s-1} - \frac{1}{s+1}\end{aligned}$$

In other words, we have

$$\begin{aligned}\mathcal{L}^{-1}(F(s)) &= \mathcal{L}^{-1}\left(\frac{2}{s-1} - \frac{1}{s+1}\right)\\ &= \mathcal{L}^{-1}\left(\frac{2}{s-1}\right) - \mathcal{L}^{-1}\left(\frac{1}{s+1}\right)\\ &= 2\mathcal{L}^{-1}\left(\frac{1}{s-1}\right) - \mathcal{L}^{-1}\left(\frac{1}{s+1}\right)\\ &= 2e^{t} - e^{-t}\end{aligned}$$

EXAMPLE 9.38 Find the inverse Laplace transform of $F(s) = \frac{2s}{s^2-1}$.
Solution: Clearly, we have

$$\begin{aligned}F(s) &= \frac{2s}{s^2-1}\\ &= \frac{1}{s-1} + \frac{1}{s+1}\end{aligned}$$

In other words, we have

$$\begin{aligned}\mathcal{L}^{-1}(F(s)) &= \mathcal{L}^{-1}\left(\frac{1}{s-1} + \frac{1}{s+1}\right)\\ &= \mathcal{L}^{-1}\left(\frac{1}{s-1}\right) + \mathcal{L}^{-1}\left(\frac{1}{s-1}\right)\\ &= e^{-t} + e^{t}\end{aligned}$$

EXAMPLE 9.39 Find the inverse Laplace transform of $F(s) = \frac{s^3+s+2}{(s-1)^2(s^2+1)}$.

Solution: It is easy to verify that

$$\frac{s^3+s+2}{(s-1)^2(s^2+1)} = \frac{2}{(s-1)^2} + \frac{s}{s^2+1}$$

In other words, we have

$$\begin{aligned}
\mathcal{L}^{-1}\left(\frac{s^3+s+2}{(s-1)^2(s^2+1)}\right) &= \mathcal{L}^{-1}\left(\frac{2}{(s-1)^2} + \frac{s}{s^2+1}\right) \\
&= \mathcal{L}^{-1}\left(\frac{2}{(s-1)^2}\right) + \mathcal{L}^{-1}\left(\frac{s}{s^2+1}\right) \\
&= 2\mathcal{L}^{-1}\left(\frac{1}{(s-1)^2}\right) + \mathcal{L}^{-1}\left(\frac{s}{s^2+1}\right) \\
&= 2t\,e^t + \cos t
\end{aligned}$$

Exercises 9.3.1

1. Find the function $f(t)$ such that $\mathcal{L}(f(t)) = \frac{s+1}{s^3+s^2-6s}$.
2. Find the function $f(t)$ such that $\mathcal{L}(f(t)) = \frac{s+12}{s^2+4s}$.
3. Find the function $f(t)$ such that $\mathcal{L}(f(t)) = \frac{3s}{s^2+2s-8}$.
4. Find the function $f(t)$ such that $\mathcal{L}(f(t)) = \frac{s^2-6s+4}{s^3-3s^2+2s}$.
5. Find the function $f(t)$ such that $\mathcal{L}(f(t)) = \frac{3s^2-2s-1}{s^3-3s^2+s-3}$.
6. Find the function $f(t)$ such that $\mathcal{L}(f(t)) = \frac{s+1}{s^2+4s+3}$.
7. Find the function $f(t)$ such that $\mathcal{L}(f(t)) = \frac{10-4s}{(s-2)^2}$.
8. Find the function $f(t)$ such that $\mathcal{L}(f(t)) = \frac{s^2+2s}{(s^2+2s+4)^2}$.
9. Find the function $f(t)$ such that $\mathcal{L}(f(t)) = \frac{s^2+s-2}{(s+1)^3}$.
10. Find the function $f(t)$ such that $\mathcal{L}(f(t)) = \frac{s^3-7s^2+14s-9}{(s-1)^2(s-2)^2}$.
11. Find the function $f(t)$ such that $\mathcal{L}(f(t)) = \frac{s^4+3(s+1)^3}{s^4(s+1)^3}$.

9.3.2 Convolution

Let S be the set of all exponentially bounded functions. Define a binary operation $*$ over S by $f(t) * g(t) = \int_0^t f(x)g(t-x)\,dx$. It is easy to verify that

1. $f(t) * g(t) = g(t) * f(t)$ according to the definition of convolution and changing the variable $u = t - x$ by

$$f * g = \int_0^t f(t-x)g(x)\,dx$$

$$= -\int_t^0 f(u)g(t-u)\,du$$

$$= \int_0^t g(t-u)f(u)\,du$$

$$= g * f$$

2. $f(t) * (h(t) + g(t)) = (f(t) * h(t)) + (f(t) * g(t))$ according to the definition of convolution by

$$f * (g+h) = \int_0^t f(t-x)(g(x) + h(x))\,dx$$

$$= \int_0^t f(t-x)g(x)\,dx + \int_0^t f(t-x)h(x)\,dx$$

$$= (f * g) + (f * h)$$

3. $f(t) * (h(t) * g(t)) = (f(t) * h(t)) * g(t)$ according to the definition of convolution, changing the order of integration and replacing $v = x - u$ by

$$f * (g * h) = \int_0^t f(t-x)(g * h)(x)\,dx$$

$$= \int_0^t f(t-x)\left(\int_0^x g(x-u)h(u)\,du\right)dx$$

$$= \int_0^t \int_0^x (f(t-x)g(x-u)h(u))\,du\,dx$$

$$= \int_0^t \int_u^t (f(t-x)g(x-u)h(u))\,dx\,du$$

$$= \int_0^t \int_u^t (f(t-x)g(x-u))\, \mathrm{d}x\, h(u)\, \mathrm{d}u$$

$$= \int_0^t \left(\int_0^{t-u} f(t-u-v)g(v)\, \mathrm{d}v \right) h(u)\, \mathrm{d}u$$

$$= \int_0^t (f*g)(t-u)h(u)\, \mathrm{d}u$$

$$= (f*g)*h$$

4. $f(t)*0 = 0$ according to the definition of convolution and changing the variable $u = t - x$ by

$$f*0 = \int_0^t f(t-x)g(x)\, \mathrm{d}x$$
$$= 0$$

where $g(x) = 0$ for all $0 \le x \le t$.

The operation $*$ is called *convolution.* Let $H(s) = \frac{p(s)}{q(s)}$ be any function such that $\lim\limits_{s\to\infty} H(s) = 0$. Express $H(s) = F(s)G(s)$ such that $\mathcal{L}^{-1}(F(s)) = f(t)$ and $\mathcal{L}^{-1}(G(s)) = g(t)$ are known. The inverse Laplace transform of $H(s)$ is the convolution of $f(t)$ and $g(t)$ which will be concluded by Theorem 9.3.1.

Theorem 9.3.1 (Convolution Theorem) If $F(s) = \mathcal{L}(f(t))$ and $G(s) = \mathcal{L}(g(t))$, then $\mathcal{L}^{-1}(F(s)G(s)) = \int_0^t f(x)g(t-x)\, \mathrm{d}x$.

Proof: Let $s > x$. According to the second shifting formula, we have

$$e^{-sx}G(s) = \int_0^\infty g(t-x)e^{-st}u_x(t)\, \mathrm{d}t$$

$$= \int_x^\infty g(t-x)e^{-st}\, \mathrm{d}t \tag{9.14}$$

Using the definition of the Laplace transform, Eq.(9.14) and change of order of integration, we have

$$F(s)G(s) = \left(\int_0^\infty f(x)e^{-sx}\, \mathrm{d}x \right) G(s)$$

$$
\begin{aligned}
&= \int_0^\infty f(x)e^{-sx}\,G(s)\,\mathrm{d}x \\
&= \int_0^\infty f(x)\left(e^{-sx}\,G(s)\right)\mathrm{d}x \\
&= \int_0^\infty f(x)\left(\int_x^\infty g(t-x)e^{-st}\,\mathrm{d}t\right)\mathrm{d}x \\
&= \int_0^\infty \int_x^\infty f(x)g(t-x)e^{-st}\,\mathrm{d}t\,\mathrm{d}x \\
&= \int_0^\infty \int_0^t f(x)g(t-x)e^{-st}\,\mathrm{d}x\,\mathrm{d}t \\
&= \int_0^\infty e^{-st}\left(\int_0^t f(x)g(t-x)\,\mathrm{d}x\right)\mathrm{d}t \\
&= \mathcal{L}\left(\int_0^t f(x)g(t-x)\,\mathrm{d}x\right)
\end{aligned}
$$

In other words, we have

$$
\int_0^t f(x)g(t-x)\,\mathrm{d}x = \mathcal{L}^{-1}\left(F(s)G(s)\right)
$$

Corollary 9.3.1 If $F(s) = \mathcal{L}(f(t))$, then $\mathcal{L}\left(\int_0^t f(x)\,\mathrm{d}x\right) = \frac{F(s)}{s}$.

Proof: Let $g(t) = 1$. Hence $\mathcal{L}(g(t)) = \frac{1}{s} = G(s)$. According to Theorem 9.3.1, we have

$$
\begin{aligned}
\mathcal{L}\left(\int_0^t f(x)\,\mathrm{d}x\right) &= \mathcal{L}\left(\int_0^t f(x)g(t-x)\,\mathrm{d}x\right) \\
&= F(s)G(s) \\
&= \frac{F(s)}{s}
\end{aligned}
$$

EXAMPLE 9.40 Find the inverse Laplace transform of the function $H(s)$ defined by

$$
H(s) = \frac{9}{s^2(s^2-9)}
$$

using convolution.

Solution: Let $F(s) = \frac{3}{s^2}$ and $G(s) = \frac{3}{s^2-9}$. It is easy to verify that $f(t) = \mathcal{L}^{-1}\left(\frac{3}{s^2}\right) = 3t$ and $g(t) = \mathcal{L}^{-1}\left(\frac{3}{s^2-9}\right) = \sinh 3t$. According to convolution Theorem 9.3.1, we have

$$\begin{aligned}
h(t) &= \mathcal{L}^{-1}\left(F(s)G(s)\right) \\
&= \int_0^t f(t-x)g(x)\,\mathrm{d}x \\
&= \int_0^t 3(t-x)\sinh 3x\,\mathrm{d}x \\
&= 3t\int_0^t \sinh 3x\,\mathrm{d}x - 3\int_0^t x\sinh 3x\,\mathrm{d}x \\
&= t(\cosh 3t - 1) - t\cosh 3t + \frac{1}{3}\sinh 3t \\
&= \frac{1}{3}\sinh 3t - t
\end{aligned}$$

EXAMPLE 9.41 Find the inverse Laplace transform of the function $H(s)$ defined by

$$H(s) = \frac{9}{s^2(s^2+9)}$$

using convolution.

Solution: Let $F(s) = \frac{3}{s^2}$ and $G(s) = \frac{3}{s^2+9}$. It is easy to verify that $f(t) = \mathcal{L}^{-1}\left(\frac{3}{s^2}\right) = 3t$ and $g(t) = \mathcal{L}^{-1}\left(\frac{3}{s^2+9}\right) = \sin 3t$. According to convolution Theorem 9.3.1, we have

$$\begin{aligned}
h(t) &= \mathcal{L}^{-1}\left(F(s)G(s)\right) \\
&= \int_0^t f(t-x)g(x)\,\mathrm{d}x \\
&= \int_0^t 3(t-x)\sin 3x\,\mathrm{d}x \\
&= 3t\int_0^t \sin 3x\,\mathrm{d}x - 3\int_0^t x\sin 3x\,\mathrm{d}x
\end{aligned}$$

$$= t(1 - \cos 3t) + t\cos 3t - \frac{1}{3}\sin 3t$$
$$= t - \frac{1}{3}\sin 3t$$

EXAMPLE 9.42 Find the inverse Laplace transform of the function $H(s)$ defined by

$$H(s) = \frac{2}{s^3(s^2+1)}$$

using convolution.

Solution: Let $F(s) = \frac{2}{s^3}$ and $G(s) = \frac{1}{s^2+1}$. It is easy to verify that $f(t) = \mathcal{L}^{-1}\left(\frac{2}{s^3}\right) = t^2$ and $g(t) = \mathcal{L}^{-1}\left(\frac{1}{s^2+1}\right) = \sin t$. According to convolution Theorem 9.3.1, we have

$$h(t) = \mathcal{L}^{-1}(F(s)G(s))$$
$$= \int_0^t f(t-x)g(x)\,\mathrm{d}x$$
$$= \int_0^t (t-x)^2 \sin x\,\mathrm{d}x$$
$$= t^2 - 2\int_0^t (t-x)\cos x\,\mathrm{d}x$$
$$= t^2 - 2\int_0^t \sin x\,\mathrm{d}x$$
$$= t^2 - 2 + 2\cos t$$

EXAMPLE 9.43 Find the inverse Laplace transform of the function $H(s) = \frac{1}{s^2(s-a)}$.

Solution: Let $F(s) = \frac{1}{s^2}$ and $G(s) = \frac{1}{s-a}$. Clearly, $f(t) = \mathcal{L}^{-1}(F(s)) = t$ and $g(t) = \mathcal{L}^{-1}(G(s)) = e^{at}$. According to convolution Theorem 9.3.1, we have

$$h(t) = \mathcal{L}^{-1}(F(s)G(s))$$
$$= \int_0^t f(t-x)g(x)\,\mathrm{d}x$$
$$= \int_0^t (t-x)e^{-ax}\,\mathrm{d}x$$

$$= -\frac{1}{a}(t-x)e^{-ax}\Big|_0^t - \frac{1}{a}\int_0^t e^{-ax}\,\mathrm{d}x$$

$$= \frac{1}{a}t + \frac{1}{a^2}e^{-at} - \frac{1}{a^2}$$

$$= \frac{1}{a^2}(at + e^{-at} - 1)$$

EXAMPLE 9.44 Find the inverse Laplace transform of the function $H(s) = \frac{1}{s(s-a)}$.

Solution: Let $F(s) = \frac{1}{s}$ and $G(s) = \frac{1}{s-a}$. Clearly, $f(t) = \mathcal{L}^{-1}(F(s)) = 1$ and $g(t) = \mathcal{L}^{-1}(G(s)) = e^{at}$. According to convolution Theorem 9.3.1, we have

$$h(t) = \mathcal{L}^{-1}(F(s)G(s))$$

$$= \int_0^t f(t-x)g(x)\,\mathrm{d}x$$

$$= \int_0^t e^{-ax}\,\mathrm{d}x$$

$$= -\frac{1}{a}e^{-ax}\Big|_0^t$$

$$= \frac{1}{a} - \frac{1}{a}e^{-at}$$

$$= \frac{1}{a}\left(1 - e^{-at}\right)$$

EXAMPLE 9.45 Solve the integral equation $y(t) = \cos t + \int_0^t y(x)\cos(t-x)\,\mathrm{d}x$.

Solution: Let $\mathcal{L}(y(t)) = Y(s)$. Applying the Laplace transform to the given integral equation, we have

$$Y(s) = \mathcal{L}(y(t))$$

$$= \mathcal{L}\left(\cos t + \int_0^t y(x)\cos(t-x)\,\mathrm{d}x\right)$$

$$= \mathcal{L}(\cos t) + \mathcal{L}\left(\int_0^t y(x)\cos(t-x)\,\mathrm{d}x\right)$$

$$= \frac{s}{s^2+1} + Y(s)\left(\frac{s}{s^2+1}\right)$$
$$= \left(\frac{s}{s^2+1}\right)(1+Y(s))$$

In other words, we have

$$Y(s) = \frac{s}{s^2-s+1}$$
$$= \frac{s-\frac{1}{2}}{\left(s-\frac{1}{2}\right)^2+\frac{3}{4}} + \frac{\frac{1}{2}}{\left(s-\frac{1}{2}\right)^2+\frac{3}{4}}$$

The required solution of the integral equation is given by

$$y(t) = \mathcal{L}^{-1}(Y(s))$$
$$= \mathcal{L}^{-1}\left(\frac{s-\frac{1}{2}}{\left(s-\frac{1}{2}\right)^2+\frac{3}{4}} + \frac{\frac{1}{2}}{\left(s-\frac{1}{2}\right)^2+\frac{3}{4}}\right)$$
$$= \mathcal{L}^{-1}\left(\frac{s-\frac{1}{2}}{\left(s-\frac{1}{2}\right)^2+\frac{3}{4}}\right) + \mathcal{L}^{-1}\left(\frac{\frac{1}{2}}{\left(s-\frac{1}{2}\right)^2+\frac{3}{4}}\right)$$
$$= \mathcal{L}^{-1}\left(\frac{s-\frac{1}{2}}{\left(s-\frac{1}{2}\right)^2+\frac{3}{4}}\right) + \frac{1}{\sqrt{3}}\mathcal{L}^{-1}\left(\frac{\frac{\sqrt{3}}{2}}{\left(s-\frac{1}{2}\right)^2+\frac{3}{4}}\right)$$
$$= e^{\frac{t}{2}}\cos\frac{\sqrt{3}\,t}{2} + \frac{1}{\sqrt{3}}e^{\frac{t}{2}}\sin\frac{\sqrt{3}\,t}{2}$$

EXAMPLE 9.46 Solve the integral equation $y(t) = t + \int_0^t y(x)\sin(t-x)\,\mathrm{d}x$.

Solution: Let $\mathcal{L}(y(t)) = Y(s)$. Applying the Laplace transform to the given integral equation, we have

$$Y(s) = \mathcal{L}(y(t))$$
$$= \mathcal{L}\left(t + \int_0^t y(x)\sin(t-x)\,\mathrm{d}x\right)$$
$$= \mathcal{L}(t) + \mathcal{L}\left(\int_0^t y(x)\sin(t-x)\,\mathrm{d}x\right)$$
$$= \frac{1}{s^2} + Y(s)\left(\frac{1}{s^2+1}\right)$$

In other words, we have

$$Y(s) = \frac{s^2+1}{s^4}$$
$$= \frac{1}{s^2} + \frac{1}{s^4}$$

The required solution of the integral equation is given by

$$y(t) = \mathcal{L}^{-1}(Y(s))$$
$$= \mathcal{L}^{-1}\left(\frac{1}{s^2} + \frac{1}{s^4}\right)$$
$$= \mathcal{L}^{-1}\left(\frac{1}{s^2}\right) + \mathcal{L}^{-1}\left(\frac{1}{s^4}\right)$$
$$= t + \frac{t^3}{6}$$

Exercises 9.3.2

1. Find the function $f(t)$ such that $\mathcal{L}(f(t)) = \frac{1}{(s-1)^2}$.
2. Find the function $f(t)$ such that $\mathcal{L}(f(t)) = \frac{1}{s(s-1)}$.
3. Find the function $f(t)$ such that $\mathcal{L}(f(t)) = \frac{1}{s(s-a)^2}$.
4. Find the function $f(t)$ such that $\mathcal{L}(f(t)) = \frac{1}{(s+a)^3}$.
5. Find the function $f(t)$ such that $\mathcal{L}(f(t)) = \frac{1}{s^2(s^2+9)}$.
6. Find the function $f(t)$ such that $\mathcal{L}(f(t)) = \frac{1}{(s-a)(s-b)}$ when $a \neq b$.
7. Find the function $f(t)$ such that $\mathcal{L}(f(t)) = \frac{1}{(s^2+a^2)^2}$.
8. Find the function $f(t)$ such that $\mathcal{L}(f(t)) = \frac{s}{(s^2+a^2)^2}$.
9. Find the function $f(t)$ such that $\mathcal{L}(f(t)) = \frac{s^2}{(s^2+a^2)^2}$.
10. Find the function $f(t)$ such that $\mathcal{L}(f(t)) = \frac{1}{(s+a)(s+b)}$ when $a \neq b$.
11. Find the function $f(t)$ such that $\mathcal{L}(f(t)) = \frac{s^2+a^2}{(s^2-a^2)^2}$.
12. Find $y(t)$ such that $y(t) = t + \int_0^t y(x)\cos(t-x)\,dx$.
13. Find $y(t)$ such that $y(t) = \cos 2t + \int_0^t y(x)\sin(t-x)\,dx$.
14. Find $y(t)$ such that $y(t) = \sin t + \int_0^t y(x)\sin(t-x)\,dx$.
15. Find $y(t)$ such that $y(t) = \sin t + \int_0^t y(x)\cos(t-x)\,dx$.

16. Find $y(t)$ such that $y(t) = t + \int_0^t y(x)(t-x)\,dx$.

17. Find $y(t)$ such that $y(t) = t - \int_0^t y(x)(t-x)\,dx$.

18. Find $y(t)$ such that $y(t) = t^3 + \int_0^t y(x)\sin(t-x)\,dx$.

19. Find $y(t)$ such that $y(t) = 1 - \int_0^t y(x)(t-x)\,dx$.

20. Find $y(t)$ such that $y(t) = e^t + \int_0^t y(x)e^{t-x}\,dx$.

21. Find $y(t)$ such that $y(t) = e^t + 2\int_0^t y(x)\cos(t-x)\,dx$.

22. Find $y(t)$ such that $y(t) = 3\sin 2t - \int_0^t y(x)(t-x)\,dx$.

23. Show that the differential equation $y'' + a^2y = f(t)$ with $y(0) = y'(0) = 0$ has $y(t) = \frac{1}{a}\int_0^t f(x)\sin(a(t-x))\,dx$.

24. Show that the differential equation $y'' - a^2y = f(t)$ with $y(0) = y'(0) = 0$ has $y(t) = \frac{1}{a}\int_0^t f(x)\sinh(a(t-x))\,dx$.

9.4 Applications to Differential Equations

Any linear ordinary differential equation with initial conditions at $t = 0$ can be solved by using the Laplace transform. One can apply the Laplace transform to any linear ordinary differential equation with initial condition at $t = 0$ by the following theorem. In this section, we have applied the Laplace transform to second order linear differential equations in particular.

Theorem 9.4.1 If $g(t) = f'(t)$ and $\mathcal{L}(f(t)) = F(s)$, then $\mathcal{L}(g(t)) = sF(s) - f(0)$.

Proof: According to the Laplace transform, we have $F(s) = \int_0^\infty f(t)e^{-st}\,dt$. Again, we have

$$\begin{aligned}\mathcal{L}(g(t)) &= \int_0^\infty g(t)e^{-st}\,dt \\ &= \int_0^\infty f'(t)e^{-st}\,dt\end{aligned}$$

$$= \int_0^\infty e^{-st} \, \mathrm{d}\,(f(t))$$

$$= e^{-st} f(t)\big|_0^\infty + s \int_0^\infty e^{-st} f(t) \, \mathrm{d}t$$

$$= -f(0) + sF(s)$$

$$= sF(s) - f(0)$$

Corollary 9.4.1 If $f(t)$ is exponentially bounded, then $\mathcal{L}\left(\int_0^t f(x) \, \mathrm{d}x\right) s = \mathcal{L}(f(t))$.

Proof: Let $g(t) = \int_0^t f(x) \, \mathrm{d}x$. Clearly, $g(0) = 0$. Again, $g'(t) = f(t)$. According to the definition of the Laplace transform and Theorem 9.4.1, we have

$$\mathcal{L}(f(t)) = \mathcal{L}(g'(t))$$

$$= s\mathcal{L}(g(t)) - g(0)$$

$$= s\mathcal{L}(g(t))$$

$$= s\mathcal{L}\left(\int_0^t f(x) \, \mathrm{d}x\right)$$

Theorem 9.4.2 If $g(t) = f''(t)$ and $\mathcal{L}(f(t)) = F(s)$, then $\mathcal{L}(g(t)) = s^2 F(s) - sf(0) - f'(0)$.

Proof: According to the Laplace transform, we have $F(s) = \int_0^\infty f(t) e^{-st} \, \mathrm{d}t$. Again, we have

$$\mathcal{L}(g(t)) = \int_0^\infty g(t) e^{-st} \, \mathrm{d}t$$

$$= \int_0^\infty f''(t) e^{-st} \, \mathrm{d}t$$

$$= \int_0^\infty e^{-st} \, \mathrm{d}\,(f'(t))$$

$$= e^{-st} f'(t)\Big|_0^\infty + s\int_0^\infty e^{-st} f'(t)\,\mathrm{d}t$$

$$= -f'(0) + s\,(sF(s) - f(0))$$
$$= s^2F(s) - sf(0) - f'(0)$$

Theorem 9.4.3 If $g(t) = f^{(n)}(t)$ and $\mathcal{L}(f(t)) = F(s)$, then $\mathcal{L}(g(t)) = s^nF(s) - s^{n-1}f(0) - s^{n-2}f'(0) - \cdots - f^{(n-1)}(0)$.

Proof: If $n = 1$, then the given formulae is reduced to $\mathcal{L}(f'(t)) = sF(s) - f(0)$, which agrees with Theorem 9.4.1. Assume that the given formula is valid for $(n-1)$th derivative of $f(t)$. In other words, we have

$$\mathcal{L}(f^{(n-1)}(t)) = \int_0^\infty f^{(n-1)}(t)\,e^{-st}\,\mathrm{d}t$$

$$= s^{n-1}F(s) - s^{n-2}f(0) - s^{n-3}f'(0) - \cdots - f^{(n-2)}(0) \tag{9.15}$$

According to the definition of the Laplace transform and Eq.(9.15), we have

$$\mathcal{L}(f^{(n)}(t)) = \int_0^\infty f^{(n)}(t)\,e^{-st}\,\mathrm{d}t$$

$$= e^{-st}f^{(n-1)}(t)\Big|_0^\infty + s\int_0^\infty f^{(n-1)}(t)\,e^{-st}\,\mathrm{d}t$$

$$= -f^{(n-1)}(0) + s\left(s^{n-1}F(s) - s^{n-2}f(0) - s^{n-3}f'(0) - \cdots - f^{(n-2)}(0)\right)$$
$$= s^nF(s) - s^{n-1}f(0) - s^{n-2}f'(0) - \cdots - f^{(n-1)}(0)$$

EXAMPLE 9.47 Find the solution of the differential equation $y'' + y = 2$ with the initial conditions $y(0) = 0$ and $y'(0) = 2$ using the Laplace transform.

Solution: Let $\mathcal{L}(y(t)) = Y(s)$. According to Theorem 9.4.2, we have

$$\frac{2}{s} = \mathcal{L}(2)$$
$$= \mathcal{L}(y'' + y)$$

$$= \mathcal{L}(y'') + \mathcal{L}(y)$$

$$= s^2Y(s) - sy(0) - y'(0) + Y(s)$$

$$= s^2Y(s) - 2 + Y(s)$$

$$= (s^2 + 1)Y(s) - 2$$

In other words, we have

$$\begin{aligned}Y(s) &= 2\left(\frac{s+1}{s(s^2-1)}\right)\\ &= 2\left(\frac{1-s}{s^2+1}+\frac{1}{s}\right)\\ &= \frac{2(1-s)}{s^2+1}+\frac{2}{s}\\ &= \frac{2}{s^2+1}-\frac{2s}{s^2+1}+\frac{2}{s}\end{aligned} \tag{9.16}$$

Applying the inverse Laplace transform to Eq.(9.16), we have

$$\begin{aligned}y(t) &= \mathcal{L}^{-1}(Y(s))\\ &= \mathcal{L}^{-1}\left(\frac{2}{s^2+1}-\frac{2s}{s^2+1}+\frac{2}{s}\right)\\ &= \mathcal{L}^{-1}\left(\frac{2}{s^2+1}\right)-\mathcal{L}^{-1}\left(\frac{2s}{s^2+1}\right)+\mathcal{L}^{-1}\left(\frac{2}{s}\right)\\ &= 2\mathcal{L}^{-1}\left(\frac{1}{s^2+1}\right)-2\mathcal{L}^{-1}\left(\frac{s}{s^2+1}\right)+2\mathcal{L}^{-1}\left(\frac{1}{s}\right)\\ &= 2\sin t-2\cos t+2\end{aligned}$$

EXAMPLE 9.48 Find the solution of the differential equation $y''-y'-2y=2$ with the initial conditions $y(0)=1$ and $y'(0)=0$ using the Laplace transform.

Solution: Let $\mathcal{L}(y(t))=Y(s)$. According to Theorem 9.4.2, we have

$$\begin{aligned}\frac{2}{s} &= \mathcal{L}(2)\\ &= \mathcal{L}(y''-y'-2y)\\ &= \mathcal{L}(y'')-\mathcal{L}(y')-2\mathcal{L}(y)\\ &= s^2Y(s)-sy(0)-y'(0)-(sY(s)-y(0))-2Y(s)\\ &= (s^2-s-2)Y(s)-sy(0)+y(0)\\ &= (s^2-s-2)Y(s)-s+1\end{aligned}$$

In other words, we have

$$Y(s)=\frac{s^2-s+2}{s(s^2-s-2)}$$

$$= \frac{s^2 - s + 2}{s(s-2)(s+1)}$$

$$= -\frac{1}{s} + \frac{2}{3}\left(\frac{1}{s-2}\right) + \frac{4}{3}\left(\frac{1}{s+1}\right) \tag{9.17}$$

Applying the inverse Laplace transform to Eq.(9.17), we have

$$\begin{aligned}
y(t) &= \mathcal{L}^{-1}(Y(s)) \\
&= \mathcal{L}^{-1}\left(-\frac{1}{s} + \frac{2}{3}\left(\frac{1}{s-2}\right) + \frac{4}{3}\left(\frac{1}{s+1}\right)\right) \\
&= -\mathcal{L}^{-1}\left(\frac{1}{s}\right) + \frac{2}{3}\mathcal{L}^{-1}\left(\frac{1}{s-2}\right) + \frac{4}{3}\mathcal{L}^{-1}\left(\frac{1}{s+1}\right) \\
&= \frac{2}{3}\mathcal{L}^{-1}\left(\frac{1}{s-2}\right) + \frac{4}{3}\mathcal{L}^{-1}\left(\frac{1}{s+1}\right) - \mathcal{L}^{-1}\left(\frac{1}{s}\right) \\
&= \frac{2}{3}e^{2t} + \frac{4}{3}e^{-t} - 1 \\
&= \frac{1}{3}\left(2e^{2t} + 4e^{-t} - 3\right)
\end{aligned}$$

EXAMPLE 9.49 Find the solution of the differential equation $y'' - y' - 2y = 2t$ with the initial conditions $y(0) = 1$ and $y'(0) = 0$ using the Laplace transform.

Solution: Let $\mathcal{L}(y(t)) = Y(s)$. According to Theorem 9.4.2, we have

$$\begin{aligned}
\frac{2}{s^2} &= \mathcal{L}(2t) \\
&= \mathcal{L}(y'' - y' - 2y) \\
&= \mathcal{L}(y'') - \mathcal{L}(y') - 2\mathcal{L}(y) \\
&= s^2Y(s) - sy(0) - y'(0) - (sY(s) - y(0)) - 2Y(s) \\
&= (s^2 - s - 2)Y(s) - sy(0) + y(0) \\
&= (s^2 - s - 2)Y(s) - s + 1
\end{aligned}$$

In other words, we have

$$\begin{aligned}
Y(s) &= \frac{s^3 - s^2 + 2}{s^2(s^2 - s - 2)} \\
&= \frac{s^3 - s^2 + 2}{s^2(s-2)(s+1)} \\
&= -\frac{1}{2}\left(\frac{1}{s}\right) - \frac{1}{s^2} + \frac{1}{2}\left(\frac{1}{s-2}\right)
\end{aligned} \tag{9.18}$$

Applying the inverse Laplace transform to Eq.(9.18), we have

$$\begin{aligned}
y(t) &= \mathcal{L}^{-1}(Y(s)) \\
&= \mathcal{L}^{-1}\left(-\frac{1}{2}\left(\frac{1}{s}\right) - \frac{1}{s^2} + \frac{1}{2}\left(\frac{1}{s-2}\right)\right) \\
&= -\frac{1}{2}\mathcal{L}^{-1}\left(\frac{1}{s}\right) - \mathcal{L}^{-1}\left(\frac{1}{s^2}\right) + \frac{1}{2}\mathcal{L}^{-1}\left(\frac{1}{s-2}\right) \\
&= \frac{1}{2}\mathcal{L}^{-1}\left(\frac{1}{s-2}\right) - \frac{1}{2}\mathcal{L}^{-1}\left(\frac{1}{s}\right) - \mathcal{L}^{-1}\left(\frac{1}{s^2}\right) \\
&= \frac{1}{2}e^{2t} + \frac{1}{2} - t \\
&= \frac{1}{2}(e^{2t} - 2t + 1)
\end{aligned}$$

EXAMPLE 9.50 Find the solution of the differential equation $y'' + 2y' + y = t\,e^{-t}$ with the initial conditions $y(0) = 0$ and $y'(0) = 1$ using the Laplace transform.

Solution: Let $\mathcal{L}(y(t)) = Y(s)$. According to Theorem 9.4.2, we have

$$\begin{aligned}
\frac{1}{(s+1)^2} &= \mathcal{L}\left(te^{-t}\right) \\
&= \mathcal{L}(y'' + 2y' + y) \\
&= \mathcal{L}(y'') + 2\mathcal{L}(y') + \mathcal{L}(y) \\
&= s^2Y(s) - sy(0) - y'(0) + 2(sY(s) - y(0)) + Y(s) \\
&= (s^2 + 2s + 1)Y(s) - 1 \\
&= (s+1)^2Y(s) - 1
\end{aligned}$$

In other words, we have

$$\begin{aligned}
Y(s) &= \left(\frac{1}{(s+1)^2}\right)\left(1 + \frac{1}{(s+1)^2}\right) \\
&= \frac{1}{(s+1)^2} + \frac{1}{(s+1)^4} \qquad (9.19)
\end{aligned}$$

Applying the inverse Laplace transform to Eq.(9.19), we have

$$\begin{aligned}
y(t) &= \mathcal{L}^{-1}(Y(s)) \\
&= \mathcal{L}^{-1}\left(\frac{1}{(s+1)^2} + \frac{1}{(s+1)^4}\right)
\end{aligned}$$

$$= \mathcal{L}^{-1}\left(\frac{1}{(s+1)^2}\right) + \mathcal{L}^{-1}\left(\frac{1}{(s+1)^4}\right)$$
$$= t\,e^{-t} + \frac{1}{6}t^3\,e^{-t}$$
$$= t\left(1 + \frac{t^2}{6}\right)e^{-t}$$

EXAMPLE 9.51 Find the solution of the differential equation $y'' - 2y' + y = e^t$ with the initial conditions $y(0) = 0$ and $y'(0) = 1$ using the Laplace transform.
Solution: Let $\mathcal{L}(y(t)) = Y(s)$. According to Theorem 9.4.2, we have

$$\begin{aligned}
\frac{1}{s-1} &= \mathcal{L}\left(e^t\right) \\
&= \mathcal{L}(y'' - 2y' + y) \\
&= \mathcal{L}(y'') - 2\mathcal{L}(y') + \mathcal{L}(y) \\
&= s^2Y(s) - sy(0) - y'(0) - 2(sY(s) - y(0)) + Y(s) \\
&= (s^2 - 2s + 1)Y(s) - 1 \\
&= (s-1)^2Y(s) - 1
\end{aligned}$$

In other words, we have

$$Y(s) = \left(\frac{1}{(s-1)^2}\right)\left(1 + \frac{1}{s-1}\right)$$
$$= \frac{1}{(s-1)^2} + \frac{1}{(s-1)^3} \tag{9.20}$$

Applying the inverse Laplace transform to Eq.(9.20), we have

$$\begin{aligned}
y(t) &= \mathcal{L}^{-1}(Y(s)) \\
&= \mathcal{L}^{-1}\left(\frac{1}{(s-1)^2} + \frac{1}{(s-1)^3}\right) \\
&= \mathcal{L}^{-1}\left(\frac{1}{(s-1)^2}\right) + \mathcal{L}^{-1}\left(\frac{1}{(s-1)^3}\right) \\
&= t\,e^t + \frac{1}{2}t^2\,e^t \\
&= t\left(1 + \frac{t}{2}\right)e^t
\end{aligned}$$

EXAMPLE 9.52 Find the solution of the system of differential equations

$$x' + 2x + y = t$$
$$y' - 3x - 2y = e^{2t}$$

which satisfy the conditions $x(0) = y(0) = 0$.

Solution: Let $X(s) = \mathcal{L}(x(t))$ and $Y(s) = \mathcal{L}(y(t))$. According to the definition of the Laplace transform, we have

$$\begin{aligned}\frac{1}{s^2} &= \mathcal{L}(t)\\ &= \mathcal{L}(x' + 2x + y)\\ &= \mathcal{L}(x') + 2\mathcal{L}(x) + \mathcal{L}(y)\\ &= sX(s) - x(0) + 2X(s) + Y(s)\\ &= (s+2)X(s) + Y(s) \qquad (9.21)\end{aligned}$$

and

$$\begin{aligned}\frac{1}{s-2} &= \mathcal{L}\left(e^{2t}\right)\\ &= \mathcal{L}(y' - 3x - 2y)\\ &= \mathcal{L}(y') - 3\mathcal{L}(x) - 2\mathcal{L}(y)\\ &= sY(s) - y(0) - 3X(s) - 2Y(s)\\ &= (s-2)Y(s) - 3X(s) \qquad (9.22)\end{aligned}$$

Clearly, the solutions of Eqs.(9.21) and (9.22) are given by

$$\begin{aligned}X(s) &= \frac{1}{s^2(s+2)} - \frac{3}{s^2(s^2-1)(s+2)} - \frac{1}{(s-2)(s^2-1)}\\ &= \frac{1}{s^2(s+2)} + \frac{1}{s(s^2-1)} - \frac{2}{s^2(s^2-1)} - \frac{1}{s^2(s+2)} - \frac{1}{(s-2)(s^2-1)}\\ &= \frac{1}{s(s^2-1)} - \frac{2}{s^2(s^2-1)} - \frac{1}{(s-2)(s^2-1)}\end{aligned}$$

and

$$\begin{aligned}Y(s) &= \frac{3}{s^2(s^2-1)} + \frac{s+2}{(s-2)(s^2-1)}\\ &= \frac{3}{s^2(s^2-1)} + \frac{s}{(s-2)(s^2-1)} + \frac{2}{(s-2)(s^2-1)}\end{aligned}$$

In other words, we have

$$\begin{aligned}x(t) &= \mathcal{L}^{-1}(X(s))\\ &= \cosh t - 1 - 2(\sinh t - t) + \frac{1}{3}(\cosh t - e^{2t} + 2\sinh t)\\ &= 2t - 1 - \frac{4}{3}\sinh t + \frac{4}{3}\cosh t - \frac{1}{3}e^{2t}\end{aligned}$$

and

$$\begin{aligned} y(t) &= \mathcal{L}^{-1}(Y(s)) \\ &= -\frac{1}{3}(\sinh t + 2\cosh t - 2\,e^{2t}) - \frac{2}{3}(\cosh t - e^{2t} + 2\sinh t) \\ &\quad + 3(\sinh t - t) \\ &= -3t + \frac{4}{3}\sinh t - \frac{4}{3}\cosh t + \frac{4}{3}e^{2t} \end{aligned}$$

EXAMPLE 9.53 Find the solution of the system of differential equations

$$\begin{aligned} x' + 2x + y &= e^{2t} \\ y' - 3x - 2y &= t \end{aligned}$$

which satisfy the conditions $x(0) = y(0) = 0$.

Solution: Let $X(s) = \mathcal{L}(x(t))$ and $Y(s) = \mathcal{L}(y(t))$. According to the definition of the Laplace transform, we have

$$\begin{aligned} \frac{1}{s-2} &= \mathcal{L}\left(e^{2t}\right) \\ &= \mathcal{L}(x' + 2x + y) \\ &= \mathcal{L}(x') + 2\mathcal{L}(x) + \mathcal{L}(y) \\ &= sX(s) - x(0) + 2X(s) + Y(s) \\ &= (s+2)X(s) + Y(s) \end{aligned} \tag{9.23}$$

and

$$\begin{aligned} \frac{1}{s^2} &= \mathcal{L}(t) \\ &= \mathcal{L}(y' - 3x - 2y) \\ &= \mathcal{L}(y') - 3\mathcal{L}(x) - 2\mathcal{L}(y) \\ &= sY(s) - y(0) - 3X(s) - 2Y(s) \\ &= (s-2)Y(s) - 3X(s) \end{aligned} \tag{9.24}$$

Clearly, the solution of Eqs.(9.23) and (9.24) are given by

$$\begin{aligned} X(s) &= \frac{1}{s^2-1} - \frac{1}{s^2(s^2-1)} \\ &= \frac{1}{s^2-1} - \frac{1}{s^2-1} + \frac{1}{s^2} \\ &= \frac{1}{s^2} \end{aligned}$$

and

$$\begin{aligned}
Y(s) &= \frac{1}{s-2} - \frac{s+2}{s^2-1} + \frac{s+2}{s^2(s^2-1)} \\
&= \frac{1}{s-2} - \frac{s}{s^2-1} - \frac{2}{s^2-1} + \frac{1}{s(s^2-1)} + \frac{2}{s^2(s^2-1)} \\
&= \frac{1}{s-2} - \frac{2}{s^2} - \frac{1}{s}
\end{aligned}$$

In other words, we have

$$\begin{aligned}
x(t) &= \mathcal{L}^{-1}(X(s)) \\
&= t
\end{aligned}$$

and

$$\begin{aligned}
y(t) &= \mathcal{L}^{-1}(Y(s)) \\
&= e^{2t} - 2t - 1
\end{aligned}$$

Exercises 9.4

1. Find the solution of the differential equation $y'' + y = 2\cos t$ with $y(0) = 2$ and $y'(0) = 0$ using the Laplace transform.
2. Find the solution of the differential equation $y'' - 4y' + 3y = 2$ with $y(0) = 2$ and $y'(0) = 0$ using the Laplace transform.
3. Find the solution of the differential equation $y'' + 2y' + y = e^{-2t}$ with $y(0) = 0$ and $y'(0) = 0$ using the Laplace transform.
4. Find the solution of the differential equation $y'' + 2y' - 3y = 2\sinh 2t$ with $y(0) = 0$ and $y'(0) = 4$ using the Laplace transform.
5. Find the solution of the differential equation $y'' + 4y = 2(\sin 2t + 2\cos 2t)$ with $y(0) = 1$ and $y'(0) = 2$ using the Laplace transform.
6. Find the solution of the differential equation $y'' + y = 2\sin 3t$ with $y(0) = 0$ and $y'(0) = 0$ using the Laplace transform.
7. Find the solution of the differential equation $y'' + y = 2\cos 3t$ with $y(0) = 0$ and $y'(0) = 0$ using the Laplace transform.
8. Find the solution of the differential equation $y'' + y = 3\sin 2t$ with $y(0) = 0$ and $y'(0) = 0$ using the Laplace transform.
9. Find the solution of the differential equation $y'' + y = 3\cos 2t$ with $y(0) = 0$ and $y'(0) = 0$ using the Laplace transform.
10. Find the solution of the differential equation $y'' + y = 2t$ with $y(0) = 0$ and $y'(0) = 0$ using the Laplace transform.
11. Find the solution of the differential equation $y'' + 3y' + 2y = e^{-t}$ with $y(0) = 0$ and $y'(0) = 0$ using the Laplace transform.

12. Find the solution of the differential equation $y'' + 2y' - 3y = 6\,e^{-2t}$ with $y(0) = 2$ and $y'(0) = 4$ using the Laplace transform.
13. Find the solution of the differential equation $y'' + y = \sin t$ with $y(0) = 0$ and $y'(0) = 0$ using the Laplace transform.
14. Find the solution of the differential equation $y'' + y = \cos t$ with $y(0) = 0$ and $y'(0) = 0$ using the Laplace transform.
15. Find the solution of the system of differential equations

$$x' + 2x - 4y = 4t$$

$$y' - 3x + y = -(1 + t)$$

which satisfy the conditions $x(0) = y(0) = 1$.
16. Find the solution of the system of differential equations

$$20y' + 5x' + 15y = 0$$

$$2y' + 6x' + 4x = 2$$

which satisfy the conditions $x(0) = y(0) = 0$.
17. Find the solution of the system of differential equations

$$x' + x + y = e^t$$

$$2x' + y' + y = \cos t$$

which satisfy the conditions $x(0) = y(0) = 2$.
18. Find the solution of the system of differential equations

$$2x'' + 3y' = 4$$

$$2y'' - 3x' = 0$$

which satisfy the conditions $x(0) = y(0) = x'(0) = y'(0) = 0$.
19. Find the solution of the system of differential equations

$$x'' + y' + 3x = 15e^{-t}$$

$$2y'' - 4x' + 3y = 15 \sin 2t$$

which satisfy the conditions $x(0) = y(0) = x'(0) = y'(0) = 0$.

9.5 Laplace Transform of Periodic Functions

Periodic functions appear in many practical problems, and in most cases they are more complicated than single cosine or sine functions. This justifies the topic of the present section, which is a systematic approach to the transformation of periodic functions. Let $f(t)$ be a function which is defined for all positive t and has the period $l > 0$. In other words, $f(t + l) = f(t)$. If $f(t)$ is piecewise continuous on an interval of length l, then its Laplace transformation exists, and we can write the integral from zero to infinity as the series of integrals over successive periods.

Theorem 9.5.1 If $f(t)$ is a periodic function with period $l > 0$, then

$$\mathcal{L}(f(t)) = \frac{1}{1-e^{-ls}} \int_0^l e^{-su} f(u)\, du$$

Proof: According to the definition of the Laplace transform, we have

$$\begin{aligned}
\mathcal{L}(f(t)) &= \int_0^\infty e^{-st} f(t)\, dt \\
&= \sum_{i=0}^{\infty} \int_{il}^{(i+1)l} e^{-st} f(t)\, dt \\
&= \sum_{i=0}^{\infty} \int_0^l e^{-s(u+li)} f(u+li)\, du \\
&= \sum_{i=0}^{\infty} e^{-sli} \int_0^l e^{-su} f(u)\, du \\
&= \sum_{i=0}^{\infty} e^{-sli} \int_0^l e^{-su} f(u)\, du \\
&= \frac{1}{1-e^{-sl}} \int_0^l e^{-su} f(u)\, du
\end{aligned}$$

EXAMPLE 9.54 Find the Laplace transform of the periodic function $f(t)$ defined over the period of length $2a$ by

$$f(t) = \begin{cases} k & 0 < t < a \\ -k & a < t < 2a \end{cases}$$

Solution: Clearly, $l = 2a$. Hence according to Theorem 9.5.1, we have

$$\begin{aligned}
\mathcal{L}(f(t)) &= \frac{1}{1-e^{-sl}} \int_0^l e^{-st} f(t)\, dt \\
&= \frac{1}{1-e^{-2sa}} \int_0^{2a} e^{-st} f(t)\, dt
\end{aligned}$$

$$= \frac{1}{1-e^{-2sa}}\left(\int_0^a e^{-st} f(t)\,\mathrm{d}t + \int_a^{2a} e^{-st} f(t)\,\mathrm{d}t\right)$$

$$= \frac{1}{1-e^{-2sa}}\left(\int_0^a k e^{-st}\,\mathrm{d}t - \int_a^{2a} k e^{-st}\,\mathrm{d}t\right)$$

$$= \frac{1}{1-e^{-2sa}}\left(-\frac{k}{s} e^{-st}\Big|_0^a + \left(\frac{k}{s}\right) e^{-st}\Big|_a^{2a}\right)$$

$$= \frac{1}{1-e^{-2sa}} \frac{k}{s}\left(1 - 2e^{-st} + e^{-2sa}\right)$$

$$= \frac{1}{1-e^{-2sa}}\left(\frac{k}{s}(1-e^{-sa})^2\right)$$

$$= \frac{k}{s}\left(\frac{1-e^{-sa}}{1+e^{-as}}\right)$$

EXAMPLE 9.55 Find the Laplace transform of the periodic function $f(t)$ defined over the period of length $2a$ by

$$f(t) = \begin{cases} t & 0 < t < a \\ 2a - t & a < t < 2a \end{cases}$$

Solution: Clearly, $l = 2a$. Hence according to Theorem 9.5.1, we have

$$\mathcal{L}(f(t)) = \frac{1}{1-e^{-sl}} \int_0^l e^{-st} f(t)\,\mathrm{d}t$$

$$= \frac{1}{1-e^{-2sa}} \int_0^{2a} e^{-st} f(t)\,\mathrm{d}t$$

$$= \frac{1}{1-e^{-2sa}}\left(\int_0^a e^{-st} f(t)\,\mathrm{d}t + \int_a^{2a} e^{-st} f(t)\,\mathrm{d}t\right)$$

$$= \frac{1}{1-e^{-2sa}}\left(\int_0^a t e^{-st}\,\mathrm{d}t + \int_a^{2a} (2a-t) e^{-st}\,\mathrm{d}t\right)$$

$$= \frac{1}{1-e^{-2sa}}\left(-\frac{a}{s} e^{-sa} + \frac{1}{s}\int_0^a e^{-st}\,\mathrm{d}t + \frac{a}{s} e^{-sa} - \frac{1}{s}\int_a^{2a} e^{-st}\,\mathrm{d}t\right)$$

$$= \frac{1}{1-e^{-2sa}}\left(\frac{1}{s}\int_0^a e^{-st}\,\mathrm{d}t - \frac{1}{s}\int_a^{2a} e^{-st}\,\mathrm{d}t\right)$$

$$= \frac{1}{1-e^{-2sa}}\left(\frac{1}{s^2}\left(1-2e^{-st}+e^{-2sa}\right)\right)$$

$$= \frac{1}{1-e^{-2sa}}\left(\frac{1}{s^2}\left(1-e^{-sa}\right)^2\right)$$

$$= \frac{1}{s^2}\left(\frac{1-e^{-sa}}{1+e^{-sa}}\right)$$

EXAMPLE 9.56 Find the Laplace transform of the periodic function $f(t)$ defined over the period of length $\frac{2\pi}{a}$ by

$$f(t) = \begin{cases} \sin at & 0 < t < \frac{\pi}{a} \\ 0 & \frac{\pi}{a} < t < \frac{2\pi}{a} \end{cases}$$

Solution: Clearly, $l = \frac{2\pi}{a}$. Hence according to Theorem 9.5.1, we have

$$\mathcal{L}(f(t)) = \frac{1}{1-e^{-\frac{2\pi s}{a}}}\int_0^l e^{-st}f(t)\,\mathrm{d}t$$

$$= \frac{1}{1-e^{-\frac{2\pi s}{a}}}\int_0^{\frac{2\pi}{a}} e^{-st}f(t)\,\mathrm{d}t$$

$$= \frac{1}{1-e^{-\frac{2\pi s}{a}}}\left(\int_0^{\frac{\pi}{a}} e^{-st}f(t)\,\mathrm{d}t + \int_{\frac{\pi}{a}}^{\frac{2\pi}{a}} e^{-st}f(t)\,\mathrm{d}t\right)$$

$$= \frac{1}{1-e^{-\frac{2\pi s}{a}}}\left(\int_0^{\frac{\pi}{a}} \sin at\, e^{-st}\,\mathrm{d}t\right)$$

$$= \left(\frac{1}{1-e^{-\frac{2\pi s}{a}}}\right)\left(\frac{a}{s^2+a^2}\right)\left(1+e^{-\frac{s\pi}{a}}\right)$$

$$= \left(\frac{a}{s^2+a^2}\right)\left(\frac{1}{1-e^{-\frac{\pi s}{a}}}\right)$$

EXAMPLE 9.57 Find the Laplace transform of the periodic function $f(t)$ defined over the period of length $\frac{2\pi}{a}$ by

$$f(t) = \begin{cases} 0 & 0 < t < \frac{\pi}{a} \\ \cos at & \frac{\pi}{a} < t < \frac{2\pi}{a} \end{cases}$$

Solution: Clearly, $l = \frac{2\pi}{a}$. Hence according to Theorem 9.5.1, we have

$$\mathcal{L}(f(t)) = \frac{1}{1-e^{-\frac{2\pi s}{a}}}\int_0^l e^{-st}f(t)\,\mathrm{d}t$$

$$= \frac{1}{1-e^{-\frac{2\pi s}{a}}}\int_0^{\frac{2\pi}{a}} e^{-st}f(t)\,\mathrm{d}t$$

$$= \frac{1}{1-e^{-\frac{2\pi s}{a}}}\left(\int_0^{\frac{\pi}{a}} e^{-st}f(t)\,\mathrm{d}t + \int_{\frac{\pi}{a}}^{\frac{2\pi}{a}} e^{-st}f(t)\,\mathrm{d}t\right)$$

$$= \frac{1}{1-e^{-\frac{2\pi s}{a}}}\int_{\frac{\pi}{a}}^{\frac{2\pi}{a}} \cos at\, e^{-st}\,\mathrm{d}t$$

$$= \left(\frac{1}{1-e^{-\frac{2\pi s}{a}}}\right)\left(\frac{-a}{s^2+a^2}\right)e^{-\frac{\pi s}{a}}\left(1+e^{-\frac{s\pi}{a}}\right)$$

$$= -\left(\frac{a}{s^2+a^2}\right)\left(\frac{e^{-\frac{\pi s}{a}}}{1-e^{-\frac{\pi s}{a}}}\right)$$

$$= -\left(\frac{a}{s^2+a^2}\right)\left(\frac{1}{e^{\frac{\pi s}{a}}-1}\right)$$

$$= \left(\frac{a}{s^2+a^2}\right)\left(\frac{1}{1-e^{\frac{\pi s}{a}}}\right)$$

EXAMPLE 9.58 Find the period and the Laplace transform of the periodic function $f(t)$ defined by

$$f(t) = \begin{cases} \sin at & 0 < t < \frac{\pi}{a} \\ -\sin at & \frac{\pi}{a} < t < \frac{2\pi}{a} \end{cases}$$

Solution: Clearly, the period $l = \frac{\pi}{a}$. Hence according to Theorem 9.5.1, we have

$$\mathcal{L}(f(t)) = \frac{1}{1-e^{-\frac{\pi s}{a}}}\int_0^l e^{-st}f(t)\,\mathrm{d}t$$

$$= \frac{1}{1-e^{-\frac{\pi s}{a}}}\int_0^{\frac{\pi}{a}} e^{-st}f(t)\,\mathrm{d}t$$

$$= \frac{1}{1-e^{-\frac{\pi s}{a}}}\int_0^{\frac{\pi}{a}} e^{-st}\sin at\,\mathrm{d}t$$

$$= \left(\frac{1}{1-e^{-\frac{\pi s}{a}}}\right)\left(\frac{a}{s^2+a^2}\right)(1+e^{-\frac{s\pi}{a}})$$

$$= \frac{a}{s^2+a^2}\left(\frac{1+e^{-\frac{s\pi}{a}}}{1-e^{-\frac{\pi s}{a}}}\right)$$

$$= \frac{a}{s^2+a^2}\left(\frac{e^{\frac{s\pi}{2a}}+e^{-\frac{s\pi}{2a}}}{e^{\frac{s\pi}{2a}}-e^{-\frac{\pi s}{2a}}}\right)$$

$$= \frac{a}{s^2+a^2}\left(\frac{\cosh\frac{\pi s}{2a}}{\sinh\frac{\pi s}{2a}}\right)$$

$$= \frac{a}{s^2+a^2}\coth\frac{\pi s}{2a}$$

Exercises 9.5

1. Find the Laplace transform of $f(t) = \sin at$ using $f(t)$ as a periodic function.
2. Find the Laplace transform of $f(t) = \cos at$ using $f(t)$ as a periodic function.
3. Find the Laplace transform of the periodic function $f(t)$ of period of length l defined by $f(t) = \frac{kt}{l}$ for $0 \leq t \leq l$.
4. Find the Laplace transform of the periodic function $f(t)$ of period of length 2 defined by $f(t) = t$ for $0 \leq t \leq 2$.
5. Find the Laplace transform of the periodic function $f(t)$ of period of length 2 defined by $f(t) = 2 - t$ for $0 \leq t \leq 2$.
6. Find the Laplace transform of the periodic function $f(t)$ of period of length 2 defined by $f(t) = t - 2$ for $0 \leq t \leq 2$.
7. Find the Laplace transform of the periodic function $f(t)$ of period of length π defined by $f(t) = t^2$ for $0 \leq t \leq \pi$.
8. Find the Laplace transform of the periodic function $f(t)$ defined over the period of length $\frac{2\pi}{a}$ by

$$f(t) = \begin{cases} 0 & 0 < t < \frac{\pi}{a} \\ \sin at & \frac{\pi}{a} < t < \frac{2\pi}{a} \end{cases}$$

9. Find the Laplace transform of the periodic function $f(t)$ defined over the period of length $\frac{2\pi}{a}$ by

$$f(t) = \begin{cases} \cos at & 0 < t < \frac{\pi}{a} \\ 0 & \frac{\pi}{a} < t < \frac{2\pi}{a} \end{cases}$$

10. Find the length of the period and the Laplace transform of the periodic function $f(t)$ defined by

$$f(t) = \begin{cases} \cos at & 0 < t < \frac{\pi}{a} \\ -\cos at & \frac{\pi}{a} < t < \frac{2\pi}{a} \end{cases}$$

11. Find the Laplace transform of the periodic function $f(t)$ defined over the period of length 4 by

$$f(t) = \begin{cases} 1 & 0 < t < 2 \\ 0 & 2 < t < 4 \end{cases}$$

12. Find the Laplace transform of the periodic function $f(t)$ defined over the period of length 4 by

$$f(t) = \begin{cases} 0 & 0 < t < 2 \\ t & 2 < t < 4 \end{cases}$$

13. Find the Laplace transform of the periodic function $f(t)$ defined over the period of length 2 by

$$f(t) = \begin{cases} 2 - t & 0 < t < 1 \\ 0 & 1 < t < 2 \end{cases}$$

Bibliography

Birkhoff, Garrett and Gian-Carlo Rota, *Ordinary Differential Equations*, John Wiley & Sons, Singapore, 1989.

Biswal, P.C., *Integral Transform*, Sengage, New Delhi, 2008.

Brauer, Fred and John A. Nohel, *The Quantitative Theory of Ordinary Differential Equations*, Dover Publications, New York, 1989.

Braun, Martin, *Differential Equations and Their Applications*, Spinger Verlag, New York, 1978.

Coddington, Earl A. and N. Levinson, *Theory of Ordinary Differential Equations*, McGraw-Hill, New York, 1955.

Ince, E.L., *Ordinary Differential Equations*, Dover Publications, New York, 1956.

Simmons, George F., *Differential Equations with Applications and Historical Notes*, McGraw-Hill, New York, 1972.

Snyder, Martin Avery, *Chebyshev Methods in Numerical Approximation*, Prentice-Hall, Englewood Cliffs, New Jersey, 1966.

Tenenbaum, Morris and Harry Pollard, *Ordinary Differential Equations*, Dover Publications, New York.

Index